SHAKESPEARE'S
TRAGIC PRACTICE

SHAKESPEARE'S TRAGIC PRACTICE

BY

BERTRAND EVANS

CLARENDON PRESS · OXFORD
1979

Oxford University Press, Walton Street, Oxford OX2 6DP

OXFORD LONDON GLASGOW
NEW YORK TORONTO MELBOURNE WELLINGTON
KUALA LUMPUR SINGAPORE JAKARTA HONG KONG TOKYO
DELHI BOMBAY CALCUTTA MADRAS KARACHI
NAIROBI DAR ES SALAAM CAPE TOWN

Published in the United States by
Oxford University Press, New York

British Library Cataloguing in Publication Data

Evans, Bertrand
    Shakespeare's tragic practice.
    1. Shakespeare, William – Tragedies
    I. Title
    822.3′3   PR2983   79–40409

ISBN 0–19–812094–X

Printed in Great Britain
by W & J Mackay Limited, Chatham

TO CAROL AND PEG

Let me have way ...
To find this practice out.
Measure for Measure, v. i. 238–9

Preface

IN the Preface to *Shakespeare's Comedies* (1960), I blithely asserted that 'a similar account of the management of awarenesses in the histories and tragedies is nearly finished at the present time'. Alas (to paraphrase Gloucester of *King Lear*), I have learned more since. Here, nearly twenty years later, are the tragedies; the histories, not yet.

I have approached the tragedies much as I did the comedies. I have studied Shakespeare's habit of creating and exploiting discrepant, or unequal, awarenesses as a means of producing various dramatic effects. But, as before, I have been concerned not with exploring this particular aspect of dramaturgy for itself only, but with illuminating the plays as total works. I have wanted to see how each tragedy looks when it is examined through my special lens. So this book is about the tragedies primarily, and not about a dramatic device.

I have found this approach more difficult when applied to the tragedies because of one large difference. In the comedies the dramatist typically opens a gap between the audience's awareness and the participants' unawareness and then proceeds to exploit it for such incidental effects as it can be made to yield—and in the comedies these effects are the be-all and the end-all of the method. But in the tragedies a participant's unawareness is likely to prove responsible for much more than passing effects: it is likely to be intimately related to the eventual catastrophe. Thus, to take a single example, when we watch Romeo lament over the seemingly dead body of Juliet and then drink poison, the dramatist has elevated us to a position from which we take note of the effects that arise as the gap between our awareness and Romeo's unawareness is exploited; yet, potent as are these effects, they are of secondary importance: what truly matters is that Romeo, *because he is unaware of what we know*, kills himself and so leaves Juliet with no choice but to follow him.

In studying the dramatist's use of awareness–unawareness gaps in the tragedies, then, I have had to be mindful always both of the incidental dramatic effects created by exploitation and of the

vii

relation of a protagonist's unawareness to the tragic outcome.

In treating the tragedies, I have found it useful also to shift my emphasis somewhat. Whereas in studying the comedies I centred on the discrepancy itself—i.e. the gap that the dramatist opens between the audience's and the participants' awareness of a situation—in the tragedies I have tended to emphasize the *means* by which the discrepancy is created. Most often, in both the comedies and the tragedies, it is created by the deceptive acts—the 'practices'—perpetrated by certain participants on others: Sir Toby's practice on Sir Andrew is designed to keep the gull's purse open by giving assurances that Olivia is thus to be won; Iago's practice on Roderigo is a deadly version of the same game. Since, by and large, and again just as much in the tragedies as in the comedies, it is the busy practisers who, through their practices, actually propel the action of any given play, it seemed appropriate to focus upon them and their devices first of all, and thus to stalk the unwary protagonists rather as the villains stalk them.

One result of this kind of stalking is that some of the tragedies, and especially their heroes, have come to appear markedly different from their traditional images. It has been no part of my intent to shock readers, but in fact I have sometimes been shocked myself by what has emerged here. In particular I had not foreseen that *Hamlet*, *Antony and Cleopatra*, and *Coriolanus*, and their respective heroes, would come out quite as they do in my analyses. Nor, for that matter, was I prepared to see *King Lear*, *Macbeth*, and *Othello* quite as they emerged as I proceeded. I was much better prepared for how *Romeo and Juliet* looks here because of an earlier study I had made of it ('The Brevity of Friar Laurence', *P.M.L.A.* Sept. 1950); my chapter in this book rests partly on that essay.

If, in short, any of my conclusions should cause the hair of my Shakespearian colleagues to rise on end like quills upon the fretful porpentine, I protest only that this effect formed no part of my intent.

As in *Shakespeare's Comedies*, I have here made no attempt to acknowledge debts, to the variety and extent of which there are truly no limits; every Shakespearian knows that no Shakespearian is an island but is a part of all that he has met, and perhaps we may let it stand at that. All references are to *The Complete Plays and Poems of William Shakespeare*, edited by Neilson and Hill, Houghton Mifflin Company, 1942. My citations of Holinshed and Plutarch are from Geoffrey Bullough's monumental *Narrative and Dramatic Sources of*

Shakespeare, Routledge and Kegan Paul; Columbia University Press, 1957-75.

To the Regents of the University of California I am indebted for allowances of time so generous that this book should have been completed years earlier.

Finally—and here again I echo remarks from my study of the comedies—I hope that this book may be received as a humble chapter in the endless search for the complex secrets of Shakespeare's incredible hold upon the mind that comes near enough to be seized.

Berkeley, California   B.E.

Table of Contents

CHAPTER I

For Practice' Sake: *Titus Andronicus*

'I'll find some cunning practice out of hand . . .'

I T is a truth universally acknowledged as unfortunate that a commentator who would study Shakespeare's tragedies chronologically is obliged to begin with *Titus Andronicus*, a play so little satisfying that a chain of critics from the seventeenth to the twentieth century would deny it a place in the canon despite the evidence of Meres and the First Folio.

But in part because of its very crudeness it offers a special advantage as starting-point for a study of the particular aspect of Shakespeare's dramatic method—his use of discrepant or unequal awarenesses between character and character and between character and audience—with which we shall be concerned here. For especially in this play the deceptive devices or 'practices' by which most of these gaps between awarenesses are brought about not only exist in great abundance but stand out nakedly as objects of primary interest, as if their mere exhibition justified their presence; hence they may be examined more easily than in later tragedies in which problems of character and a wealth of other dramatic elements make it more difficult to get at them.

Though here is not the place to debate questions of authorship, it seems appropriate to remark that later chapters will repeatedly evince the close similarities that exist in the dramatist's handling of practices and awarenesses in *Titus Andronicus* and the subsequent tragedies. Though indeed the management of awarenesses in this first attempt shares in the general crudeness of the play, yet even in their apprentice state the techniques and the very emphasis given them bear unmistakable likenesses to those of *Hamlet* and *Othello*. One notable difference does exist, however, which may best be identified by reference to Shakespeare's habit not in the tragedies but in the comedies. In the comedies the opposition of our awareness to the participants' unawareness results typically in incidental, passing effects that have little to do with determining outcomes, whereas in the later tragedies the unawareness of protagonists is an integral part of the movement towards catastrophe. Thus *Titus Andronicus*, in this single respect, more resembles the comedies than the later tragedies: the unawareness of participants is exploited for passing effects only and serves not at all in the determination of outcomes. The general slaughter that ends the tragedy is at best only remotely related to any of the many practices that have entertained the audience during five acts.

The relevant statistics of scenes and participants in *Titus Andronicus* form in themselves a persuasive argument not only for Shakespeare's authorship but for *Titus* as archetype of what was to follow in the management of awarenesses. In nine of the fourteen scenes the audience holds exploitable advantage over one or more of the participants; and of the twelve principal characters eleven are sometimes unaware of facts known to us. Aaron alone, the chief instigator and perpetrator of the practices that at some time victimize all other participants, never appears in a situation about which he knows less than we. It is noteworthy that in his first try at tragedy Shakespeare gives audience and arch-villain the advantage in awareness, leaving the 'good' persons to stumble about in varying degrees of darkness; with few exceptions (*Hamlet*, most notably), the same disposition holds in the later tragedies: the powers of darkness hold the light, and the disciples of light hold the dark. In the comedies, in contrast, we typically share the highest vantage point not with villains, but with disguised heroines, the proprietresses of all key secrets and the prime movers of main and subordinate actions alike.

Foreshadowing the pattern of later tragedies, in *Titus* it is the hero who first falls below our vantage point, and he does so even before the dramatist has supplied our awareness by some such device as soliloquy or aside. At first Titus is in as good position as we to see that Saturninus is unprincipled and vicious. Before Titus arrives on the scene, Saturninus says or does nothing to give us advantage; hence we gain nothing simply by being present while Titus is absent. The remarks that first show us what Saturninus is are spoken in Titus's hearing as well as ours:

> *Marc.* Titus, thou shalt obtain and ask the empery.
> *Sat.* Proud and ambitious tribune, canst thou tell?

2

Tit.   Patience, Prince Saturninus.
Sat.   Romans, do me right.
          Patricians, draw your swords, and sheathe them not
          Till Saturninus be Rome's Emperor.
          Andronicus, would thou were shipp'd to hell
          Rather than rob me of the people's hearts!
                                                (I. i. 201–7)

Though it advises us plainly enough, this sudden and dramatically unprepared display of Saturninus's villainous propensity advises Titus not at all—for, in perfect innocence, having won the unsolicited voices of the populace, he promptly bestows them upon Saturninus as 'your Emperor's eldest son'. Thereafter Saturninus's protestations of gratitude, together with his blatantly condescending choice of Titus's daughter as his Empress, cannot erase our abruptly gained insight: Saturninus is a thoroughly undeserving recipient of Titus's favour. Our impression is confirmed immediately when the dramatist tenders his first overt and private prompting of our awareness through Saturninus's 'aside' as he eyes the swarthy captive Tamora: 'A goodly lady, trust me, of the hue / That I would choose, were I to choose anew.' Turning to her, he then confides that 'he comforts you / Can make you greater than the Queen of Goths', and, as a stage direction informs us, 'courts Tamora in dumb show'.

Thus early we have gained a potently exploitable advantage over Titus, having learned that Saturninus is spiteful in nature and more disposed towards Tamora's charms than Lavinia's. When, therefore, Titus reacts to Bassianus's seizure of Lavinia as his own betrothed— a detail not earlier revealed to us—by summarily slaying his own son Mutius, who attempts to aid the abduction, and repudiates all his family for having compromised his honour, we recognize and deplore his haughty and obstinate blindness. Superficially, this blindness anticipates that of Marcus Brutus in the next tragedy but one; but Brutus's blindness is the effect of a deep-seated and well-marked trait of character, his soul's addiction to honour. Titus's blindness results from no such cause, but exists, together with that rashness which always prompts his actions, rather as a theatrical device than as a trait of character. We shall note often hereafter that much of what passes for dramatic effect throughout the play proceeds from just such abrupt, unprepared, unpredictable action— as when Aaron suddenly demands a severed hand and Titus as

3

suddenly obliges with one. Unlike Brutus, Titus is blind and rash merely because it is theatrically expedient for him to be so.

Near the end of the first scene is initiated what then seems certain to become the enveloping practice of the play. Just after Bassianus has seized Lavinia and Titus has slain his son for preventing her rescue, Saturninus bluntly denounces all the Andronici—'the Emperor needs her not, / Nor her, nor thee, nor any of thy stock'— and claims Tamora as his bride. Returning soon after—having briefly withdrawn with Tamora to 'consummate our spousal rites'— he continues to vent his rage upon his brother Bassianus and the Andronici for their abduction of Lavinia—though we know, certainly, from his earlier 'aside', that he is delighted with the turn of events that gave him Tamora rather than Lavinia; hence his seeming fury throughout this portion of the scene is itself an incidental practice on Bassianus and the Andronici, who never suspect the truth. And it is just here that Tamora initiates the first major practice of the play, urging Saturninus aside to 'Dissemble all your griefs and discontents' and assuring him that 'I'll find a day to massacre them all'. The first act thus closes with all the Andronici ignorant of the peril in which they stand, and with Titus especially, elated at his apparent return to favour, jubilantly inviting Saturninus and his new Empress 'To hunt the panther and the hart with me' on the morrow.

But though the end of Act I promises that Tamora's practice on the Andronici, with Saturninus serving as a kind of silent partner, will direct the course of subsequent events and constitute the main action of the play, Aaron's soliloquy that opens Act II abruptly redirects the plot and obliges us to entertain different expectations. Present on stage during most of Act I, his dark hue making him conspicuous first among the crowd of prisoners and next in the company of the Emperor and his bride, Aaron has spoken never a word, and his role, other than as it serves the purpose of spectacle—a very considerable purpose in *Titus Andronicus*—has remained unpredictable. But then, after a prelude studded with imagery as spectacular as his own shining skin, he announces the second major practice of the play—one that quickly reduces Tamora's, announced only moments before, to relative insignificance:

> Then, Aaron, arm thy heart and fit thy thoughts
> To mount aloft with thy imperial mistress,
> And mount her pitch, whom thou in triumph long
> Hast prisoner held, fett'red in amorous chains

4

And faster bound to Aaron's charming eyes
Than is Prometheus tied to Caucasus.

(II. i. 12–17)

'This goddess, this Semiramis, this nymph', this Tamora, we learn, belongs body and soul to Aaron, not Saturninus; and she 'will charm Rome's Saturnine / And see his shipwreck and his commonweal's'. With these remarks Aaron suddenly replaces Tamora as the prime mover of action to come; and, with the same stroke, Saturninus replaces Titus as the primary victim of intended harm: in the abruptly shifted perspective, the destruction of the Andronici is to be but incidental to the ruin of Saturninus.

However (even so abrupt are the shifts that occur throughout the play), we have barely readjusted our vision to accommodate the announcement of Aaron's all-encompassing new practice when the direction changes again. The instigators, though not the devisers, of the next practice are Chiron and Demetrius, Tamora's wayward sons, who, being smitten by lust for Lavinia as sudden as was Saturnine's for their mother, voice their readiness to endure a thousand deaths to achieve her. Lending his masterly touch to the cause of these mere apprentices, Aaron dictates that the 'dainty doe' shall be enticed during the hunt to where 'The woods are ruthless, dreadful, deaf, and dull'—and there the boys, by turns, shall 'revel in Lavinia's treasury'.

The summary announcement, in quick succession, of a third major practice leaves large questions without clear answers: Just what relation has the latest practice to those previously announced— Tamora's on the Andronici and Aaron's on Saturninus? Though Tamora, not yet advised of the plot against Lavinia, will surely approve it as a means of injuring the hated Andronici, why should Aaron, whose stated intent is to destroy Saturninus 'and his commonweal', seize on a chance to strike at Titus, who, if apprised of the facts, might prove his best ally against Saturninus? Though the new twist returns the drama to its theme of revenge on the Andronici, it does so as it were by accident: the initial motive for revenge on Titus was Tamora's, yet Tamora has had no part in the new device against Lavinia and as yet knows nothing of it; the device itself is invented by Aaron, though Aaron's announced target is Saturninus, not Titus; and the immediate impetus is given by the boys' lust for Lavinia, the source of which is unrelated to either Tamora's grudge against Titus or Aaron's against Saturninus. We are presumably to imagine that

the two enveloping practices, Tamora's and Aaron's, are still somehow operative and only await implementation; meanwhile we are to
be entertained by whatever passing chance may provide opportunity
for sensationalism.

In II. ii, in quick succession, Shakespeare exploits the various
unawarenesses of virtually his entire cast, who enter by ones and
twos for the hunt. First comes the unwitting, exuberant Titus,
crying 'The hunt is up, the morn is bright and grey, / The fields are
fragrant and the woods are green'—adjectives that clash noisily in our
minds against those we have just heard Aaron use in instructing
Tamora's boys how to surprise Lavinia: 'The woods are ruthless,
dreadful, deaf, and dull'. Titus is ignorant both of Tamora's threat
that she will find a day 'to massacre them all' and of the imminent
ambush that awaits his daughter. Next enter all the royalty:
Saturninus, who presumably remembers Tamora's promise to get
revenge on the Andronici but is ignorant of her private liaison with
her lover Aaron and of Aaron's promise to destroy him; Tamora,
who is aware of both the major secrets, being chief proprietor of that
which involves Titus and equal partner in that which involves
Saturninus, but who remains ignorant of her sons' and Aaron's plot
to rape Lavinia (a detail, however, of potential delight, not grief, to
the Empress); Bassianus and Lavinia, ignorant of the plots against
Saturninus and Titus, and, more pertinently, of the ambush
designed for Lavinia; and, finally, the boys, Chiron and Demetrius,
keenly aware of their plot against Lavinia—'. . . we hunt not, we,
with horse nor hound, / But hope to pluck a dainty doe to ground'—
but ignorant of all else, including their mother's private liaison with
Aaron (a detail that, however, we should hardly expect to shock them
if they knew of it). Brief and non-violent, but bloated with impending
danger, the scene is dramatically superior to more typical scenes in
the play where violence merely erupts without warning and for its
own sake.

But for skill in handling multiple plays and crossplays of awareness
it is the next scene (II. ii) that stands foremost in the tragedy even
despite its gross offences against the staunchest dramatic stomach.
The scene begins startlingly, inexplicably, with Aaron burying gold
that 'must coin a stratagem'. What stratagem, we are not told,
though we are promised that it 'will beget / A very excellent piece of
villany'; doubtless the later Shakespeare would have made Aaron
explicate his intent at this point, even as he makes Iago label his

6

devices for us in advance. Tamora enters, unaware of Aaron's immediate purposes, and attempts, with voluptuous language as gross
in its way as are the many physical atrocities of the play in theirs, to
entice her lover into sexual games— '. . . each wreathed in the other's
arms, / Our pastimes done'—but to no avail: 'Vengeance is in my
heart, death in my hand', Aaron retorts, and 'Blood and revenge are
hammering in my head'. He then tells that her sons will kill
Bassianus and 'make pillage' of Lavinia's chastity, and forces a letter
on her: '. . . give the King this fatal-plotted scroll'. Instructing her
to 'Be cross' with Bassianus, just entering with Lavinia, he runs off to
bring Chiron and Demetrius 'To back thy quarrels, whatsoe'er they
be'. Chancing upon Aaron and Tamora, Bassianus and Lavinia have
discovered the secret of her unfaithfulness to Saturninus, which
Bassianus threatens to disclose to his brother; but death at the hands
of Tamora's sons summarily blocks his purpose. At the very end of
the play Saturninus goes to his own death never having learned of
his wife's infidelity.

Practices and action are inseparable during most of this scene: in
being 'cross' with Bassianus and Lavinia, as Aaron had ordered,
Tamora practises upon these two; when her sons arrive, she
practises not only on Bassianus and Lavinia but simultaneously on
Chiron and Demetrius in telling them monstrous lies:

> They told me, here, at dead time of the night,
> A thousand fiends, a thousand hissing snakes,
> Ten thousand swelling toads, as many urchins,
> Would make such fearful and confused cries
> As any mortal body hearing it
> Should straight fall mad or else die suddenly.
>
> (II. iii. 99–104)

They threatened to bind her, she cries, 'Unto the body of a dismal
yew', and, what is worse, had called her 'foul adulteress, / Lascivious
Goth'. Responding quickly to this supposed insult to their mother,
the boys stab Bassianus. They are doubly unaware in doing so:
unaware that in fact Bassianus and Lavinia had made no such fierce
threats as the mother claims, and unaware that the epithets 'foul
adulteress' and 'Lascivious Goth' are quite true. Like Saturninus, the
boys will eventually go to their own deaths ignorant of Tamora's
private games with Aaron; but here, with disarming simplicity and
candour, they prevent their mother's prompt murder of Lavinia by
asserting the obvious: she should first be raped.

7

As the scene lurches to its next complications, the inexperience of the dramatist in managing multiple practices and exploiting the unawareness of numerous persons simultaneously becomes increasingly evident. So far as we had been informed, Aaron's scheme encompassed only the murder of Bassianus and the rape of Lavinia; but also, mysteriously, he had buried a bag of gold and presented Tamora with a 'fatal-plotted scroll' to be given eventually to Saturninus. Left unexplained at the time of their occurrence, these details begin to make sense only when Aaron enters with two of Titus's sons and notifies us of his intent to have them found trapped in the pit with Bassianus's body. 'My sight is very dull', remarks one son, falling into the pit to which Aaron has guided him; and, while Aaron runs off, the other son, too, tumbles in. There they remain, ignorant that the very villain who had enticed them into their dire plight had also contrived the murder of Bassianus, whose body shares the pit, and had also helped Tamora's sons devise Lavinia's ambush. Only thereafter, belatedly, does Aaron inform us fully about his practice: 'Now will I fetch the King to find them here, / That he thereby may have a likely guess / How these were they that made away his brother.' Guided to the pit by the busy Moor, and easily persuaded by the evidence of the 'fatal-plotted scroll' given him by Tamora and the bag of gold conveniently retrieved by Aaron from beneath the elder-tree, Saturninus—who has now become enwrapped with more layers of ignorance than Titus himself—orders Titus's sons confined in prison until 'we have devis'd / Some never-heard-of torturing pain for them'.

Here, then, just before the end of Act II, the practices of Aaron and Tamora have generated as great a volume of dramatic force as the play is to achieve—dramatic force, that is to say, as distinct from the mere exhibition of horrors, of which the most shocking are yet to come. Mainly this force derives from the difference between the participants' and our awareness of crucial facts: the private relations of Tamora and Aaron, their arrangement of Bassianus's murder so that Titus's sons will appear guilty, and their deception of Saturninus both as to their relations and as to the murder. Action moves with startling rapidity from the moment Aaron guides the sons to where they fall into the pit, through the moment of their discovery of the body, on through the entrance of the main group of persons, the presentation of the scroll, the 'discovery' of the gold by Aaron, the accusation of Titus's sons, the raging of Saturninus, and the pleas of

Titus. Yet throughout the mad rush of these spectacular incidents on the stage, perhaps what adds most power to the scene is not what occurs before our eyes but our awareness of what, during these very moments, is taking place elsewhere—the rape of Lavinia by the same villains who killed Bassianus. The concluding scene of Act II exhibits a ravished and mutilated Lavinia and features the long and anguished—yet elaborately allusion-ornamented—speech of her uncle Marcus, who finds her abandoned by her persecutors; but though the scene affords a sight of unprecedented horror, it seems dramatically less forceful than the preceding scene of multiple practices and exploitable gaps between awarenesses.

It is Lavinia's tongueless and handless predicament that creates the next such gaps. We share with Tamora, Aaron, and Tamora's sons, and, more poignantly, with the victim herself, knowledge of where the guilt lies; the remaining Andronici are ignorant of the truth. Though he drags out the secret until Act IV, Shakespeare actually exploits it very little for dramatic effect, giving instead nearly the whole third act to the shocking sounds of rage and grief among the family of Titus: Titus hurls himself before the procession of judges, senators, and tribunes on their way to execute his sons; Lucius, banished, sword waving, joins his father in lamentation; Marcus leads and exhibits the mutilated Lavinia; and in all some one hundred and fifty lines are spent upon the family's sorrow and anger. But clearly this prolonged, outright exhibition rather loses than gains force by its very excesses. As Titus himself insists, his grief had attained its zenith before he saw the maimed Lavinia—'My grief was at the height before thou cam'st'—and though thereafter he attempts new heights of lamentation, the very directness of his renewed outpouring defeats its effect. Soon afterwards, Shakespeare learned to convey the experience of profound grief by means of oblique, not head-on, expression: 'O thou untaught,' mumbles old Montague at sight of dead Romeo in the tomb of the Capulets, 'what manners is in this, / To press before thy father to a grave?'

Unprepared as it is, Aaron's next improvised practice on Titus comes as a welcome respite from the extended and yet essentially undramatic outpouring of grief: '. . . chop off your hand / And send it to the King; he for the same / Will send thee hither both thy sons alive'. This spur-of-the-moment practice is as wretchedly motivated as any to be found in all Shakespeare: (1) We do not know, and will never learn, the basis of Aaron's grudge against Titus; for his sole purpose, earlier announced, was to ruin Saturninus; (2) We are not advised, until after Titus has obliged by cutting off a hand, that it is merely Aaron's impromptu practice; (3) We are not advised, and are never to learn, whether Saturninus had any part in the deception or ever learns of it. (A minor and similarly incidental practice occurs within the frame of Aaron's practice when Titus tricks Marcus and Lucius into withdrawing so that he can oblige Aaron by amputating his own hand before either can contribute his.) But, grossly contrived as it is, the incident serves to rescue the scene from the tedium of incessant lamentation already stretched too far. Our respite, however, is brief: after the sons' heads and Titus's hand are presented Titus by an ambiguous messenger—Saturninus's, or only Aaron's?—the outcries resume with redoubled volume. Thus Marcus itemizes his brother's griefs:

> . . . die, Andronicus.
> Thou dost not slumber; see, thy two sons' heads,
> Thy warlike hand, thy mangled daughter here,
> Thy other banish'd son with this dear sight
> Struck pale and bloodless; and thy brother, I,
> Even like a stony image, cold and numb.
>
> (III. i. 254–9)

Yet all the howls together, which continue without further respite for us to the end of the scene, plus the horrendous spectacle itself— 'Come, brother, take a head; / And in this hand the other will I bear. / Lavinia . . . bear thou my hand, sweet wench, between thy teeth'—rather lose than gain in force by their excess; if a little more than a little is by much too much, then a great deal more than a little is intolerable. For aside from its excess the scene has nothing dramatic to offer: practice for the sake of practice only (Aaron's tricking Titus out of his hand, only to return it with the heads of the sons it was to save) is only momentarily diverting, not dramatic. What is perhaps most damaging to the total effect is the dramatist's mismanagement of the relation of our awareness to Titus's. At the end of the scene we remain uninformed as to what part, if any, Saturninus played in the woes that now afflict the protagonist. The Emperor certainly had no part in the rape and mayhem committed on Lavinia; he had no reason to question the evidence supplied by Aaron that Titus's sons murdered Bassianus, and therefore the beheading of the sons must be taken as execution, not persecution. The dramatist never hints of collusion among Saturninus, Tamora, and Aaron; we know only that Saturninus, like Titus, is a victim of the Moors' treachery. Yet the wails of the Andronici against Saturninus and Rome dominate the middle portion of the drama. Titus orders Lucius to join the Goths and raise an army, and Lucius's ending line expresses determination 'To be reveng'd on Rome and Saturnine'. But for what? Neither Saturninus nor Rome even knows about the injuries inflicted on the Andronici by Tamora and Aaron.

The final scene of Act III merely continues the misdirected lamentation of the Andronici. Curiously, however, near the end of this scene, when Marcus has struck at a fly on his plate and been scolded for useless killing by the half-mad Titus, he likens the fly to 'the Empress' Moor', whereupon Titus, too, strikes at the body: 'There's for thyself, and that's for Tamora.' Here Titus's rage is directed at the proper targets, Tamora and Aaron, the authors of all his woes; yet the difficulty is that Titus could not have learned, between the preceding scene and this, that these, not Saturninus, are his persecutors. Though Lavinia and Bassianus had indeed stumbled upon the truth that Aaron is 'the Empress' Moor', yet Bassianus was slain immediately afterwards and Lavinia's tongue was severed before she could reveal any secrets. Thus, though here the dramatist has corrected one fault, he has made a worse: formerly, we could hardly sympathize with Titus's rage against Saturninus, knowing that it was misdirected; now, when it is directed at the right target, sympathy is blocked by our realization that Titus cannot possibly know the facts against which he complains.

Indeed, Titus does not learn until Act IV, when Lavinia reveals the identities of her attackers, that any injuries have been done his family by Tamora and her sons. Apparently he had not even entertained suspicion of their guilt, and is surprised, like Marcus, when it becomes known: 'What! what! The lustful sons of Tamora / Performers of this heinous, bloody act?' But next, inexplicably, he speaks as though he knows of Tamora's relations with Aaron and their mutual deception of Saturninus:

> She's with the lion deeply still in league,
> And lulls him whilst she playeth on her back,
> And when he sleeps will she do what she list.
> (IV. i. 98–100)

In any event, his discovery of the sons' guilt serves briefly to divert his mis-spent rage from Saturninus to Tamora's family, and now, for the first time, having taken a step up the ladder of awarenesses, he promptly sets a trap to catch his persecutors. In IV. ii Chiron and Demetrius receive 'the goodliest weapons' of Titus's armoury as gifts wrapped in a scroll, all the while remaining ignorant—for it is briefly villainy's turn to be unaware—that Titus has learned of their guilt. Aaron, however, reads the facts aright: 'Here's no sound jest! The old man hath found their guilt; / And sends them weapons wrapp'd about with lines / That wound, beyond their feeling, to the quick.' But then, again inexplicably, he decides to keep secret the fact that Titus knows the sons' guilt; even Tamora is not to be told: '. . . let her rest in her unrest a while'. Meanwhile the sons gloat, in their ignorance, over their supposed deception of Titus; Demetrius finds it agreeable 'to see so great a lord / Basely insinuate and send us gifts', and Chiron wishes for 'a thousand Roman dames / At such a bay, by turns to serve our lust'. Brief as it is, the dramatist's exploitation here of their ignorance that Titus knows their guilt is productive of satisfying dramatic effect.

With the arrival of the Nurse bearing 'a blackamoor Child', the secret that Tamora's sons are the last—except for her husband—to learn now comes to light. Predictably, the sons' indignation on learning of their mother's relations with Aaron is short-lived and mainly reflects their consent that Tamora may thereby lose her—and their own—royal status. When, therefore—after casually stabbing the Nurse, who might otherwise betray them all—Aaron devises yet another practice by which the Emperor will be made to accept a Goth's white child in place of the black one, the boys are not only appeased but ecstatically grateful: 'For this care of Tamora / Herself and hers are highly bound to thee.' During all this while, of course, the grateful sons remain unaware that Aaron is withholding from them his certainty that Titus knows their guilt. But that the scene is overwrought with frightful language—'I'll broach the tadpole on my rapier's point'—and sudden, casual murder, and marred by Aaron's unexplained decision to keep the boys in the dark, its combination of practices and abundance of exploitable gaps would make it one of the more effective scenes in the tragedy.

It is superior, in any event, to the next scene, which features Titus's absurd shooting of arrows into the sky and offers little but a spectacle of madness exhibited for its own sake. It is perhaps the greatest fault of *Titus Andronicus* that the largest part of the protagonist's very considerable expenditure of rage and grief is wholly misdirected. Instead of the shooting scene, we should have expected to find Titus still howling for revenge on Chiron and Demetrius for having mutilated Lavinia, and on Aaron for having cheated him of a hand. But he continues to cry for revenge on Saturninus and on Rome in general:

> Ah, Rome! Well, well; I made thee miserable
> What time I threw the people's suffrages
> On him that thus doth tyrannize o'er me.
>
> (IV. iii. 18–20)

Yet what injuries has Saturninus done him? We have not seen the Emperor for two acts and have heard of him only in Titus's ravings and in connection with the scheming of Aaron and Tamora. We must assume that he knows nothing of Lavinia's mutilation by Tamora's sons and that he had no part in Aaron's vicious practice in getting Titus to cut off his hand. Further, at this point, we have no cause to suppose otherwise than that, in executing Titus's sons for the murder of Bassianus, he believed that he was administering simple justice, for all the evidence, manufactured by Aaron, pointed to their guilt. A degree of dramatic integrity might be salvaged here had we been made to understand that Titus, in his madness, *mistakenly* supposes Saturninus responsible for the injuries done him; but in fact we know that he knows of Tamora's sons' guilt with Lavinia and of Aaron's trick in robbing him of his hand. If he believes Saturninus guilty of deliberately executing the young Andronici while aware of their innocence in Bassianus's murder, he never mentions the fact; indeed, in all his fulminations against Rome and the Emperor he mentions no specific grievances, but merely blames 'Rome' for all ills. Yet 'Rome' had tried to make him Emperor, has done him no injuries since, and presumably is ignorant that he has been injured by anyone. All his raving, therefore, is mere misdirected rant, and we can scarcely become involved sympathetically when he has neither been wronged by those he accuses nor even believes mistakenly that those he rails against have wronged him.

That Saturninus remains ignorant of Titus's sons' innocence is proved by two speeches in the last scene of Act IV. When arrows shot by Titus with messages directed variously to Jove, Mercury, and Apollo are picked up by the Emperor, he remarks that however wildly the Andronici 'Buzz in the people's ears', he has in fact done nothing 'But even with law against the wilful sons / of old Andronicus'. And again he shows his innocence when the Clown brings in

yet another scroll: 'May this be borne? As if his traitorous sons, /
That died by law for murder of our brother, / Have by my means
been butcher'd wrongfully!' Though Titus's ravings that constitute
the climactic portions of the play appear designed to claim our
sympathy for the protagonist and simultaneously stir our hatred of
Saturninus, the dramatist has failed to convince us that Saturninus
even deserves the title of villain. Aside from his ill-showing in the
opening scene, he has committed no injustices in his own person and
appears unaware of those committed by Tamora and Aaron. Our
lone hint that he might once have become involved in a plot against
the Andronici was given long ago, in Act I, with Tamora's 'aside' to
the effect that if he would but dissemble and seem to forgive Titus,
she would herself 'find a day to massacre them all'. But during subsequent scenes we have heard nothing of this threat, which seemed
at the time intended as the enveloping practice of the play but was
then obscured by Aaron's confidential assertion that, with Tamora,
he would destroy both Saturninus and Rome. Yet this latter threat
was then, too, rendered meaningless by the inexplicable attacks of
Aaron not upon the Emperor but upon the Andronici, and perpetrated even without Tamora's knowledge or complicity.

Near the end of Act IV, after the Emperor's angry but justifiable
outburst over the affair of the arrows and Titus's rabble-rousing
declarations against Rome, Tamora first feigns to soothe him, as she
had done in Act I, then speaks aside:

> Why, thus it shall become
> High-witted Tamora to gloze with all;
> But, Titus, I have touch'd thee to the quick.
> Thy life-blood out, if Aaron now be wise,
> Then is all safe, the anchor in the port.
>
> (IV. iii i. 34–8)

The speech is ominous enough in its sound, but ambiguous, if not
utterly meaningless, in its sense. Wherein has she touched Titus 'to
the quick'? She has just now soothed Saturninus's rage against him
and has told us of no plan of attack on him that would explain her
claim that Titus's life-blood will shortly be out. Her 'if Aaron now
be wise' implies some comprehensive practice devised between the
pair; but when was it worked out? No prior conversations that we
have overheard hinted at such a plot; further, we know that at the
moment Aaron's sole concern is for their black baby (about the fate
of whom, and, incredibly, even the birth of whom, the mother

appears totally ignorant). And, finally, against whom is the threatened plot? It can hardly be aimed at Titus, for the last clause of the sentence—'if Aaron now be wise'—is contingent upon the prior condition, that Titus's blood will already be out. And if the plot is against Saturninus, as may be implied by the final 'the anchor in the port', then we have truly been left unadvised, for no previous conferences between Tamora and Aaron have even mentioned Saturninus.

Act V begins, thus, with a plethora of practices, some clear, some muddy, ready to be activated. First, Titus has sent gifts to Tamora's sons, together with a note in Latin that they do not understand; but Aaron perceives that Titus has found the sons' guilt and intends to strike back (yet, curiously, as we have noted, Aaron, in a negative kind of practice, has decided not to tell either the boys or Tamora that Titus knows). Second, Aaron, again curiously, without conferring with Tamora or expressing intent to inform her, has devised a scheme for passing off a Goth's child as Tamora's in place of his own black baby. Third, 'High-witted Tamora' has boasted that she has touched Titus to the quick and that if Aaron will but be wise, then 'is all safe, the anchor in the port': but from this practice, or two-pronged device, we know not what to expect. And, finally, Tamora has promised Saturninus that she will 'so enchant the old Andronicus' that he will call home his son Lucius, who has gone to bring an army of Goths against Rome. At the end of Act IV we heard Tamora promise Saturninus—by all odds the least aware of the participants, having no practice of his own and being ignorant of all others—to 'be blithe again, / And bury all thy fear in my devices'.

It is thus with an awesome backlog of dramatic promises, at least in quantitative terms, that Act V is set to begin. But the difficulty is that all the pending practices are misguided or ambiguous, so that we cannot really be sure just what is to come. And indeed it turns out that Act V, from the point of view of the management of the awarenesses and the fulfilment of dramatic promises, is the most garbled and unsatisfactory portion of the tragedy. In the first scene a Goth brings Aaron, with his baby, to Lucius; and the Moor, quite gratuitously, recounts in lurid detail the crimes he has committed or caused others to commit: the begetting of the black baby, the murder of Bassianus, the rape of Lavinia, the entrapment of Titus's sons in the pit with dead Bassianus, the deception by which Titus was robbed of his hand. To this recital Lucius reacts with great consternation—though in fact most of the details were evidently known to him beforehand; for even before Aaron begins to speak Lucius calls him 'the incarnate devil / That robb'd Andronicus of his good hand' and refers to the baby as 'the base fruit' of the Empress's lust and the 'growing image' of Aaron's own 'fiend-like face'. Further, as we know, Lucius saw Lavinia after her rape, though indeed he left Rome before the names of her assailants were divulged. But, in short, the entire first scene of the final act serves no useful purpose but to make occasion for Aaron's horrendous account of dire deeds already done. Perhaps no incident in the play more amply demonstrates that the horrors of *Titus Andronicus* exist only for the sake of horror; and to confirm the point, Aaron, having finished his account of injuries done the Andronici, continues with a similarly lurid survey of his past career in general, ending only at Lucius's command to 'stop his mouth, and let him speak no more'. But Shakespeare has let him speak more than enough.

The second scene of Act V opens with a game of charades like nothing else to be found in Shakespeare. Launching a new practice, Tamora presents herself to Titus as 'Revenge', and Titus soon supplies the titles of 'Rape' and 'Murder' for her companions, Chiron and Demetrius. During this period the status of our own awareness is uncertain: who is deceiving whom? The first round of the game, Tamora mistakenly imagines, is hers:

> For now he firmly takes me for Revenge;
> And, being credulous in this mad thought,
> I'll make him send for Lucius his son;
> And, whilst I at a banquet hold him sure,
> I'll find some cunning practice out of hand
> To scatter and disperse the giddy Goths.
>
> (v. ii. 73–8)

When she urges Titus to send for Lucius and then invite 'the Empress and her sons', with the Emperor and 'all thy foes', to a banquet where 'at thy mercy shall they stoop and kneel', Titus appears to be taken in and promptly dispatches Marcus to bid Lucius to return. So sure of the success of her practice is Tamora that she agrees to let 'Rape' and 'Murder' remain as Titus's guests 'Whiles I go tell the Emperor / How I have govern'd our determin'd jest'. It is not until this point—very late by Shakespeare's typical schedule—that the muddled state of our own awareness is clarified: 'I knew them all though they suppos'd me mad,' Titus tells us aside, 'And will o'erreach them in their own devices.'

But Titus's revelation immediately occasions other doubts about the accuracy of our understanding. He is not mad at all, he blandly assures us—and then proceeds to prove his assertion by effecting a masterful seizure of Tamora's sons, reciting in cogent detail the catalogue of wrongs done him, describing in vivid detail the menu for the pending banquet, and finally, adroitly, cutting the throats of the victims who are to serve as *pièces de résistance*. Indeed, not only is he no longer mad—he appears also to be no longer stupid. From the first scene of the play until this nearly final one, Titus's head has appeared notably uninhabited, and the record of actions in which he has been involved reveals rather a succession of personal blunders than of victimization at the hands of super-subtle practisers; for though Aaron and Tamora have been ubiquitously active, it is not so much their devices as Titus's own deficiencies that account for his persistent condition of unawareness—which seems rather chronic than acute, general than local, congenital than communicated. He grandly but stupidly gave the title of Emperor to Saturninus when all Rome had chosen himself, nobly but stupidly slew his own son when Bassianus claimed Lavinia as his betrothed, eagerly but stupidly lopped off and surrendered his hand to a mere prankster who asked for it, raged at Saturninus as author of all his woes when in fact Saturninus was ignorant of any woes done him, madly shot arrows into the Emperor's courtyard with messages demanding justice of the gods, and finally dispatched his sole remaining son to bring an army of Goths to avenge him on Rome when not Rome but only Aaron and Tamora's sons have wronged him. Earlier scenes, thus, have hardly prepared us for the transformation when, in the immediate scene, Titus shrewdly penetrates 'High-witted' Tamora's disguise, evinces rare acumen in divining her purposes, subtly manœuvres her into leaving her boys to be dealt with at his pleasure, and generally demonstrates command of himself and the whole, murky situation. In short order he has sprung from the lowest to the topmost rung on the ladder of awareness, while, hardly less strangely, the clever practisers Tamora and Aaron, formerly at the top, have plummeted to the bottom. Incredibly, it is thus Titus at last who usurps the Moor's title of chief practiser in the play, and who designs and executes the practice that disposes not only of his immediate enemies but of everyone else within reach.

Titus's practice in the last two scenes is directed, as his wrath should always have been, primarily at Tamora and her faction. The banquet itself is prepared for her special benefit: 'This is the feast that I have bid her to, / And this the banquet she shall surfeit on.' Though he, too, is invited, Saturninus is not mentioned in the revenge that Titus designs; yet as late as Act IV, though he then knew who had mutilated his daughter and robbed him of a hand, Titus's fury was still directed at Saturninus and Rome in general. The shift in the direction of his anger is marked by no explicit acknowledgment that he had erred in blaming Saturninus; indeed, it is not marked at all, but simply occurs. Yet if our record of shifting awarenesses is correct, Titus should have no truer sense in the final scenes of who his enemies have been than he had in Act IV, when he still cursed Saturninus. True, in the first scene of Act V Aaron related to Lucius the various crimes committed by himself and Tamora's sons against the Andronici; but this recital cannot explain Titus's change, for, first, it presented little not already known to him and, second, Lucius, being with the Goths all the while, has had no opportunity to communicate additional facts to his father. The shift must therefore be regarded as mere dramatic expedience. Either the climactic middle scenes when Titus roared for revenge on Saturninus and Rome are structurally out of line, or those scenes are correct and the final scenes wherein he plots and executes vengeance on Tamora are at fault.

It would redeem much if we could but credit the apparent flaws of the play to the character of its hero as one who, like Othello, misjudged his enemies and recognized the truth too late. But it is as useless to probe the character of Titus as to probe the characters of Tamora and Aaron, for there is nothing, really, to probe. Like the practices that abound throughout the play, they are only devices serving the needs of plot and the ends of shock and horror. Thus Tamora functions, at the start, to enunciate dire threats of revenge on the Andronici, and, later, to hint darkly at even more encompassing deeds, with Aaron as accomplice, against Saturninus and Rome; but these threats, after serving the immediate purpose of effect, come to nothing more. She is not, in fact, involved directly in any of the harms that come to the Andronici or to Saturninus. Aaron, on the other hand, performs busily enough as 'Chief architect and plotter' of the woes visited on the Andronici; but his practices are those of a lone and unpredictable entrepreneur, undertaken for their own sake only and without regard to Tamora's more ambitious ends. Because of the excessively sentimental concern he exhibits for his child's survival, Aaron is sometimes singled out by critics as fore-shadowing Shakespeare's mature skill in deep characterization; but the black baby in fact merely represents one more object in the play's range of bizarre exhibits, and Aaron's concern is but another manifestation of the dramatist's own concern to surprise us. As for Titus himself, he serves to exhibit extremes of madness, uncontrolled grief, and thirst for revenge; his character, if it may be so called, twists and turns at the dramatist's whim and with the demands of the plot. Only so is it possible to reconcile the contradictory facts of his brilliant performance during the final two scenes with his earlier record of blindness and stupidity.

These structural flaws show most conspicuously in the final action, even though within itself that action, when we hold perhaps our most potent advantage in all the play—our awareness of the ingredients that comprise Titus's main dish—offers sights and sounds enough to glut the most capacious stomach. It is this final portion that lines up least well with the middle of the play, when Titus railed against Rome and the Emperor. What we should have expected here is a dénouement in which the protagonist meets death in his final lunge against Saturninus. But, as we have noted, the Titus of the last moments is not the erring Titus of the rest of the play. The new Titus recognizes Tamora as his enemy and, though he invites Saturninus to the feast, expresses no resentment towards him. He is right in not doing so, of course, since Saturninus has never done him any harm; but, then, neither does he apologize to the Emperor for the curses, the arrows, and the rabble-rousing speeches that preceded. As for Saturninus himself, nearly the last words that he speaks before he dies reveal what we could only surmise earlier, that he did not even know of the wrongs done by others to Titus. When Titus has slain Lavinia and likened himself grandiloquently to Virginius, the Emperor cries, 'What, was she ravish'd? Tell who did the deed.' When Titus names Tamora's sons, the Emperor orders them brought forward, obviously intending to punish them even as he had earlier executed Titus's sons when their guilt in Bassianus's death seemed certain. If the justice administered by Saturninus tends to be abrupt and final, yet we have every indication of his intent to be even-handed. He strikes Titus down at last, not from mere tyrannical malice but because Titus has just slain Tamora, and we must assume that if he were not still ignorant of the wrongs done by Tamora he would have rewarded rather than slain Titus. Lucius then kills Saturninus—but not, indeed, for any real or imagined past wrongs done the Andronici but only because the Emperor has just slain Titus: 'Can the son's eye behold his father bleed? / There's meed for meed, death for a deadly deed.'

Titus has his revenge, then, at last, but not the revenge he howled for during the middle scenes, not the revenge for which he dispatched Lucius to bring the Goths against Rome. As for the Emperor, he dies unaware both of the wrongs done him by Tamora and Aaron and of that he has himself done in slaying Titus for slaying Tamora. He dies, thus, not at all the tyrant who was denounced by Titus earlier, and not the villain that, indeed, the opening scene led us to expect. Tamora dies an instant too soon to witness the deaths of Saturninus and Titus, both of whom she had expressed intent to destroy; but there is no connection between her initial threat and their deaths. Aaron, the 'breeder of these dire events', whose wicked deeds—even after we have been obliged to witness them in the doing —are thrice recounted in ugly detail, will shortly be put to an even more brutal death. As 'Chief architect and plotter of these woes', he alone has directed or committed all the crimes against the Andronici —but, unfortunately for the integrity of the tragedy, there is no sound reason for his having committed any of them. His practices have generated virtually all the sensational effects of an otherwise essentially actionless and dull drama; and yet these practices bear no integral relationship to the ostensible theme of the tragedy, which is the fall of the hero Titus Andronicus at the hands of an ungrateful Rome.

With the others dead, Lucius inherits everything at last, as Fortinbras does at the end of *Hamlet*. But in the later tragedies, *Hamlet* included, the ordeal of blood through which the survivors have passed is instrumental in bringing about a more healthful state governed more ably and compassionately than before. Expectations of a benevolent Rome under Lucius, however, cannot be high: it was he who initially aroused Tamora's wrath by demanding her son 'That we may hew his limbs and on a pile . . . / sacrifice his flesh'; he who later reports that 'Alarbus' limbs are lopp'd, / And entrails feed the sacrificing fire'; he who orders Aaron's baby slain: 'First hang the child, that he may see it sprawl'; and he, finally, who orders Aaron set breast-deep in earth and starved. Rome, thus, will hold to the same brute level as before—that exhibited when Titus slew his son Mutius for standing in his way, when Tamora's sons ravished

Lavinia, when Aaron stole Titus's hand, when Titus cut the throats of Tamora's sons and Lavinia, with her stumps, held a basin to catch the blood.

Presumably Shakespeare's first tragedy was meant to exploit Rome's historic reputation as a city of blood and ruthless might; hence, merely because he is Rome's Emperor, Saturninus is to be identified automatically with tyranny and oppression, and Titus, as an out-of-favour hero, is to be recognized as the wronged and idealistic rebel at war with the injustices associated with the very name of Rome. Titus's ravings during most of II, III, and IV partly carry out this intent. But the great fault of the tragedy is that in the actual translation of the theme into action, all the 'ruthless might' of oppressive Rome is reduced to the gratuitous practices of one wicked wretch who is not even a Roman, but a Moor, and all of whose fierceness dissolves in his bizarre concern for a spectacular but essentially irrelevant black baby. It is, at last, only by this rabid alien, not Rome, that Titus has been tormented and against whom Lucius has brought an army of Goths—and of whom neither the Emperor nor Rome knew anything at all.

CHAPTER II

Fate as Practiser: *Romeo and Juliet*

'Some consequence yet hanging in the stars
Shall bitterly begin his fearful date.'

UNLESS one consciously recalls that several comedies—*Love's Labour's Lost*, *The Two Gentlemen of Verona*, *The Taming of the Shrew*, and (perhaps) *A Midsummer Night's Dream*—separate them in time, one finds the tragic progression between *Titus Andronicus* and *Romeo and Juliet* unbelievable. Examined chronologically, as though one followed on the heels of the other, they appear too far apart to permit any sort of bridging: even Shakespeare could not have moved directly from the one to the other.

With the full poetic and dramatic growth that occurred between them we are not concerned here; it must suffice to state the obvious —that in all aspects of his art the dramatist's development during the intervening three or four years was phenomenal. What is relevant immediately is the growth of his skill in managing the awarenesses of participants and audience; perhaps no difference in the basic workmanship of the two plays is more noteworthy or conspicuous. In the briefest terms, the management of awarenesses in *Titus Andronicus* produces only incidental theatrical effects, whereas in *Romeo and Juliet* it is so integral to the tragedy that it comes very near to *being* the tragedy.

More than any other of Shakespeare's, not even excepting *Othello*, *Romeo and Juliet* is a tragedy of unawareness; and more than any other of Shakespeare's, not even excepting *Macbeth*, it is a tragedy of Fate. Fate is the controlling practiser, and the entire action of the play represents her at work in the details of her housekeeping. The Prologue that boldly announces the business of the play stands without parallel not only as the most vital expository passage in the tragedy, but, in terms of its precise revelation of what will follow, as the most informative such passage in all Shakespeare. The Prologue provides us with an instant and inclusive guide that remains valid until the end; it notifies us of two essential facts: that the lovers' deaths will be the means by which Fate ends the feud of the rival households, and that the details of Fate's progress to that end will constitute the very action of the play.

Because of the Prologue we hold advantage over all participants *even at the moment the action commences*, and we retain our advantage to the end of the play—and beyond, for there is no Horatio to tell 'th' yet unknowing world / How these things came about'. Friar Laurence tries, but lamely, for he does not know enough of what we know. *Romeo and Juliet* is not the only play in which a gap between the awarenesses of audience and participants is maintained during most of the action; but it is the only one in which we hold advantage steadily, without interruption of any kind or degree, throughout every scene in the entire action, and over all participants—main, minor, and unnamed—during every moment. In other plays we share great central secrets variously with bright-eyed heroines, villains, professional fools, old dukes of dark corners. In *Romeo and Juliet* we share only secondary secrets, and those with only a select few persons—the lovers, the Nurse, the Friar; but the one all-encompassing secret of Fate's announced intent we share with no one, even at the end. Perhaps it is this fact, as much as any, that gives the tragedy its uniquely powerful hold upon our minds and emotions and makes this tragic experience peculiarly excruciating and enduring.

By advising us of what the end must be and what the role of the lovers is with respect to this end, the Prologue equips us to recognize that each subsequent incident in the 'two hours' traffic' represents a step in the process of Fate's working; Fate, in getting to its end, will waste no moves. We are to see these incidents not by their own light only, as the participants do, but also by the light of the Prologue. We are to recognize successive incidents as links in the chain that leads to catastrophe. Because the Prologue's pronouncement is all-inclusive, we shall need no asides, soliloquies, or other expository devices to keep us privately informed; the Prologue's warning, if we attend it faithfully, will suffice to keep us apprised of the fatal undercurrent even when surface movements may seem to contradict it. No participant's vision will vie with ours; when, for an instant, one or another actor catches a sudden glimpse of the truth that we see steadily, and expresses dire misgivings—as Romeo often does—he will have nothing new to teach us, as characters in other plays often have, but his remarks will produce instead a flash of irony unseen by himself. *Romeo and Juliet* is peculiarly abundant in the repetition of such flashes, perhaps exceeding even *Othello* in this respect. Though certain participants know more than others, and thus hold advantage, yet none knows the great secret; thanks to the words of the Prologue, the action of the play is spread before us like a map on a table.

Throughout the first act our advantage derives solely from the Prologue. Because they begin the action, it operates first over Sampson and Gregory, Abraham and Balthasar, the rival servants whose seemingly merely comic antics actually initiate the fatal march of events. Though the lovers have not yet met or, presumably, heard of each other, the pattern of their destruction is begun with Sampson's thumb-biting, followed a moment later by his 'Yes, better, sir.' From this start everything proceeds, even to the dismal end that is marked by the Prince's 'A glooming peace this morning with it brings.' For upon this 'Yes, better, sir' the servants fall to the fray; the well-intentioned but unwitting Benvolio, acting quite in character, strives to beat down their swords, crying more wisely than he dreams, 'You know not what you do'; Tybalt mistakes the intent of Benvolio's sword drawn against 'heartless hinds' and draws his own; the Citizens join in the mêlée; the families enter; and at last the Prince unwittingly sets the seal of doom upon lovers who are not yet acquainted:

> If ever you disturb our streets again,
> Your lives shall pay the forfeit of the peace.
>
> (I. i. 103–4)

The full significance of this brief pronouncement we shall note hereafter; but even at the moment, thanks to the Prologue that has alerted us, we can recognize that all the persons we have so far seen— servants, Benvolio, Tybalt, Citizens, families, and Prince—have unknowingly acted out the first movement in Fate's tragedy.

The second movement begins without a missed beat in the cadence: Romeo, fashionably lovestruck and self-consciously exhibiting the traditional symptoms of his malady, laments to Benvolio, who, well-meaning as always, extends a fatal hand:

> *Ben.* Be rul'd by me; forget to think of her.
> *Rom.* O, teach me how I should forget to think.
> *Ben.* By giving liberty unto thine eyes;
> Examine other beauties.
>
> (Ibid. 231–4)

It is Benvolio's second unknowing contribution to the chain; and Fate, quick to seize the chance, immediately capitalizes on the entrance of Capulet's illiterate servant carrying invitations to a feast. Unaware that Romeo is a Montague—for it is in the prevailing climate of unawareness that incidents occur throughout the action— he solicits his aid in identifying the names of those invited. Romeo discovers Rosaline's name on the list, and Benvolio eagerly repeats his urging:

> Go thither; and with unattainted eye
> Compare her face with some that I shall show,
> And I will make thee think thy swan a crow.
> (I. ii. 90–2)

It is at Benvolio's insistence that Romeo, then, goes to the feast— not, indeed, to seek new and richer beauty but 'to rejoice in splendour of mine own'.

Already Fate's deadly chain is lengthening; several persons have unknowingly added their links and gone their ways: the testy servants, Benvolio not once but twice, Tybalt, the Prince, the illiterate servant. Even Rosaline, whom we have not met and will not meet, has contributed, for it is to view her that Romeo decides to attend the Capulets' feast. In the second and third scenes of Act I we are advised that three young people will attend this feast for their private purposes: Romeo's we already know, and Paris's and Juliet's we soon learn. Says Capulet to Paris, 'Hear all, all see, / And like her most whose merit most shall be.' And Lady Capulet to Juliet, 'This night you shall behold him at our feast; / Read o'er the volume of young Paris' face / And find delight writ there with beauty's pen.' The Prologue has said nothing of Paris, and we are not to learn until III. iv, when Capulet—unaware that Juliet is already married— makes a 'desperate tender' of his daughter's hand, of the deadly use that Fate will make of him. But the paralleling of Paris's interest in the ball with that of the doomed lovers suggests a fatal tie; each of the three will come to the ball with open eyes and private purpose, and the Prologue has advised us that for two, at least, the meeting is a step towards death.

It is not to be forgotten, certainly, that when we first see Juliet we already know that she is doomed; and as though to make sure that we shall not forget, the dramatist boldly exploits the advantage that the Prologue has given us. About sixty of the one hundred lines in Juliet's introductory scene are given to the Nurse, and these,

combining tender sentiment with gross jest, set two ideas in sharp relief
—the extreme youth of Juliet and her keen anticipation of imminent
joy: 'The pretty wretch left crying and said, "Ay,"' says the Nurse,
not once but four times: 'It stinted and said, "Ay."' It is the very
heavy emphasis on the joyousness of the present and the immediate
future that most prods our recollection of the Prologue's ominous
word; through her coarse jest, four times repeated, and her recital of
details about that day when Juliet learned to stand high-lone, the
Nurse pricks the sensitive area of our mind, reminding us that this
rosebud just ready to burst into bloom is already blighted.

Similar exploitation occurs in the first portion of the scene immediately following that in which we are introduced to Juliet. Here
the dramatist pairs the Nurse with the bawdy Mercutio, whose wit,
as brilliant as it is bawdy, dominates these moments. As with the
Nurse in the preceding scene, so here more than half the lines are
given to Mercutio, who, if he epitomizes one idea beyond any other,
personifies a buoyant zest for life; he is, indeed, the very spirit of life,
and so his vital presence functions as reminder of its opposite, death,
and the threat of the Prologue accordingly looms over the moments of
hilarity. In the remainder of the scene, after Mercutio's exuberant
Queen Mab aria, this threat suddenly invades the occasion, for here
the dramatist injects heavy-handed and direct exploitation. Romeo
remarks that ''tis no wit to go' as the revellers move to enter
Capulet's house. 'I dream'd a dream to-night,' he adds, and
continues:

> . . . my mind misgives
> Some consequence yet hanging in the stars
> Shall bitterly begin his fearful date
> With this night's revels . . .
>
> (I. iv. 106–09)

Without the Prologue, these lines would function as foreshadowing;
with it, they do not foreshadow but sharply remind, as do all others
of like tenor throughout the play, and especially in those moments
when the fullest dramatic effect depends on our remembering. Foreshadowings give hints of outcomes; but when an outcome has
already been told, we have nothing to learn from such a line as
Romeo's 'my mind misgives . . .', which gives voice to his own
fleeting, imperfect glimpse of the peril that we know well. Nor is his
premonition here quite accurate: he fears that 'Some consequence'
will just begin its operation 'With this night's revels'. But we know

that Fate began its march when one servant bit his thumb at another and that the progress towards death has already begun.

Such reminders, not foreshadowings, prick our sore awareness steadily in the closing scene of Act I during the lovers' first meeting. In the seventeen lines of their initial conversation, the force of their sudden love, intense as religious devotion, is figured forth by the motif of their exchange: 'profane ... holy shrine ... pilgrims ... pilgrim ... devotion ... saints ... pilgrims ... holy palmers' ... saints ... holy palmers ... pilgrim ... prayer ... saint ... pray ... faith ... Saints ... prayers' ... prayer's ... sin ... sin ... Sin ... trespass ... sin'. Such is the predominant vocabulary of their exchange, the very sound of which marks the brief encounter as being a moment of transcendent passion; and these lines are made even more potent by the special setting in which, because of the Prologue, we hear them: by its very intensity this sudden love bids us remember that the lovers are doomed, and the sounds echo as in a tomb.

During this scene especially it is evident not only that the participants are ignorant of what we know—that they are doomed—but that their very unawareness is the channel through which Fate advances. Thus, in the instant before Romeo speaks the lines that forever seal his devotion to Juliet, that commit his heart irretrievably to her, this seemingly casual exchange occurs:

> *Rom.* (To a Serving-man.) What lady's that which doth enrich
> the hand
> Of yonder knight?
> *Serv.* I know not, sir.
>
> (I. v. 43–5)

The 'lady' is Juliet. And why, indeed, should the servant not know his own master's daughter? Perhaps, being busy, he simply tosses off the easiest answer, without even glancing in Juliet's direction; or perhaps he is not a regular servant to Capulet, but a transient hired for the special occasion. Whatever the cause, he does not identify the Capulet daughter, and thus, like the earlier illiterate servant who sought a Montague's aid in identifying the names of guests for the Capulet ball, he contributes his link, unknowingly, to Fate's chain: Romeo falls in love with Juliet before he knows her name. Immediately after the Servant has rushed on about his business—never to learn what he has done—Romeo goes into his aria from which there will be no recovery: 'O, she doth teach the torches to burn bright.'

The same unawareness afflicts Juliet, and the pair speak the entranced lines of their sonnet-duo in mutual innocence; the kisses that, at the end, seal the bond that Fate has wrought between them are exchanged in ignorance of the one fact that, known in time, could have prevented the mutual commitment of their souls. They have no sooner parted than Shakespeare points up the fact that their symbolic betrothal took place in ignorance; indeed, he points it not once but twice: Romeo asks Juliet's name of the Nurse, and, on being told, cries out, 'Is she a Capulet? / O dear account! my life is my foe's debt'; and Juliet, having learned Romeo's name, cries out, 'My only love sprung from my only hate! / Too early seen unknown, and known too late!' Of course we cannot know that Juliet's brain would have blocked the impulse of her heart had she known before; but the dramatist's emphasis in thus paralleling the lovers' reactions makes his point unmistakable.

With the lovers' meeting Fate has made a great stride, but is not content even with so much. For while the lovers are sealing their devotion another strand is also being woven into the deadly design. Recognizing a Montague, Tybalt seeks 'To strike him dead' on the spot, and when Capulet furiously overrules him, Tybalt's wrath is bottled up for a future explosion: '. . . this intrusion shall / Now seeming sweet convert to bitt'rest gall'. The incident parallels the climax of the opening scene of the play, when the angry Prince proclaimed that if ever the feud should break forth again, 'Your lives shall pay the forfeit of the peace'. Both incidents threaten future eruptions of which the lovers, in their ecstatic meeting, remain oblivious.

With the opening of Act II, we have gained a second enormous advantage over all participants except the lovers. Benvolio and Mercutio, shouting to Romeo from outside Capulet's wall, know nothing, of course, of the secret given us by the Prologue; but more immediately they are ignorant that Romeo is no longer in love with Rosaline, having found Juliet. 'I conjure thee', cries Mercutio, 'by Rosaline's bright eye', and Romeo, hidden within, surely enjoys the flash of irony; it is one of the rare occasions in the tragedy of one participant's holding sufficient advantage over another to perceive irony. Yet the dramatic effect sparked here by our recognition of Mercutio's error is incidental and unimportant; the tragic consequence of his unawareness is not to appear until later, when he duels with Tybalt because he knows nothing of Romeo's marriage to

Juliet, hence cannot conceive why Romeo refuses to fight Tybalt.

Over the unnaturally bright and shining first balcony scene, which follows, our sense of the Prologue's doom casts an oppressive shadow. Perhaps there exists no more striking example of Shakespeare's fondness, particularly in climactic scenes, for forcing two contradictory demands simultaneously on our emotions. We learn 'to love that well', insists Sonnet 73, 'Which we must leave ere long'. The voices of the lovers, transformed by passion, ring out in the moonlight as expressions of vibrant love, life, and youth, of which the lovers are themselves the embodied symbols; but the Prologue's ominous word, which all the preparatory scenes have forced into our consciousness, is of death, not life, and obliges us to view the lovers as we might view two fresh roses whose bright hues still show no sign that they have been cut from the stem. The sparkling vision of eager life that is the balcony scene is thus framed by the blackness of death, and the dramatist never allows us to forget the fact. Though Juliet speaks of danger, she imagines only Romeo's risk of being found within the Capulet orchard, knowing nothing of the doom that awaits both. Unaware of Romeo's presence, she speaks her love aloud: '. . . be but sworn my love, / And I'll no longer be a Capulet', and, more boldly still, 'Romeo, doff thy name, / And for thy name, which is no part of thee, / Take all myself.' After Romeo has made his presence known, she declares that, had she but known beforehand, no such outright declaration would have escaped her lips:

> I should have been more strange, I must confess,
> But that thou overheard'st, ere I was ware,
> My true love's passion . . .
>
> (II. ii. 102–4)

'. . . ere I was ware': again, Fate has moved through the unawareness of a participant. Though young and suddenly much in love, Juliet is not naïve; had she known of Romeo's presence, she would indeed have acted otherwise—perhaps even tried the role that Rosaline played, and that Romeo, fresh from his earlier experience of young ladies, would have expected her to play. Even as matters stand, having caught his new love unaware and heard her love confessed, Romeo at first knows no kind of response but that he had learned 'from the book'—the artificial, fashionable language of the lover disdained. Hence he would neither have sought nor gained a betrothal on this night, and there would have been no marriage on

the morrow. Fate has taken a giant stride through Juliet's unawareness of Romeo's presence, even as, earlier, Fate moved through the
unawarenesses of a succession of lesser participants—Capulet's
illiterate servant, and the house-servant who failed to identify Juliet
at the ball; ultimately, Fate will have moved through the unawareness
of every principal participant and of many minor figures in attaining
its end.

Because Romeo overheard Juliet's admission 'ere I was ware',
then, the lovers do move swiftly with a practice of their own—a
secret marriage on the very next day. It is the second-greatest
practice of the play, but of course it is circumscribed by Fate's
practice and is immediately converted to Fate's purpose. At this
precise point it is necessary to mention a vital detail of which, in
rushing their marriage date, the lovers are ignorant. This is the fact
that the feud, of which both are keenly mindful, is on the verge of
expiring. Caught up as both were at first sight of each other, they saw
nothing of the drama that suddenly flared in the Capulet ballroom
when Tybalt recognized Romeo's voice and would have struck him
with his sword had not Capulet, in a white fury, reprimanded him.
Romeo 'bears him like a portly gentleman', said Capulet then, 'And,
to say truth, Verona brags of him / To be a virtuous and well
govern'd youth'. These are the words of a father who might possibly,
if properly approached, entertain the thought of welcoming such a
young man as his son-in-law. When fiery Tybalt persisted, Capulet's
rage burst its bounds:

> What, goodman boy! I say he shall; go to!
> Am I the master here, or you? Go to!
> You'll not endure him! God shall mend my soul!
> You'll make a mutiny among my guests!
> You will set cock-a-hoop! You'll be the man!
>
> (I. v. 79–83)

This nearly violent interlude serves multiple and potent dramatic
purposes. First, it bottles up Tybalt's wrath to be exploded on
another day; second, it shows us what to expect of Capulet when he
is angered, and thus prepares us for the later moment when Juliet
crosses him by refusing to marry Paris; third, and of great importance as an underlying irony of the tragedy, it reveals that Capulet is
capable of warm regard for a Montague and perhaps even amenable
to ending the feud. Nor is Capulet's behaviour towards Romeo the
sole evidence that the feud, if not rekindled, would soon sputter out.

Both Capulet and Montague are old: 'A crutch, a crutch! Why call you for a sword?' chided Lady Capulet when, in the opening scene, her husband wished to take arms against 'Old Montague'. And Lady Montague similarly chided her own husband: 'Thou shalt not stir one foot to seek a foe.' Thereafter, in the second scene, Capulet, speaking for himself and Montague, told Paris that "'tis not hard, I think, / For men so old as we to keep the peace'—and silently assented to Paris's pregnant reply: 'And pity 'tis you liv'd at odds so long'. Ultimately, of course, the fact that the feud was near to dying a quiet death is confirmed by the readiness with which each bereaved father moves to memorialize the other's dead child.

But of all this the lovers know nothing. Their agreement to marry in haste is enfolded by layers of unawareness. It was precipitated by Juliet's being overheard 'ere I was ware', when she spoke honestly rather than coyly as she would otherwise have done; and they continue in haste because of their mutual unawareness that the delay of a week or a month, with the Prince's edict demanding amity and with the certain urging of Friar Laurence, could bring a joyous reconciliation of the families. To our own hearing, Capulet's warm defence of Romeo suggested that a Montague might well come within Juliet's free 'scope of choice' of which he boasts to Paris. More than once, Shakespeare goes out of his way to emphasize the strength of Capulet's love for his daughter—'The earth hath swallowed all my hopes but she', and 'She is the hopeful lady of my earth'; within her scope of choice, he tells Paris, 'Lies my consent and fair according voice'. Approached in the right way at the right moment, Capulet would doubtless consent to his daughter's marriage to the 'portly gentleman' of whom, he says, 'Verona brags'.

Not only the lovers, but Friar Laurence, too, is ignorant of these circumstances, and in his ignorance agrees to marry them in secrecy and haste:

> For this alliance may so happy prove
> To turn your households' rancour to pure love.
>
> (II. iii. 91–2)

With the Prologue's message fixed in our memory we cannot miss the flash of irony that accompanies this sentiment: 'pure love' will indeed unite the households, but only through the lovers' deaths; yet, given time and opportunity, the breach would heal of itself. It is necessary to consider at this point whether Fate may not be more than chief architect and practiser of the tragedy—whether Fate is not chief villain as well. The Prologue itself expresses no attitude towards Fate; we are only advised, bluntly, that 'nought . . . but their children's end' could terminate the feud. If any bias at all is detectable in the wording of the Prologue, it is modestly favourable to Fate, positing it as a benign force that uses a lesser evil (the lovers' deaths) as means to a greater good (the end of the feud). But, then, are we to take it that Fate itself is ignorant of the true status of the feud as that has been hinted by both Capulet and Montague? It would be preposterous to suppose that the dramatist means Fate to occupy a lower level of awareness than our own; hence we are obliged to suspect that Fate is malign and aware rather than benign but unaware. Fate is stubborn and absolute in its intent, and will end the feud only on its own terms: through the deaths of the lovers. Not only will it not accept, but it will high-handedly prevent a resolution by any less drastic means. 'A greater power than we can contradict', says the Friar at last, 'Hath thwarted our intent.'

Until the actual moment of the marriage, the Prologue has provided us with the one great advantage that we hold over the participants. With the marriage, however, comes into being the second major secret, which we share with the lovers, the Friar, the Nurse, and Romeo's man, Balthasar. Until the closing moments of the play all others are ignorant even that Romeo's affection has veered from Rosaline to Juliet—and their ignorance is the channel through which Fate continues to move. Earlier, Shakespeare twice exploited the general unawareness that Romeo had changed loves: first, directly after the ball, when Mercutio and Benvolio vainly sought him; second, with the same participants, just before the marriage. Directly after we have seen Romeo go to the Friar's cell to arrange the details, Mercutio complains to Benvolio,

> Why, that same pale hard-hearted wench, that Rosaline,
> Torments him so that he will sure run mad.
>
> (II. iv. 4–5)

Then, emerging from the Friar's in high good humour, Romeo exchanges wit with Mercutio, and triumphs; says Mercutio,

> Why, is not this better now than groaning for love?
> Now art thou sociable, now art thou Romeo.
>
> (Ibid. 92–4)

Nothing more than passing effect results from Mercutio's unawareness on these two occasions. But on the third occasion, which immediately follows the marriage, calamity results from it. The noonday sun itself here enlists on Fate's side: 'The day is hot, the Capulets abroad, / And, if we meet, we shall not scape a brawl, / For now, these hot days, is the mad blood stirring.' But the general irritability that comes of the heat is at most a contributing factor; the essential condition through which Fate operates is the unawareness of the participants, most notably Mercutio and Tybalt. We have just been pointedly reminded, before Tybalt enters, that Mercutio does not even know that Romeo has found a new love; of course he knows nothing of the marriage, nor does Tybalt. The two quarrel before Romeo enters, but they do not fight: 'Peace be with you, sir', says Tybalt, mildly enough, and dismisses Mercutio as Romeo approaches. So grossly does Tybalt insult Romeo that Romeo is honour-bound to fight; but Romeo, conscious of their new relationship, refuses: 'I see thou know'st me not.' Nor is it Tybalt's fiery temper that actually precipitates the fight with Mercutio. The cause is Mercutio's ignorance of Romeo's marriage; not knowing of it, he is enraged by Romeo's 'calm, dishonourable, vile submission' to Tybalt's insults. Fiery as he is, Tybalt is yet reluctant to draw his sword: 'What wouldst thou have with me?' He fights only when Mercutio threatens to strike him while his sword is still undrawn: 'Make haste, lest mine be about your ears ere it be out.' Once Mercutio has been slain, in defence, as he believes, of Romeo's honour, the new bridegroom has no choice: 'My very friend hath got this mortal hurt / In my behalf; my reputation stain'd / With Tybalt's slander,—Tybalt, that an hour / Hath been my cousin!' With this final line the dramatist again prods our awareness of the circumstances that occasion the deaths of Mercutio and Tybalt: not the noonday heat, the enmity of the families, the quarrelsome natures of Mercutio and Tybalt—though clearly all are contributory factors—but Mercutio's and Tybalt's ignorance of Romeo's marriage. So much Romeo himself recognizes; but he, too, remains unaware that Fate, operating as always through men's unawareness, is actively directing events and has just now forged yet another link in its chain.

The final lines of II. i are those of the Prince banishing Romeo. They are measured and reasoned lines, evincing a judgement not merely just but merciful, if we remember that the Prince knows less of the situation than we. 'I can discover all / The unlucky manage of this fatal brawl', Benvolio says, and proceeds to tell the story fairly enough; but Benvolio's own vision is limited: he knows nothing of Romeo's marriage, which was the cause of his refusal to fight Tybalt and the cause also of his intervention, fatal to Mercutio, when Tybalt and Mercutio assailed each other. Hence Benvolio cannot tell the whole truth, and therefore the Prince's sentence is made in ignorance of the all-important fact known to us. Even so, the Prince would presumably have pardoned Romeo—for he says 'Romeo slew him, he slew Mercutio; / Who now the price of his dear blood doth owe?'—but for the stern edict pronounced by himself after the previous outbreak of the feud in I. i: 'If ever you disturb our streets again, / Your lives shall pay the forfeit of the peace.' Having so declared, he cannot now pardon Romeo; the sentence of banishment is as lenient as circumstances allow. His decision thus confirms our earlier sense that the outbreak with which the tragedy opens was one of terrible significance: had not one servant then bitten his thumb at another, the Prince would have had no occasion to pronounce his edict, hence would not now be bound to banish Romeo. But for that thumb-biting servant, Romeo would now be a free man, even as his father rightly contends he should be: 'His fault concludes but what the law should end, / The life of Tybalt.' Once more, then, with Romeo's banishment, Fate has struck through the unawareness of participants.

The Prince's words banishing Romeo come at the end of III. i. The next scene opens on Juliet's ecstatic aria: 'Gallop apace, you fiery-footed steeds, / Towards Phoebus' lodging. . . .' She speaks thirty-five lines of the most impassioned language in the play as she anticipates the coming of night and Romeo; but she speaks them while the Prince's dire words still hang in the air: 'Let Romeo hence in haste, / Else, when he's found, that hour is his last.' Her words emerge from within a triple layer of ignorance: she does not know that Romeo has killed her cousin; that he is banished at the very threshold of their bliss; that Fate is using all that has passed in their relations towards its own end, the reconciliation of their families. No word in her aria need exploit directly the gap between her awareness and ours, for the very pitch of her ecstasy exploits it: 'Come, night, come, Romeo; come, thou day in night / . . . Come, gentle night, come, loving, black-brow'd night, / Give me my Romeo.' While these high notes continue, we are unlikely to forget the Prince's words closing the preceding scene.

In all, the period of Juliet's acute unawareness continues until

34

after line 110, far beyond the initial thirty-five line speech. It does not end until at last the Nurse's word, 'banished', penetrates her active consciousness. At the start, the Nurse's failure to communicate the facts clearly appears as a cruelly grotesque variation on her earlier, deliberate delay for comic effect in reporting to impatient Juliet the results of her interview with Romeo (II. v) about their marriage plans. By line 70 the Nurse has managed to make the fact explicit that Romeo is banished, but during the next forty lines the dreadful word remains buried in Juliet's under-consciousness— 'Some word there was . . . that murd'red me'—and, though it 'presses on my memory' during the ensuing conversation, does not break through to the surface until her brain can no longer suppress it: '"Tybalt is dead, and Romeo—banished."' In much the same way, throughout the tragedy, but most painfully in those moments when unsuspecting participants appear most joyous, the Prologue's blunt warning that the lovers are doomed burdens our own memories: so it is, for example, during the lovers' first meeting, during the first balcony scene, during the meeting at the Friar's cell before the marriage ceremony, during Romeo's high-spirited rejection of Tybalt's challenge, and during Juliet's ecstatic apostrophe to night and Romeo. During such periods we need no direct reminders of the Prologue's doom: the lovers' very joy is cue enough.

But the Prologue's dire word pains our awareness also during moments of participants' despair; more precisely, it then intensifies our feeling of their distress. For Romeo, the first desperate moments follow the killing of Tybalt, when he, like Juliet later, hears the dread word 'banishment'. 'Be merciful, say "death"; / For exile hath more terror in his look, / Much more than death,' he cries, and his distress continues unabated through about one hundred and fifty lines until the Friar can persuade him to 'Go, get thee to thy love, as was decreed', and thereafter to flee to Mantua to await fulfilment of the Friar's promise to reconcile the families, win the Prince's pardon, and 'call thee back / With twenty hundred thousand times more joy / Than thou went'st forth in lamentation'. Lifted from the pit of despair to the height of anticipated bliss, Romeo grasps the Friar's hand: 'But that a joy past joy calls out on me, / It were a grief, so brief to part with thee.' At which time is our contemplation of the lovers' plight more painful?—when they register their own anguish during moments of recognized misfortune, or when they fairly burst their lungs with ecstatic song? The tragedy

35

poses a peculiarly difficult dilemma for the emotions: we easily grieve with the lovers when they see their own miseries; but we cannot join in their rare moments of joy, knowing that it is Fate's means of killing them. Between the two extremes, the play allows no respite, for its moods are alternately those of ecstasy and despair, with no middle ground.

Such is the dilemma with which our emotions must contend during the second balcony scene, perhaps the most painful in the play before the final scene and the deaths of the lovers. On the one hand it is the time of the consummation of their loves; but on the other our consciousness of the inevitable and imminent deaths of both is here peculiarly sharp and urgent. It is a high-pitched poetic language that the dramatist makes them speak during the scene; the vowels are forward in the mouth—mainly a's, e's, and i's, so that both lovers might almost be screaming: 'Wilt thou be gone? it is not yet near day. / It was the nightingale, and not the lark, / That pierc'd the fearful hollow of thine ear . . . Let me be ta'en, let me be put to death . . . It is not day . . . It is, it is! Hie hence, be gone, away! / It is the lark that sings so out of tune, / Straining harsh discords and unpleasing sharps . . . More light and light; more dark and dark our woes!' The lovers themselves know that their plight is desperate, and the dramatist gives a bitter line to each: 'O God, I have an ill-divining soul! / Methinks I see thee, now thou art below, / As one dead in the bottom of a tomb,' cries Juliet; and Romeo, 'Dry sorrow drinks our blood.' Even so, their anguish differs from ours: though they are desperate, their desperation is not devoid of hope, for Romeo is to live in Mantua until the Friar can call him back 'With twenty-hundred thousand times more joy / Than thou went'st forth in lamentation', and so Romeo reassures Juliet, insisting that '. . . all these woes shall serve / For sweet discourses in our times to come'. But of course our own awareness holds no shred of hope, for we have heard the Prologue; and it is precisely by the admixture of the lover's expressions of hope that the dramatist blends the exquisite torture that this balcony scene inflicts upon our emotions.

What is more—and worse, if any further information can worsen that provided by the Prologue—the dramatist, with cold calculation, has placed immediately before the balcony scene a cruel new warning of the lovers' imminent peril and the vanity of their hopes; says Capulet,

on being crossed, he so ferociously berated his nephew as to bottle up his wrath to explode at a later crucial instant; just so, when Juliet crosses him and deflates his buoyant humour, he responds with a fury that drives her to embrace the desperate remedy that destroys both lovers and wins Fate's game. Yet, vital as they are in Fate's programme, these instances of Capulet's pendular swings are mainly illustrative, and preparatory for the later casual decision that a new burst of good humour causes him to make.

At this point, as we are about to enter on Fate's final movement, it is appropriate to repeat that three great secrets constitute the environment within which action proceeds. The first, told us by the Prologue, encloses everything else—participants, incidents, words— and conditions our view of all. The second, that of the marriage, we share with the lovers, the Friar, the Nurse, and Romeo's man; it is through other participants' ignorance of this secret that Fate takes the greatest number of strides. The final secret, which we share only with Juliet and the Friar, comes into being at the start of Act IV, with the Friar's offer and Juliet's acceptance of a potion that will induce the 'borrowed likeness of shrunk death'. Just as the number of those who know this secret is smallest, so is the number of those ignorant of it largest—and this latter group includes two, Romeo and his man, through whose unawareness Fate makes its final thrust.

In IV. i it is Paris who steps into a situation of which he is wholly ignorant. Few persons in Shakespeare's plays ever stand in a more unbecoming light than that in which Paris stands here, and through no fault of his own. He is an innocent pawn whose responsibility for the deaths of the lovers is nevertheless enormous. In his three principal scenes—III. iv, where Capulet promises him Juliet; IV. i, where he converses with the distraught Juliet about their imminent wedding; and V. iii, where he attempts to block Romeo's way into the Capulet tomb and is slain for his trouble—his image is unenviable in the extreme. Elsewhere, most persons who are caught in situations of which they know nothing enjoy at least our sympathy and our good wishes: we laugh happily at Orlando as he makes love to a supposed Rosalind who is in fact his very Rosalind; we grieve for the good and generous captain Antonio who mistakes 'Cesario' for Sebastian and is cruelly rebuffed; we are soon moved to pity even Shylock, who, after Portia's entrance into the Duke's court, suddenly becomes the victim and does not know it. Indeed, even villains like Richard III and Macbeth, when they do not recognize as soon as we that they are doomed, extract a measure of pity, however ill-deserved, from us. But the innocent, unblemished, and unfortunate Paris we resent merely because he exists. From the first, we can sense that his existence threatens the lives of the lovers; but for him, Juliet might indeed simply wait in Verona and Romeo in Mantua until the Friar can find time to reconcile the families, win pardon of the Prince, and make all well. But Paris is offered Juliet's hand, and merely therefore gains our involuntary resentment. His very presence is a gnawing reminder of the Prologue's decree, and his continuing unawareness of situations in which he stands provides one of the larger openings through which Fate rushes.

And yet what manner of man is this Paris? To assert that his character is as exemplary as any ever created by Shakespeare is hardly to exaggerate. When we first see him, he is earnestly voicing hope that, for the good of others, not himself, the bloody feud will end. When we next see him, he is expressing his love for Juliet and accepting Capulet's 'tender' with a manner that is modest, respectful, irreproachable. When we next see him, meeting his promised bride at the Friar's cell, his solicitude for Juliet in her grief—the cause of which he innocently mistakes—marks him for the gentleman he is; ignorant that she is married, ignorant that her tears are for Romeo, not Tybalt, aware only that she is troubled, he is chivalrous, self-effacing, kind. If Shakespeare exaggerates the quality of his nobility, it is perhaps to underscore the way of Fate's deadly working throughout the tragedy: not through conscious villainy, but through the unawareness of the well-intentioned and the good. Each word that Paris speaks to Juliet during their interview is considerate and gentle; and yet, because the time is so dreadfully wrong, each word grates us.

In the remainder of this scene, after the good Paris leaves, Juliet and the Friar conspire to deceive the Capulets, and, indeed, all Verona. Theirs is a deliberate practice, the only meditated act of deception in the play:

> Go home, be merry, give consent
> To marry Paris. Wednesday is to-morrow.
> To-morrow night look that thou lie alone;
> Let not thy nurse lie with thee in thy chamber.
> Take thou this vial . . . .
>
> (IV. i. 89–93)

It is also the only direct and determined attempt by participants to

block the course of events, which the Friar perceives is bent towards
disaster; their effort, in his words, is a 'desperate . . . execution'—
as desperate 'As that is desperate which we would prevent'. But
though they recognize the direction of events, they lack our knowledge that their adversary is no less than Fate itself, Fate moving
swiftly and inexorably onwards, striking through each opening that
is provided, in their unawareness, by principal and casual participants
alike.

And so it is with the Friar's own practice, which, designed to
thwart the progress of events, opens instead a wide new channel
through which Fate rushes. Heartened by the Friar's plan, and with
the vial in her pocket, Juliet arrives home with 'merry look' and
prostrates herself before her father: 'Henceforward I am ever rul'd
by you.' It is here that the dramatic purpose of Shakespeare's earlier
illustrations of Capulet's pendular swings of mood becomes clear.
Ignorant now of three vital details known to us, the last of which is
that Juliet shows herself submissive only because of the Friar's plan,
Capulet reacts with a surge of joy as extreme as was, earlier, his
lunatic rage against his daughter. In his joy, without an instant's
reflection, he advances by twenty-four hours the time set for the
wedding: 'Go, nurse, go with her; we'll to church tomorrow.'
Reminded that tomorrow is Wednesday and that he had himself
previously named Thursday, he will no more be swayed in his joy
than formerly in his fury. The shift of date will shortly prove disastrous for all but Fate: Juliet must now take the potion on Tuesday
night, not Wednesday, as the Friar had planned; Romeo's man will
see her 'laid low in her kindred's vault' twenty-four hours earlier
than he could otherwise have seen her, and be off with false word to
Romeo by that much sooner; and the Friar will have twenty-four hours
less time than he had expected for getting the true word to Romeo.

During the concluding scenes of Act IV all three major secrets have
immediate relevance, and our knowledge of them imposes multiple
and contradictory demands upon the emotions. The disparity
between our awareness and the participants' is here at its widest: not
one of those who appear in these scenes knows of Fate's design;
none knows of the marriage except the Friar, the Nurse, and Juliet—
who is insensible; and none but the Friar knows that Juliet is asleep,
not dead. The most urgent advantage that we hold during IV. iv
derives from our knowledge that, while all is stir and bustle down-
stairs at three o'clock on Wednesday morning, upstairs lies Juliet in

apparent death. Servants dash hither and yon, clanging pots and carrying firelogs; the Nurse and Lady Capulet echo and re-echo the roaring exuberance of Capulet as he directs frantic preparations for the wedding. He is in a towering fever of the spirit as he orders servants about, jests with the Nurse, even jokes with his wife: 'A jealous-hood, a jealous-hood!' And all the while we watch this bustle, our minds retain the image of what we last saw above-stairs: 'Romeo, I come! This do I drink to thee!' With these words, just before, Juliet drank and fell upon the bed.

The following scene, in which all those who are preparing to celebrate Juliet's wedding discover instead her seemingly dead body, confronted Shakespeare with a dramatic dilemma: on the one hand, the members of the family must be represented as undergoing intense shock and grief; on the other, since we are ourselves well aware that Juliet is not really dead, we can hardly be expected to respond as to a scene in which we know that the participants have genuine cause to grieve. In these circumstances, what should the dramatist do? Should he portray shock and grief as ably as his imagination and skill allow? If so, the very force of the scene might either dim our awareness that Juliet, after all, is alive, or, worse, appear embarrassingly overdone and excessive. Or should he defer to our knowledge that Juliet is alive, that the occasion is not real, by offering a mere token show of shock and grief? If so, his scene will falsify the emotions of the mourners, none of whom could be expected to treat the situation casually. Shakespeare would confront a problem like this one only twice more in his career, both times in the same play, *Cymbeline*: there, first, Belarius and the two princes find Imogen seemingly dead and must weep for her while we know that she lives; and, second, Imogen wakes to find the headless body of the detested Cloten stretched beside her, and, because the body is dressed in her husband's clothes, believes it to be her husband indeed.

What Shakespeare actually does with his peculiarly complex problem in *Romeo and Juliet* is to present a loud but essentially stylized and formal scene of shock and grief. The resulting effect resembles that of a play within a play, in which the participants speak like actors playing roles, as though separated from their real world and selves. Though their cries are emphatic, they are declamatory and theatrical; what the scene offers are the motions and sounds of grief—'but the trappings and the suits of woe'—without the heart.

Nineteen times in the forty lines that comprise the body of the scene, the cry of 'O!' is raised, and the same sound is repeatedly echoed in words like 'woe', 'woeful', 'cold', 'behold'. Similarly conspicuous is the repetition, in series, of exclamatory adjectives, thus: 'Accurs'd, unhappy, wretched, hateful day!' and 'Beguil'd, divorced, wronged, spited, slain!' and again, 'Despis'd, distressed, hated, martyr'd, kill'd!' These are truly but the trappings and the suits of woe; they are artificial, and resemble the false language of love used earlier by Romeo when he was playing at the fashion of being in love and knew only to speak 'by the book'. Once Romeo has felt the throb of genuine passion, as he does at first sight of Juliet—'O, she doth teach the torches to burn bright!'—he speaks another language; yet the contrast between his earlier and later manners of speaking is no more striking than that between the false scene of grief that is played when we know Juliet to be alive and the very moving scene of grief that we are to witness at the end of Act V, when the lovers are dead to us as well as to their families. If Shakespeare's solution to the dilemma of IV. v was not flawless, yet it would be exceedingly difficult even to imagine, let alone devise, any suitable alternative. The scene stands, in the tragedies, as an early example of that extraordinary dramatic shrewdness which, no less than poetic inspiration, would shortly distinguish his greatest tragedies.

Once the mourners have departed, the musicians take over the scene, and the play-within-a-play effect vanishes. Excepting Peter, presumably none of the musicians who participate in these final fifty lines knows Juliet or has particular cause to weep for her; they had been hired to play at a wedding, and when there is no wedding their business is only to 'put up our pipes and be gone'. It is all one to them: whether Juliet is asleep or dead, whether she is married to Romeo or to Paris, whether Fate has or has not doomed her; they seem, indeed, unaware of her and her world entirely. Like the Porter who answers the knock at the gate after Duncan's murder, and the knocker outside, they represent the great world going on about its regular business, oblivious of particularities. Simon Catling, Hugh Rebeck, James Soundpost: in their loose trade-talk and saucy jocularity they are reminders of the widest gap of all, in the tragedies —that which separates the private world of the principals, a world which altogether stops at times, and agonizes over its troubles, from the great impersonal world that never stops, but goes on its way unknowing, hence unperturbed and seemingly indifferent. At this point we are only a step or two away from 'th' yet unknowing world' that engulfs the deadly game between Hamlet and Claudius.

In *Romeo and Juliet* this cold reminder of the great world's indifference is strategically placed, for it stands just before the opening of Act V, where, in quick succession, Fate takes its final strides. These final steps, like those that have preceded, are made not only through but because of participants' unawareness. Here the three major secrets continue to comprise the heavy burden of awareness that we ourselves must bear: the secret of Fate's purpose, known to us alone; that of the marriage, shared with the lovers, the Nurse, and the Friar; and that of the potion, shared only with the Friar while Juliet sleeps. All three provide advantages that figure potently in the dramatic effects of the final scenes. Thus, to start, Act V opens with a brilliant flare of irony when Romeo's words exploit the gap between his ignorance and our knowledge of the potion and what has just happened in Verona: 'If I may trust the flattering truth of sleep,' says the hero, relaxing as best he can in Mantua, 'My dreams presage some joyful news at hand.' His dream, superficially false, is in fact true: the 'joyful news' that should be at hand is that Juliet sleeps and that Romeo can come to carry her off to Mantua; but the news that actually arrives is Balthasar's false report:

> Her body sleeps in Capel's monument,
> And her immortal part with angels lives.
>
> (v. i. 18–19)

Balthasar's 'sleeps' is cruelly precise, as we alone recognize. Unaware of their truth, Romeo rejects both the happy promise of his dream and the literal sense of 'sleeps', and, grown suddenly to full manhood after only four days of drastically maturing experience, seeks immediate means to join his wife in death: 'Well, Juliet, I will lie with thee tonight.' And once again a cruel flash attends his bitter play on 'lie with thee': if he but knew the truth, he might yet indeed lie with Juliet, and joyfully.

But identification of incidental effects that rise from the difference between the participants' and our awareness is of minor importance here; what matters is how the participants' unawareness contributes to catastrophe. Balthasar's action is the immediate case in point. Balthasar saw Juliet 'laid low in her kindred's vault' and quite dutifully sped to Mantua and Romeo: 'Since you did leave it for my office, sir'. The action, like the man, is impeccably true. It is noteworthy that in the quarrel of the servants with which the play opens, Balthasar, though present, speaks not at all, though each of the others speaks repeatedly. When we first hear directly of him, it is from Romeo as he arranges with the Nurse for his marriage: 'I warrant thee, my man's as true as steel.' Next Friar Laurence, bidding Romeo 'Sojourn in Mantua' until the families can be reconciled, advises that 'I'll find out your man, / And he shall signify from time to time / Every good hap to you that chances here'. Next Balthasar, having the testimony of his own eyes, brings word that Juliet is 'laid low'. His final action is to remain in the graveyard, despite Romeo's threat to rend him joint by joint, in order to serve his master in any way he can. The dramatist's characterization of Balthasar is brief but conspicuous, and it underscores the fact that the final disaster comes not through villainy or any manner of reprehensible conduct, or even personal flaw, but through the unavoidable unawareness of many persons. Because it is on receipt of Balthasar's false report that Romeo rushes to Verona and kills himself, Balthasar's share in the total responsibility for the catastrophe seems to loom larger than that of others. Yet it is no more than one of many equal links, neither more nor less crucial than the action of the illiterate servant who innocently inquired of Romeo what names were listed on Capulet's invitation to his banquet; no more important than the words of another servant at the feast who, to Romeo's inquiry for Juliet's name, replied, 'I know not, sir.' Because of the climactic occasions of their actions, Tybalt, Mercutio, Capulet, and Balthasar all appear to bear greater responsibility than others; but in the swift process of Fate's deadly working their actions are only equal to those of the least conspicuous members of the fatal chain—only equal, let us say, to the mere fact of Rosaline, whom we never meet, never see, never hear to speak, who, in truth, may never even have known that Romeo loved her—yet for whose sake Romeo first entered the Capulet house and perceived Juliet: 'I'll go along no such sight to be shown, / But to rejoice in splendour of mine own.' None could be more innocent than Rosaline of the lovers' deaths; yet none is more guilty. In Fate's chain, the links are of a size.

During the scene in which Balthasar delivers his false report, Shakespeare allows our own awareness to lag: we do not yet know what has happened to the brother whom Friar Laurence had told Juliet he would send to Mantua with word of the 'desperate remedy'. It is the only such lapse in the play, and it is of little significance, for if we do not yet know why the true word failed to arrive, we know well enough that the false word has come. We are left to suppose, briefly, that Balthasar, rushing to Mantua on the instant he saw Juliet 'laid low', naturally arrived hours earlier than an elderly friar who knew no need for haste. We learn of Friar John's detention at the opening of v. ii, and with this reinforcement of our knowledge we gain a double advantage over Friar Laurence. He knows nothing, of course, of the true nature of the force that he has vainly sought to thwart; but in the scene with Friar John we hold also an immediate and agonizing advantage, having heard Balthasar's false report and seen Romeo buy poison. On learning of his own messenger's detention, the Friar knows a need for haste, but cannot know how urgent is the need:

> Within this three hours will fair Juliet wake.
> She will beshrew me much that Romeo
> Hath had no notice of these accidents.
>
> (v. ii. 25–7)

At this point, the worst misfortune that he can foresee is that Juliet will wake and scold him because Romeo is not beside her as had been promised; he can see nothing in the situation that cannot be readily remedied: '. . . I will write again to Mantua, / And keep her at my cell till Romeo come'.

Here, when we ourselves can clearly foresee the few remaining steps in Fate's course, it should be noted that the failure of the true word to reach Romeo in Mantua is not due simply to the critically oft-maligned accident of Friar John's detention; without question, Balthasar, riding swiftly with his false word, would have reached Romeo ahead of the Friar even if there had been no detention. But neither is the disaster due only to Balthasar's speed: it is due most of all to the loss of twenty-four hours of precious time. Juliet was to have been married on Thursday, and this fact was in Friar Laurence's mind when he told her that he would send a brother to Mantua. But, as we have noted, Juliet's cheerful and submissive manner when she returned from the Friar's cell with the potion prompted Capulet—blind to every aspect of the situation—to advance the date to Wednesday. Therefore Juliet was obliged to drink off her potion twenty-four hours earlier than planned, Balthasar saw her 'laid low' twenty-four hours earlier than he could otherwise have done, and Friar John had twenty-four hours less time than he would otherwise have had in which to reach Romeo. Had Balthasar not seen Juliet 'dead' at the time he did, Friar John would have had time to be detained, be freed, and be sent again to Romeo; for even as the actual schedule stands, Friar John reports back to Friar Laurence three hours before Juliet is to wake, leaving ample time for Friar Laurence himself to have carried Juliet safely to his cell, where she could have remained while Friar John made a leisurely trip to Mantua to inform Romeo. But because of Balthasar's false word, Romeo has already slain himself in the Capulet tomb when Friar Laurence arrives to carry Juliet away. Had Capulet not advanced the wedding date by twenty-four hours, Friar John could have returned from his detention with twenty-seven hours to spare before Juliet woke, and many hours before Romeo could possibly have been misinformed by Balthasar.

The so-called 'accident of Friar John's detention' has long been advertised as the single great structural defect of *Romeo and Juliet* considered as tragedy. But in fact the detention is a mere trifle, a veritable nothing in the total scheme of Fate's campaign. There might have been no Friar John at all, no effort by Friar Laurence to reach Romeo before Juliet woke; the Friar might simply have planned, from the start, to be at Juliet's side when she woke. And indeed he *is* at her side, in spite of all, and might have borne her to safety but for the fact that Romeo had already come to the tomb and slain himself—a fact with which Friar John's detention had no connection. Thus, in the final analysis, not the Friar's abortive effort to reach Romeo but Balthasar's success in reaching him makes Fate's final link.

We must return briefly now to the last movement of the tragedy, which begins with the arrival of Paris at the tomb, intent on strewing flowers on Juliet's 'bridal bed'. Here as elsewhere Paris's shining goodness is rendered ugly in our view by the very fact that he is present. All that he knows, as he enters, is that Juliet, who was to have been his bride, is dead. But we know that she is not dead, only sleeping and ready to wake; that at this moment her husband Romeo, thinking her dead and prepared to kill himself, is rushing from Mantua; that at this moment also Friar Laurence, knowing nothing of Romeo, is approaching the tomb to be with Juliet when she wakes; and, finally, that all these elements are integral parts of Fate's deadly design. Entering the scene at such a moment, Paris, however noble his intent, shows as a mere irrelevance, a brash intruder whose gentlest utterances jar unpleasantly against the multiple awarenesses that pack our minds:

> Sweet flower, with flowers thy bridal bed I strew,—
> O woe! thy canopy is dust and stones—
> Which with sweet water nightly I will dew,
> Or, wanting that, with tears distill'd by moans.
>
> (v. iii. 12–15)

At Romeo's entrance the dramatist exploits the gaps that separate the awarenesses of the two men and that separate ours from both; says Paris, all but entirely mistaken about everything,

> This is that banish'd haughty Montague,
> That murder'd my love's cousin, with which grief,
> It is supposed, the fair creature died;
> And here is come to do some villanous shame
> To the dead bodies.
>
> (Ibid. 49–53)

He is unaware that there was any connection between Romeo and Juliet other than that Romeo killed Tybalt; he knows nothing of the love that sprang up full-blown three days before at the Capulet feast, and of the marriage that followed. In the next moment he dies, totally ignorant, like Mercutio and Tybalt before him, of the situation in which he dies. As for Romeo, he of course knows much more than Paris, and, in the moment before killing him, marks the distance between his own and Paris's understanding of the situation. Romeo, indeed, knows all the salient facts of the immediate situation except for the crucial one: that Juliet sleeps. It is his unawareness of this detail on which Shakespeare concentrates his exploitative powers from Romeo's entrance at line 20 until his dying gasp at line 120—a period of one hundred lines during which Paris's challenge, the fight, and Paris's death appear rather as incidentals, extraneous to the central issue, hence jarring to our sensibilities. In the confusion of our emotions we are likely to wish Romeo to kill Paris abruptly, with one blow, so that he can get on into the tomb and Juliet; a prolonged duel seems vexatious and irrelevant. Yet a shrewder insight, if we were capable of it at the moment, might remind us that any and every delay, just here, is of utmost value and should be welcomed—for we know that the Friar is even now running towards the churchyard and that Juliet is near to waking.

Romeo's great final speech, after Paris's death, makes an extraordinary assault upon the emotions. A gorgeous speech, rich with image, symphonic in combinations of vowel and consonant, a sumptuous banquet for both the eye and the ear, it nevertheless contradicts its own bid for a response of sheer pleasure by means of cruelly pointed reminders of the gap between Romeo's awareness and ours. Seven times in the passage Romeo speaks the dread word 'Death', most notably in the course of wondering that Juliet's lips and cheeks should still be crimson, her beauty still fresh: 'Why art thou yet so fair?' For us the answer is plain enough: the potion is losing its power, and Juliet's young blood is again stirring; in another moment she will wake. Repeatedly, insistently, the dramatist prods and turns the knife in our wound with Romeo's unaware words, down to his dying gesture: 'Thus with a kiss I die'—whereupon, instantly, as the dramatist has conditioned us to expect, 'Enter Friar Laurence, with lantern, crow, and spade.' And in the next moment the Friar notices Juliet: 'The lady stirs.'

Romeo, like Mercutio, Tybalt, and Paris before him, dies not only unaware but because he is unaware. He is the only hero in Shakespeare's tragedies to die so. A moment later Juliet, too, dies by her own hand; but she does not die unknowing, and no truth that she might learn could save her. If she does not know why Romeo is dead beside her, his lips still warm, neither does she question why; for her the only relevant fact is that he is dead.

From this point on the dramatist's focus is on the very breadth of the gulf that, with each passing scene, has further separated the participants' awareness from our own. In successive lines, from the mouths of those who approach the tomb, two key questions are asked and asked again: What has happened? Why has it happened? Astonished, the Watchmen note Juliet's warmth in death—'Juliet, dead before, / Warm and new kill'd'. They marvel at the presence of Romeo and Paris dead in the tomb, and at the incongruous conduct of the Friar, who 'trembles, sighs, and weeps'. 'What should it be', cries Capulet, 'that they so shriek abroad?' 'What fear is this which startles in our ears?' demands the Prince. The first question is answered soon enough, for they behold the bodies in the tomb; but the second question—*how* this dire calamity came about—they attempt in vain to answer. If there is a single theme that runs through the closing hundred lines, it is that the gulf between our awareness and the unknowing world's ignorance is too wide ever to be bridged. 'We see the ground whereon these woes do lie', says the first Watchman, 'But the true ground of all these piteous woes / We cannot without circumstance descry.' The Prince's first order is to 'Search, seek, and know how this foul murder comes.' His second is the same: 'Seal up the mouth of outrage for a while, / Till we can clear these ambiguities, / And know their spring, their head, their true descent.' But of course the 'true descent' they will never discover, for it is composed of the many hidden details of Fate's workings that began when the rival servants quarrelled on the street some five days earlier; we alone witnessed the forging of all the links, and we are powerless to advise the Prince.

The best answer that any participant can provide is the Friar's, and it, though lengthy and detailed, is woefully incomplete. The Friar's forty-line recapitulation of events that we have watched has been much criticized as merely repetitious and tedious. But of course it is neither repetitious nor boring for any of the survivors who hear it, but fresh and astonishing. Further, it is vitally important for our final experience of the tragedy that we be present to see and hear the reactions of the survivors as they first glimpse truths that we have known and been pained by from the outset. How wide is the gulf that has grown between the actors and ourselves is partially suggested by the fact that Montague, for example, never knew even of his son's infatuation with Rosaline—let alone of any subsequent events by which at last his body was brought to lie in the Capulet tomb; it is thus no wonder that Montague, staring at his dead son, should utter a seeming irrelevance that marks the totality of his own and all Verona's ignorance of what has happened: 'O thou untaught! what manners is in this, / To press before thy father to a grave?' Wisest and best-informed among them, the Friar himself can tell only in general terms how the disaster came about. He reveals two of the great secrets, that of the marriage and that of the potion, and his words have terrible significance for certain of those present—Capulet, the Prince, and Balthasar. For Capulet the most chilling disclosure is that it was for Romeo, not Tybalt, that Juliet pined: 'You, to remove that siege of grief from her, / Betroth'd and would have married her perforce / To County Paris.' Hearing these words, and remembering his own command that Juliet marry Paris or 'hang, beg, starve, die in the streets', Capulet must blanch to his very soul. So, too, must the Prince on hearing what part his edict played: 'I married them; and their stol'n marriage-day / Was Tybalt's dooms-day, whose untimely death / Banish'd the new-made bridegroom from this city.' But no doubt the deadliest shaft is reserved for loyal and innocent Balthasar, who could not have known until this instant what contribution to catastrophe was his: 'Then gave I her, so tutor'd by my art, / A sleeping potion; which so took effect / As I intended, for it wrought on her / The form of death.' Even as he speaks, the Friar himself knows nothing of Balthasar's role; it is only when the Friar's speech is done that Romeo's man blurts out the chilling words: 'I brought my master news of Juliet's death.'

Far from being a repetitious exercise best deleted on the stage, the Friar's speech is an indispensable part of the total experience of the tragedy; not to be present when some key participants learn how their acts resulted in the pile of bodies in the Capulet tomb—Romeo's, Juliet's, Tybalt's, Paris's—would be to miss too much. Of course many who played roles no less significant in Fate's progress are not present to hear: Mercutio, Tybalt, and Paris died without learning anything. Others who contributed links may or may not be present among the throng at the tomb, but in any event none would be able to recognize his own role from the Friar's sketchy account of what happened. Rosaline would not learn that but for her Romeo would not have attended the Capulet ball; the illiterate servant who sought Romeo's aid in reading names of those invited would never guess the connection between his request and the catastrophe; the servant who, at the ball, failed to identify Juliet as a Capulet and so left Romeo to commit his heart forever to a 'loathed enemy' before he was aware would never learn of his contribution; and as for the rival servants whose guarded quarrelling and thumb-biting initiated Fate's action, they would never learn, however long the Friar might talk, of the part they played.

The Prince promises 'more talk of these sad things', but we know that the gulf between us and the unknowing world will not be closed. Fate's practice, to which we alone have been privy, has buried its own tracks, and at the end we are left with knowledge unshared and unsharable, with pressures of awareness that will not be relieved. In the end Horatio's account will make Hamlet's story known to the world, and Lodovico's account will advise Venice of what happened on Cyprus. *Romeo and Juliet* is unique among the tragedies in leaving its burden of awareness upon us.

CHAPTER III

The Blind and the Blinded: *Julius Caesar*

'Why dost thou show to the apt thoughts of men
The things that are not?'

THE contrasts between *Romeo and Juliet* and *Julius Caesar* as
tragedies are as striking as those between any two of Shakespeare's
tragedies. The basic materials of the first are those of romance, of
fiction; those of the second are Roman history. The spirit of the first
is that of youth—fresh, vibrant, passionate, poignant; that of the
second is adult fact and cold reality. The protagonists of the first are
pawns, incidentals to Fate's purpose; the protagonist of the second is
directed by the choices that his own character mandates.

But of particular relevance here is the contrast in the dramatist's
use of practices and the discrepant awarenesses that they create. In
Romeo and Juliet Fate's enveloping practice dominates and directs
the entire action, and because of it we hold advantage over all
participants at all times, from the first moment to the last—and
beyond; two additional practices, the secret marriage and the use
of the potion, create broad discrepancies that are intensively
exploited over long periods of time. In *Julius Caesar* no enveloping
practice exists, and we hold exploitable advantage over only three
major participants during portions of only four of the seventeen
scenes in the play. During no scenes do multiple discrepancies exist.
Moreover, the dramatist's method sometimes differs drastically from
his more typical way in that he occasionally denies us special
advantage, as though he were content to keep us abreast of or even
slightly behind the participants' awareness of their situations. In
short, *Julius Caesar* makes less use of practices and discrepant
awarenesses than almost any other play of Shakespeare's.

It is possible, certainly, to take a contrary view and to consider
that, just as we hold advantage throughout *Romeo and Juliet* because
the Prologue has told us all, so also here we hold continuous

advantage because, like the Elizabethan audiences, we already know the main outline and the end of *Julius Caesar*: Caesar will be assassinated; the conspirators will be destroyed; Brutus will die; Octavius and Antony will inherit all. Further, the tragedy abounds with portents and prophecies that we recognize as true because of our historical knowledge, whereas the participants lack such advantage, though they hear the prophecies just as we do. But in this play as elsewhere (with the partial exception of *Troilus and Cressida*, where he openly trades on our knowledge of what name and fame Cressida would hold afterwards) Shakespeare prefers not to presume upon extra-dramatic knowledge held by the audience. It can be stated as a principle of his method that he rarely relies upon, presumes upon, or exploits advantages that we happen to hold by virtue of living at a later date than the action depicted. Rather, when he wishes us to enjoy an advantage over participants in particular situations, he takes pains to give us that advantage by dramatic means. His scenes are regularly built on the assumption that we will view the action by the light provided by the play, and by that light only. As to the prevalence of portents and prophecies in *Julius Caesar*, it is to be noted that they are made in the presence of participants, and not privily to us after the fashion of the Prologue in *Romeo and Juliet*. It is probable that we set more store by these advance warnings than the participants do; certainly when participants refuse to heed prophecies, as Caesar does just before he is murdered, we are expected to recognize our advantage in awareness and to view the scene by its light. But in such cases we need not rely on historical knowledge for advantage.

Julius Caesar anticipates *Hamlet* in leaving a measure of ambiguity about the status of our awareness, and thus again it differs sharply from *Romeo and Juliet*, where there is never a degree of ambiguity. The principal problems about our advantage, or lack of it, concern Caesar and Brutus, with secondary ones that concern Antony and Cassius. Just how early, for example, do we gain advantage over either Caesar or Brutus? And how long does our advantage over Brutus endure? The earliest possible moment in the case of Caesar comes in i. ii, when the Soothsayer bids him 'Beware the Ides of March'. The dramatist gives heavy weight to the warning by having the Soothsayer shout it a second time, directly to Caesar's face. Caesar dismisses both the man and his warning abruptly: 'He is a dreamer; let us leave him. Pass.' Does the advantage that we gain here derive simply from our historical knowledge of Caesar's assassination on the ides of March? Or does it derive from the fact that, as playgoers of a certain sophistication, we know that when soothsayers speak, they speak the truth? However it comes about, we do gain at this point our first insight into the flaw in Caesar's character that will shortly destroy him: it is his overweening pride that makes him vulnerable.

When we next see Caesar, our advantage has changed from generality to particular fact, for while he has been engaged, offstage, with being thrice offered and thrice refusing a crown, 'every time gentler than other', we have been watching Cassius whet Brutus against him. During nearly a hundred lines Cassius's forked tongue has been busy, so that when Caesar returns, flushed with fury, we know not only that his vanity and his pomposity invite assault but that he is now in fact marked for murder. Yet our advantage is not absolute, for Caesar is not totally blind to the danger of Cassius: 'Yond Cassius has a lean and hungry look, / He thinks too much; such men are dangerous.' But if not blind, he is blinded, in part by the reassuring words of Antony—'Fear him not, Caesar; he's not dangerous; / He is a noble Roman, and well given'—but mainly by his own pride, which repeatedly shows itself in his pronouncements. Much of his ostentation consists in his stubborn refusal to admit the idea of danger into his mind even when it beats at the door: 'Would he were fatter! but I fear him not. / Yet if my name were liable to fear, / I do not know the man I should avoid / So soon as that spare Cassius.' Thus he continues for a dozen lines that end in a veritable eruption of pomposity: 'I rather tell thee what is to be fear'd / Than what I fear; for always I am Caesar.' It is surely not accidentally that, directly after Caesar has thus puffed himself up to the dimensions of an immortal god, Shakespeare punctures him with a mortal pin: 'Come on my right hand, for this ear is deaf.'

Self-blinding ostentation continues to insulate Caesar's awareness from danger up to the last instant before his death. Before we see him again, we have been shown the perfervid preparations of Cassius, the blindly reasoned decision of Brutus to join the conspirators, and the meeting of 'the faction' where the final details of Caesar's murder are fixed. All that we have been shown are hard facts, which give us certainty of immediate action; the last lines that we hear before we next see Caesar are those of Brutus and Ligarius, on their way to Caesar's house:

Bru.   What it is, my Caius,
I shall unfold to thee as we are going
To whom it must be done.
Lig.   Set on your feet,
And with a heart new-fir'd I follow you,
To do I know not what; but it sufficeth
That Brutus leads me on.   (Thunder)
Bru.   Follow me, then.
(II. i. 329–34)

It is with the words and footsteps of the approaching assassins still echoing in our ears that we are introduced to the following scene— the first in the play upon which the dramatist concentrates his full exploitative power. Our advantage over Caesar here is towering; we have been advised already not only by the coldly agreed decisions of the conspirators but by thunder, lightning, and other baldly ominous events, and now we are further warned by Calpurnia's prophetic dreams of Caesar's imminent death. The storm, the freakish incidents of the night— '. . . ghosts did shriek and squeal about the street'—and the details of Calpurnia's dream all are laid as open to Caesar's understanding as to ours; he lacks only our sure knowledge of the conspirators' immediate purpose. But Caesar is vain and ostentatious not only publicly but privately, in his own house, alone with his wife—presumably, indeed, if he were quite alone he would put on a show to impress himself; therefore he spurns all warnings and insulates himself from awareness of danger as completely as if he were in fact totally ignorant of it:

Danger knows full well
That Caesar is more dangerous than he.
We are two lions litter'd in one day,
And I the elder and more terrible;
And Caesar shall go forth.
(II. ii. 44–8)

Clear-sighted Calpurnia succinctly defines the fact of the matter: 'Alas, my lord, / Your wisdom is consum'd in confidence.' When at last Caesar yields, it is not because he is aware of danger but because Calpurnia begs him to stay. A subtle distinction is here made between his private and his public face: alone with Calpurnia, he agrees to have Antony report that he is sick; but when Decius arrives and is instructed by Calpurnia to 'Say he is sick', Caesar instantly repudiates his private agreement: 'Shall Caesar send a lie? /

Have I in conquest stretch'd mine arm so far, / To be afeard to tell greybeards the truth? / Decius, go tell them Caesar will not come.' So, too, when Decius—who shrewdly knows on which of Caesar's strings to play—urges him to ignore dreams and come to the Senate lest it be said that 'Caesar is afraid', he quickly brushes off Calpurnia's renewed pleas: 'Give me my robe, for I will go.'

Upon the instant that this line is spoken, Shakespeare brings in the conspirators, who have come to usher Caesar to his death; it is one of the less subtle strokes in a generally unsubtle play in its exploitation of the gap between our awareness and Caesar's, and in fact from this moment to the end of the scene the dramatist's exploitation is anything but subtle. Thus Caesar greets his assassins with a burst of goodwill and jollity: 'Now, Cinna; now, Metellus. What, Trebonius! / I have an hour's talk in store for you; / Remember that you call on me to-day; / Be near me, that I may remember you.' Trebonius's reply is baldly exploitative: 'Caesar, I will; (aside) and so near will I be, / That your best friends shall wish I had been further.' In the interim between his departure from his house and his arrival at the Capitol our advantage over Caesar is again bolstered, first by Artemidorus, who tells us that he knows the names of the assassins and will so place himself as to be able to warn Caesar, and next by the Soothsayer, who advises us that he will once again attempt to put Caesar on his guard. At the entrance to the Capitol the Soothsayer vainly reasserts his earlier warning. Artemidorus tries next; the letter that he wishes Caesar to read immediately is, we know, more direct than the Soothsayer's prophetic argument, for it contains the names of those who even now begin to swarm about the unprotected Caesar. But again, predictably, Caesar's propensity for public ostentation serves him ill: to Artemidorus's insistence that his letter be read first because it 'touches Caesar nearer', Caesar replies grandly, 'What touches us ourself shall be last serv'd.' Thus, self-blinded, he spurns his last chance.

The same repeatedly illustrated passion for show prevents him from noting that the conspirators are pressing tightly about him, isolating him from his friends. Trebonius draws Antony aside; others crowd about, kneeling at his feet, shouting their pleas that Publius Cimber be repealed from banishment. Caesar's final speech, of some fifteen lines, uttered even as the assassins draw their daggers, is a veritable monument to pride. 'I am constant as the northern star,' he boasts, 'Of whose true-fix'd and resting quality / There is no fellow in the firmament.' So he continues, his mind sealed by pride against danger, until the daggers silence him.

In each scene with Caesar, Shakespeare has exploited the gap between his awareness and ours for incidental theatrical effect; great, glaring flashes of irony accompany his utterances on the street, in the privacy of his house with Calpurnia, and at the Capitol. The dramatist's handling of the dialogue and his manipulation of incidents to produce these effects are anything but subtle; it would be necessary to go back to such an early work as *Richard III* to find examples of equally bold and obvious exploitative technique— though indeed we shall note such effects again in abundance in *Othello*. Nor can it be doubted that these flashes contribute to the dramatic interest of the scenes in which they appear; almost, it may be asserted, they *are* the dramatic interest, and *Julius Caesar* would be drab without them. Nevertheless, it is not only for passing effects that Shakespeare uses Caesar's unawareness: it serves also as the means to Caesar's destruction, even as, in the preceding tragedy, it was made the means to the destruction of the lovers. But whereas, as we have noted, Romeo and Juliet were powerless to discern and effectively oppose the invisible force that was bent on ending them, Caesar, as we have noted also, bears a mind insulated against the approach of his assassins; prideful ostentation is the self-erected barrier that blinds him to reality. Thus Caesar's unawareness, unlike that of the lovers, derives from his own character. It owes nothing to any subtlety in the practice mounted by the conspirators against him, for indeed their practice is inept and crude, and might easily have been anticipated and thwarted by one less self-blinded than he.

The task of defining and tracing the relation of our awareness to Brutus's is much more difficult than that of marking the obvious gap between our awareness and Caesar's. Of what situations, exactly, or aspects of them, is Brutus actually unaware? When and how do we gain an advantage over him? Does his unawareness, like Caesar's, derive from some element, virtuous or vicious, in his own nature? And, above all, is there a causal relation between this character-caused unawareness—if such it is—and his destruction?

On the quintessential point, the question whether Brutus was 'right' in assassinating Caesar, the dramatist remains ambiguous to the end; or, in any event, he gives us no such certainty as we have when Othello kills Desdemona and when Macbeth kills Duncan. On the one hand, Shakespeare very quickly establishes a strong case against Caesar: he wants to be crowned, as Casca makes evident; he wants an heir, presumably to the throne that he aspires to make perpetual, for in his first scene he orders Calpurnia to stand in Antony's way, '. . . for our elders say, / The barren, touched in this holy chase, / Shake off their sterile curse'. And indeed Caesar shows himself everywhere to be a monster of egotism and show, wholly dictatorial and peremptory in his relations with his fellow Romans. Further, Brutus fears that, once crowned, he will worsen: 'He would be crown'd; / How that might change his nature, there's the question.' From what we are shown of it, his nature is intolerable while he yet lacks a crown; if Brutus is right, and crowning would make him even worse, then, on the surface, it appears that our own view and Brutus's view of the necessity for his assassination are the same.

But the total play also presents evidence for the other side of the case: Antony's extolling arguments are mightily persuasive; so is the will left by Caesar, bequeathing, in effect, all his works to the people; and, in the end, Rome under the rule of the coldly calculating, mutually distrustful, mutually jealous triumvirs appears at best to be no better off than it would have been under Caesar and at worst to be in worse circumstances.

If the dramatist does not, then, unmistakably distinguish our awareness from Brutus's on the question of the necessity for Caesar's death, he does, however, sharply differentiate it from the hero's on the main dramatic, as opposed to political or historical, question. We learn early, whereas Brutus never learns, that he is the wrong man for the task that he undertakes, and indeed the tragedy as a whole is essentially a demonstration of this fact. From time to time we hold other advantages over Brutus, but they are secondary and concerned rather with the creation of incidental effects than with the outcome. We do not gain our main advantage abruptly, at one stroke, as we do, for example, over Othello after hearing but a single speech of Iago's even before we are introduced to the Moor; though our advantage over Brutus starts early, it builds gradually during the first two acts.

It begins, actually, with our perception of the character and motives of Cassius. After only two of his speeches during the long conversation of the two men in I. ii we recognize his intent first to sound Brutus out and then to whet him against Caesar. But we do not here gain full advantage, for Brutus quickly signifies his own sense of Cassius's purpose: 'Into what dangers would you lead me, Cassius, / That you would have me seek into myself / For that which is not in me?' Brutus's unguarded choice of the word 'fear'— 'I do fear, the people / Choose Caesar for their king'—gives the eager Cassius his first break, whereupon he launches into two long and vicious tirades against Caesar, to which Brutus responds with some show of awareness: 'What you would work me to, I have some aim.' Next we hear Caesar, with more astuteness than he displays elsewhere, describe Cassius as one who does not sleep of nights, looks lean and hungry, reads and observes much, looks 'quite through the deeds of men', smiles as if he mocked himself for smiling, is suspicious by nature, and is therefore 'very dangerous'. Still, our advantage over Brutus in understanding Cassius does not become dramatically useful until Cassius's soliloquy that ends the scene:

> Well, Brutus, thou art noble; yet, I see,
> Thy honourable metal may be wrought
> From that it is dispos'd; therefore it is meet
> That noble minds keep ever with their likes;
> For who so firm that cannot be seduc'd?
> (I. ii. 312–16)

Earlier we heard Brutus's own statement about himself:

> If it be aught toward the general good,
> Set honour in one eye and death i' th' other,
> And I will look on both indifferently;
> For let the gods so speed me as I love
> The name of honour more than I fear death.
> (Ibid. 85–9)

With Cassius's final lines of the scene, promising that he will have letters thrown in at Brutus's window as if they came from sundry Romans all pointing at Caesar's ambition and urging Brutus to act, the distinction between our awareness and Brutus's becomes vivid. Brutus, we see, is ignorant of Cassius's dishonesty and ignorant that their two creeds are incompatible.

Yet it should be noted that at the same time we gain advantage over Brutus with respect to Cassius's qualities, we also gain advantage over Cassius with respect to those of Brutus. That Cassius has grossly misjudged his man is evinced by the first lines of his soliloquy: 'Well, Brutus, thou art noble; yet, I see, / Thy honourable metal may be wrought / From that it is dispos'd.' Cassius's own crooked nature recognizes the enterprise upon which he is engaged as a dishonourable one, and he mistakenly believes that Brutus's honour can be corrupted to accommodate the crooked route that must be travelled. Not only Cassius, but the other conspirators as well recognize the dubious nature of their undertaking, and it is precisely therefore that they wish to include Brutus, who, as Casca says, '. . . sits high in all the people's hearts / And that which would appear offence in us, / His countenance, like richest alchemy, / Will change to virtue and to worthiness'. But in desiring Brutus's participation because his reputation will put a face of honour upon the assassination and in assuming, at the same time, that Brutus's 'honourable metal may be wrought / From that it is dispos'd', Cassius is self-deceived; the course of the tragedy, beginning even before the assassination and lasting to the end, confirms our own sense, which is already truer than Cassius's, that Brutus's honour is inflexible and cannot be 'wrought'.

Against the exhibition of the conspirators' dark and sinister plottings, the dramatist immediately sets Brutus's high-minded and brooding soliloquy on the necessity for Caesar's death. It is a noble exercise in idealistic logic, and it confirms all that earlier indications have led us to expect of Brutus's absolute dedication to honour; indeed, Brutus is not merely dedicated, but addicted to honour: 'I know no personal cause to spurn at him, / But for the general.' Thus he reasons for a total of twenty-five lines, never deviating from the theme of personal devotion to the general cause, the good of Rome. But at the same time that it further stresses Brutus's nobility, the speech also evinces a certain blindness to reality; speaking directly after we have heard Cassius and his fellows darkly plotting to ensnare him in order to give respectability to their enterprise, Brutus, in his total unawareness of how differently Cassius conceives this enterprise, shows himself to be the naïve dupe of his own nobility. His speech is barely finished when the boy Lucius enters with a paper 'thus seal'd up' that he had found by a window, and we are reminded of Cassius's words: 'I will this night, / In several hands, in at his window throw / . . . . Writings all tending to the great opinion / That Rome holds of his name.' The nobly reasoned soliloquy is thus wedged between proofs, before and after, that the realities of conspiracy are not at all as Brutus supposes them to be.

Brutus, then, makes up his mind to kill Caesar as result of Cassius's urging—of that face-to-face urging that we witnessed, and of the dishonest urging represented by forged letters thrown in at the window. These latter Brutus gullibly assumes to have come from citizens at random. 'Such instigations have been often dropp'd / Where I have took them up', he remarks as he reads the latest of them. And thereupon he commits himself irrevocably to the cause:

'Speak, strike, redress!' Am I entreated
To speak and strike? O Rome, I make thee promise,
If the redress will follow, thou receivest
Thy full petition at the hand of Brutus!

(II. i. 55–8)

It is true that, as he tells Cassius, Brutus had thought much about the danger of Caesar to Rome even before Cassius sounded him; it is therefore conceivable that, prodded only by patriotism, he would ultimately have reasoned his way to the conclusion that Caesar must be removed. It is even possible, though unlikely, that he would eventually have moved beyond mere philosophizing and initiated a conspiracy to kill Caesar. But the hard fact remains that, as Shakespeare represents the case, Brutus joins the conspiracy as a direct result of Cassius's campaign to win him—'Since Cassius first did whet me against Caesar, / I have not slept'—and in ignorance of Cassius's having practised on him with forged letters.

At the same time, of course, the dramatist's handling of the crucial point of Brutus's enlistment in the conspiracy leaves no doubt that the hero succumbs not because Cassius's powers of persuasion are peculiarly potent but because Brutus's own character makes him peculiarly susceptible. Cassius's two tirades against Caesar in I. ii reek of selfishness; envy, bitterness, private grudge all show themselves only too plainly in both the words and the venomous emphases of the speaker. Perhaps it is not to go too far to suppose that almost any man other than Brutus would have marked Cassius for a sore loser and turned his back on him; but Brutus is blinded by his very honesty, which makes him incapable of suspecting another's capability of dishonesty. He falls victim to Cassius's practice because his own honesty limits his vision of the world's way. It comes as a shock to him that, according to Lucius, the conspirators have arrived at his door with 'their hats . . . pluck'd about their ears / And half their faces buried in their cloaks'. It had not occurred to him that this conspiracy, being directed to an honourable end, should need secrecy and furtiveness: 'O Conspiracy, / Sham'st thou to show thy dangerous brow by night, / When evils are most free?'

A second shock to the purity of his own impulses occurs shortly after, when Cassius proposes that the conspirators swear an oath of fidelity to their 'resolution'. To Brutus an enterprise such as this requires no swearing: '. . . what other oath / Than honesty to honesty engag'd / That this shall be, or we will fall for it?' He continues on this plane for twenty-five lines, and the speech is as noble as any that Shakespeare gives to one whose every thought is noble. But, standing where it does, delivered amid a gang of conspirators whose caps are pulled over their faces, and whose leader has tricked Brutus into joining them by throwing false letters in at windows, the speech serves most of all to fortify our sense that Brutus's qualities are wrong for this kind of enterprise.

Then, in quick succession, Brutus twice more shows his unfitness for his chosen task. Presumably the dramatist here intends us to recognize his hero's errors of judgement at the moment they are committed, though some doubt exists that we are sufficiently advised to do so. First, Brutus abruptly overrules the recommendation made by Cassius and seconded by all the others that Cicero be invited to join them. Metellus Cimber's urging is the most persuasive: '. . . for his silver hairs / Will purchase us a good opinion / And buy men's voices to commend our deeds'. Brutus's lame rebuttal is only that Cicero 'will never follow anything / That other men begin'—but even this is sufficient to end the matter. Later, in the market-place, when Antony easily turns the citizens against Brutus's faction, we are in position to recognize how badly Brutus erred in leaving Cicero out, for then those silver hairs might have given Antony some competition. But at the instant Brutus's decision is made we have only the conspirators' arguments, chiefly Cimber's, to advise us that Brutus is wrong.

In the second instance, where Brutus's decision is of even greater moment, we are again insufficiently informed to be positive of the error at the moment it is made. Cassius argues that Antony must die with Caesar, but Brutus, in another of his nobly impassioned speeches, again overrules him: 'For he can do no more than Caesar's arm / When Caesar's head is off'. Again it is only later, in the market-place, when Antony's oration converts the citizenry into a rabid mob, that we can recognize with certainty how very faulty was Brutus's decision. Though at the time the decision is made we are sufficiently acquainted with Brutus to expect his nobility to misguide him, and though we are sufficiently acquainted with

Cassius to suspect that his judgement in such matters is superior, still, Brutus's speech is eloquent and moving, and Cassius makes only a token resistance to it. Had Cassius argued with conviction, we might have been made sure that Brutus was wrong, and our advantage would accordingly have been unmistakable; but Cassius has too much at stake to risk losing the newly won Brutus by adamantly opposing his wishes at this point. Yet what is most needed here to make our advantage certain is fuller knowledge of Antony. We cannot know whether it is dangerous to let him live, for Shakespeare has as yet given us no reason to think him any more than 'a limb of Caesar'. We saw him first half-naked, dressed 'for the course', seemingly wholly subservient to Caesar: 'When Caesar says "Do this", it is performed.' We saw him next when, after the games, he replied to Caesar's 'Yond Cassius has a lean and hungry look' by blandly reassuring his master: 'Fear him not, Caesar; he's not dangerous.' Antony speaks this categorical assessment immediately after we have seen Cassius plotting Caesar's murder and seducing Brutus to join in the conspiracy. We learn at the same time that Antony, unlike Cassius, with whom Caesar contrasts him, is a lover of plays; and though we have not yet heard Caesar's masterful description of him as one who 'revels long o' nights', the implication is evident: Antony is a playboy, and no threat to the conspiracy. By giving us but one hint of Antony's latent power, Shakespeare could have ensured our advantage at the moment that Brutus overrules Cassius's argument for killing Antony with Caesar; we could then have perceived more clearly the double excellence of the superb 'Let's carve him as a dish fit for the gods' speech—both for its marvellously conceived and noble sentiments and also for its rich irony as the speech of a hero whose very nobility causes him to mistake his situation utterly. What finally lulls the doubts of Cassius and the others is Brutus's brilliantly persuasive misjudgement of Antony: 'If he love Caesar, all that he can do / Is to himself —take thought and die for Caesar.' When he adds that Antony 'is given / To sports, to wildness, and much company', he further clouds our own awareness, since up to this time we have had no evidence to the contrary. And of course his words silence the opposition, when Trebonius speaks for all: 'There is no fear in him; let him not die; / For he will live, and laugh at this hereafter.' Doubtless the dramatist intended all these lines, both those by which Brutus misleads and the conspirators express their acquiescence, to exploit the gap between the participants' and our awareness; but the effect is compromised by our uncertainty about Antony.

The relation of our awareness to Brutus's in regard to Antony's capabilities does not cease to be ambiguous with the decision to kill only Caesar. Directly after the assassination, Antony sends a servant to Brutus and Cassius with this soothing message:

> Brutus is noble, wise, valiant, and honest;
> Caesar was mighty, bold, royal, and loving;
> Say I love Brutus, and I honour him;
> Say I fear'd Caesar, honour'd him, and lov'd him.
> If Brutus will vouchsafe that Antony
> May safely come to him, and be resolv'd
> How Caesar hath deserv'd to lie in death,
> Mark Antony shall not love Caesar dead
> So well as Brutus living; but will follow
> The fortunes and affairs of noble Brutus
> Thorough the hazards of this untrod state
> With all true faith.
>
> (III. i. 126–37)

Again, we are given no reason to doubt the truth of this message or to take it other than as Brutus does. We have had no private 'aside' from Antony, and, as we do not yet know the man well enough to distrust him, we are obliged to assume that his intent is just what his words make explicit. If we do gain any advantage over Brutus at this point, we do so because of Cassius's warning: '. . . but yet have I a mind / That fears him much, and my misgiving still / Falls shrewdly to the purpose'. We have already had some indication of the truth of Caesar's observation that Cassius 'looks quite through the deeds of men', and if we had faith earlier in the superiority of his judgement to Brutus's, perhaps we should also have faith here. In that event, we do hold advantage over the hero at this point, just before Antony enters, and should not take the servant's words at face value as Brutus does.

The basic question, however, continues unresolved throughout the hundred lines between Antony's entrance and his soliloquy that ends the scene. The content of his speeches to the conspirators during this period leaves no hold to be taken whether he is or is not perfectly sincere:

> I doubt not of your wisdom.
> Let each man render me his bloody hand.
> First, Marcus Brutus, will I shake with you;

Next, Caius Cassius, do I take your hand;
Now, Decius Brutus, yours; now yours, Metellus;
Yours, Cinna; and, my valiant Casca, yours;
Though last, not least in love, yours, good Trebonius.
Gentlemen all,—alas, what shall I say?

(Ibid. 183–90)

Seen in the perspective of the whole play, each salutation is a biting irony uttered by Shakespeare's supreme master of irony. But in the immediate context we cannot know that these words are ironical at all. Repeatedly, as the scene proceeds, Antony professes himself overcome by his emotions—'Sway'd from the point by looking down on Caesar'. To Brutus, his display of overflowing feeling presents high recommendation of sincerity and perfect faith. Have we ourselves been prepared to see more clearly than Brutus here?

Without Cassius we would surely find ourselves as much deceived by Antony as Brutus is, for we have as yet learned nothing about him to give us clear advantage. Nothing that he has said earlier has prepared us to discern the deep current of irony that, as we learn shortly afterwards, flows through his being. If we do hold advantage, it must yet derive from our confidence in the perspicacity and shrewdly suspicious nature of Cassius, whose mind 'fears him much'. When Antony, agonizing over Caesar's body, speaks of the conspirators as hunters 'Sign'd in thy spoil, and crimson'd in thy Lethe', Cassius hotly cuts him off with a demand for a forthright commitment: 'But what compact mean you to have with us? / Will you be prick'd in number of our friends; / Or shall we on, and not depend on you?' But Antony's reply is so honest-seeming that it disarms even Cassius: he was, he says, 'Sway'd from the point by looking down on Caesar. / Friends am I with you all and love you all.' Moved by this show of affection, Brutus promptly makes the gravest error of his tragic course—aside, certainly, from his initial error in joining in the conspiracy at all; to Antony's plea for permission to speak in Caesar's funeral, he answers eagerly with the one response compatible with his nature: 'You shall, Mark Antony.' At this crucial point we need to be in a position to recognize the enormity of his error, and to recognize it instantly, with no uncertainty. But again our sole adviser is the crooked Cassius, speaking privately to Brutus: 'You know not what you do. Do not consent / That Antony speak in his funeral. / Know you how much the poeple may be mov'd / By that which he will utter?' It is the fourth time Cassius has directly warned Brutus, and is the most pointed of his warnings against Antony; yet, as before, he lamely yields when Brutus reassures him. 'I know not what may fall; I like it not,' he says, and says no more. If our awareness exceeds Brutus's here, it is not because of what the dramatist has shown or told us of Antony, whose dominant image as a playboy, fixed earlier, still stands; it is because, lacking Brutus's native blindness, we set a higher estimate on Cassius's shrewdness.

Ten lines later whatever doubts have lingered with us are un-equivocally resolved and vanish utterly, for immediately after the conspirators have left him alone, Antony rises full-grown, to the stature of a giant:

> O, pardon me, thou bleeding piece of earth,
> That I am meek and gentle with these butchers!
> (Ibid. 254–5)

He continues with another twenty of the most thunderous lines in Shakespeare, building to a climax that curdles the blood: 'Cry "Havoc!" and let slip the dogs of war.' At this speech hindsight recasts for us the blurred earlier moments when Brutus asserted that Antony need not die, being only a limb of Caesar; when he allowed Antony to come in to the company of assassins and assure them of his love; and, above all, when he agreed to let Antony speak in Caesar's funeral. But why did Shakespeare wait until this moment to let us see what Antony is? It was certainly not his usual way, in tragedy, history, or comedy. In *Richard III* we are told boldly, in advance of critical scenes, just what Richard is, and, scene by scene thereafter, are provided with pertinent information to remind us what he is, so that, unfailingly, we can enjoy perfect advantage over his successive dupes. In *Othello*, Iago spreads his plans out before us like a map on a table, so that we invariably see clearly what, in succession, Roderigo, Cassio, Othello, and Desdemona do not see at all. In the comedies we are regularly provided beforehand with the bright heroines' secrets—Rosalind's masquerade in the Forest of Arden, Viola's in Illyria, Portia's in Venice—so that we can relish the one-sided hilarity of the great comic scenes. In *Much Ado About Nothing* we know at the time Claudio denounces her that Hero is innocent, and are not left, like the bridegroom, to learn the truth later. But such examples in the plays are endless. Why, then, did the dramatist deviate from his usual way by allowing us to remain virtually as unaware as Brutus of Antony's formidable talents?

Had he followed his typical practice, he would have advised us at latest by the time of Brutus's refusal to kill Antony with Caesar. Thus the ambiguity of our awareness during the key scenes that build directly towards the climactic peak of Antony's oration would have been clarified. Possibly Shakespeare simply trusted to our historical knowledge of Antony's prowess to keep us sure of the truth even while the play itself represents him as but a reveller subservient to Caesar; but, as we have noted, it was not Shakespeare's way to presume on external knowledge. Further, it is unlikely that, at this point in his career, he merely failed to recognize that his own neglect to advise us clearly had left our advantage uncertain. It must be, then, that the ambiguity was by design. But what design? Perhaps the dramatist preferred to keep our awareness as uncertain as Brutus's lest the hero's prestige be diminished; if we saw clearly, at the moment he commits them, how monumental Brutus's errors are, would not his heroic stature suffer? Later, with Othello, Shakespeare took no such precaution, but quickly opened an enormous gap between the Moor's perception and ours of Iago—with the result that, to many spectators in the theatre, the Moor looks like a fool. By allowing Antony to deceive us almost as badly as he deceives Brutus, the dramatist at least partly protected his hero from the same fate.

Another possible explanation is that we are denied early knowledge of Antony's prodigious talents because the dramatist wished to shoot him suddenly into our view like a meteor after the high dramatic peak of the assassination, when the tragedy so desperately needs something phenomenal to prevent a disastrous decline of interest. Possibly Shakespeare, in hiding Antony's talents for so long, was influenced by the example of his own Prince Hal, who would have been much in his mind at this very time. It was Hal's practice to 'imitate the sun, / Who doth permit the base contagious clouds / To smother up his beauty from the world, / That when he please again to be himself / Being wanted, he may be more wond'red at.' His sudden change, argues Hal, 'Shall show more goodly and attract more eyes / Than that which hath no foil to set it off'. At a later point the rebel Vernon replies to Hotspur's scornful reference to Hal—'The nimble-footed madcap Prince of Wales'—with discomfiting words: 'I saw young Harry with his beaver on, / His cuisses on his thighs, gallantly arm'd, / Rise from the ground like feathered Mercury, / And vaulted with such ease into his seat / As if an angel dropp'd down from the skies / To turn and wind a fiery Pegasus.'

Just so, Antony's formidable powers, having been long held in obscurity, suddenly loom before us in the tremendous 'Cry Havoc!' soliloquy. That speech is followed immediately by what we now easily recognize as an ominous declaration; commanding Octavius's servant to accompany him, he sets off for the market-place: 'There shall I try, / In my oration, how the people take / The cruel issue of these bloody men.' It accords with Shakespeare's usual manner that just after hearing this declaration, we are shown Brutus, naïvely assuring the people that 'public reasons shall be rendered / Of Caesar's death'. It is the first time that we have held unqualified advantage over Brutus with respect to Antony, and, what is more, the gap widens as the scene proceeds. Shakespeare gives Brutus a curiously cold and formal speech embellished with obvious rhetorical devices; rhetoric aside, the message is the same by which Brutus had persuaded himself to enter the conspiracy: 'Not that I lov'd Caesar less, but that I lov'd Rome more'. Earlier we learned of Brutus's perfect confidence in the power of simple truth to affect others as it affected himself; and, indeed, momentarily simple truth suffices to win the citizens, who had begun by howling, 'We will be satisfied! Let us be satisfied!' After he has spoken, they all cry, 'Live, Brutus! live, live!' Brutus must be carried home in triumph, given a statue with his ancestors, and, at the peak of their frenzy: 'Let him be Caesar.'

But in the opening scene we witnessed the fickleness of the populace, and only a moment before Brutus speaks we heard Antony vow to try the 'cruel issue of these bloody men' in his oration. At this point, therefore, we are well prepared for the play's boldest exploitation of the gap between our awareness and the hero's. With all the citizens shouting his praises and protesting their loyalty to him, Brutus actually begs them to remain in the market-place to give Antony a hearing: '. . . let me depart alone, / And, for my sake, stay here with Antony'. And he concludes: 'I do entreat you, not a man depart . . . till Antony have spoke.' So he departs, leaving the lambs to the wolf, blindly confident that his cause has been vindicated and that all hearts are his own forever; as for ourselves, we hardly need to hear Antony's oration in order to know its outcome. So smoothly does he blend emotion with mocking irony that after only a score of lines he has reversed the minds of his hearers: 'Caesar has had great

wrong', and 'There's not a nobler man in Rome than Antony.'
Another twenty, and the citizens are so aroused that he can scarcely
quiet them in order to continue. Another twenty-five, during which
he adroitly exhibits Caesar's body, and the hearers have become a
frenzied mob: 'Seek! Burn! Fire! Kill! Slay!' In their mindless fury
they forget even that they have not heard Caesar's will, and again
Antony must quiet them. When they have heard it, they dash off in
all directions to dismember Rome. 'Mischief, thou art afoot,' exults
Antony: 'Take thou what course thou wilt!' Elsewhere in Shakespeare's plays we are conscious of quick, passing flashes of irony;
but Antony's tremendous oration is one massive block of irony, as
hard as granite. Perhaps the dramatist erred by denying us early
sight of Antony's prodigious powers by which we might have had
clear advantage over Brutus when he commits his fatal errors and
thus have enjoyed the effects of an exploitable gap over a longer
period. But more likely his decision to surprise us was better
calculated, for Antony's performance, coming with startling suddenness and concentrated within a few moments at dead centre of the
play, truly achieves the mesmerizing effect of a meteor.

After the high peak of Antony's oration, the direction of *Julius
Caesar* is inevitably downwards. If there is a fault in the tragic
structure as a whole, it is that Shakespeare did his work too well in
the first three acts, leaving the remaining two only to move routinely
to a foregone conclusion.

Between the time Brutus departs from the Forum, sure of his
success with the citizens, and his next appearance in the camp at
Sardis, the result of Antony's speech, anticipated by us even before
Antony spoke, becomes known to Brutus also. Though in the somewhat disarrayed final scenes our knowledge is surer than his that the
end must be the death of all his hopes and the triumph of Antony
and Octavius, we never again hold perfect advantage over him, for
he, too, is aware of the probability of this end. The Ghost of Caesar
advises him bluntly, and Cassius warns him of the significance of
the ravens, crows, and kites that 'Fly o'er our heads and downward
look on us'. Our edge over Brutus before the last battle thus marks
only the difference between probability and certainty. 'O, that a man
might know / The end of this day's business ere it come!' he
laments; but his expectation is close to our own. Brutus and Caesar
are the only persons in the play over whom we hold significant
advantage during any considerable period of the play. Our advantage

over Caesar ends, of course, when the conspirators' knives strike him, and our advantage over Brutus ceases, for all practical purposes, as soon as he learns that Antony has reversed the citizens' minds. Brutus and Caesar are also the only persons in the play whose own natures blind them or make them peculiarly vulnerable to deception. Antony and Cassius, the other principal figures, do not blind themselves and are all but invulnerable to deception. What remains here, then, is to review briefly the history of the awareness of each of these men in relation to our own.

Of all major persons, Antony consistently occupies the highest level of awareness; indeed, as result of Shakespeare's failure to characterize him fully during the opening scenes, he may be said to stand above our own level until his soliloquy at the end of III. i. Very briefly, at two points, he does see less clearly than we, but in neither instance is he more than a bystander to the main action. On the first occasion, we hear Antony say of Cassius, 'Fear him not, Caesar; he's not dangerous'—directly after we have watched Cassius skilfully whetting Brutus's mind against Caesar; here, however, it is not Antony but Caesar whose unawareness is exploited, for it is Caesar whose life is at stake. Though at the time Antony speaks the line we do not yet know him well enough to form any opinion of his acuteness, yet in retrospect his gross misjudgement of Cassius and categorical dismissal of him as a danger appears as one of the more curious facts of the play; for, considering his later brilliance, it seems incredible that he should err so badly. It is tempting to conjecture wildly here, to question whether Antony does truly underestimate Cassius, or whether he deliberately lulls Caesar with an eye to his own ambitions should Caesar be eliminated. No hard evidence exists for so drastic a view; but Antony's later acumen in all things is hard to reconcile with such a blunder. What is more, the exhilaration of spirit and the alacrity with which he seizes the reins to command the situation immediately after Caesar's death might serve to strengthen a suspicion that he is not wholly desolated by Caesar's assassination. It goes without saying that the emotions he displays with the conspirators and the citizens after the assassination and at the funeral are part and parcel of his oratorical showmanship and as profoundly ironic as his repeatedly uttered and shrewdly placed mockeries of Brutus and Cassius as 'honourable men'.

On the second occasion of his possible unawareness Antony is again a bystander, Caesar the central figure. Antony enters the

Capitol with Caesar and is present during the first twenty-five lines
that precede the conspirators' final manœuvrings and the assassination itself. He speaks not at all during this period, and his departure
is marked only by Cassius: 'Trebonius knows his time; for, look you,
Brutus, / He draws Mark Antony out of the way.' It is curious that
Shakespeare keeps the talkative Antony totally silent during this
period; possibly he wished merely to avoid drawing particular
attention to him. On the other hand, when the facts of his silence and
his convenient exit just before the murder are added to the fact of his
earlier gross misjudgement of Cassius, it becomes still more tempting
to think that Shakespeare, at the very least, meant to leave the way
open to varying interpretation. Doubtless we are meant to assume
that Antony is wholly ignorant of the impending murder; yet
ignorance is incompatible with the highly circumspect character of
Antony that becomes apparent immediately afterwards. Perhaps we
are meant to suppose that Antony simply stands by and lets matters
take their course; if that were so, then Brutus and Cassius are not so
much his enemies as his fools, even as Macbeth is Fate's fool and as
Hotspur is Hal's 'factor'.

With Cassius as with Antony no scene gives us unqualified
advantage that is dramatically exploitable. Cassius, like Antony, sees
everything about him, and Shakespeare takes care with him as with
Antony to see that he is not exposed at a disadvantage. In relation to
one matter of first importance perhaps we do hold brief advantage;
in relation to two or three others it appears that we share our unawareness rather than our awareness with Cassius. First is the
difference between Cassius's and our understanding of Brutus's
honour. We have already noted how Brutus mistakes Cassius's
character and purposes in simply assuming that Cassius is ruled,
like himself, by a passion for the good of Rome. But we have already
noted also that just as Brutus's honour blinds him to Cassius's
crookedness, so Cassius's crookedness blinds him to Brutus's honour.
Cassius values Brutus as a conspirator for his unsullied reputation:
'. . . he sits high in all the people's hearts', he says in seconding the
choice. But Cassius is clearly incapable of discerning the quality of
Brutus's honour, and so supposes that his 'honourable metal may
be wrought / From that it is dispos'd'. The fact is, and certainly we
mark the fact long before Cassius does, that Brutus's honour is not
only undefiled but undefilable; in the whole course of events from
his entrance into the conspiracy until his death Brutus never

71

deviates from the narrow path of honour. He will have none killed but Caesar; he trusts Antony's honour as he does his own and therefore allows him to take part in Caesar's funeral as a proper and fitting act; he tells the citizens the exact truth about the assassination; and, in the midst of war, he will tolerate no dishonourable means of supporting the army: 'For I can raise no money by vile means. / By heaven, I had rather coin my heart / And drop my blood for drachmas than to wring / From the hard hands of peasants their vile trash / By any indirection.' It appears, then, that villains can make errors, too: Cassius discovers too late that the very quality that makes Brutus seem so useful a colleague in a dishonourable enterprise also renders him unfit for it.

Less certain is our advantage over Cassius with respect to Antony; perhaps, indeed, we should reverse the case and consider Cassius's advantage over us with respect to Antony. Shakespeare tells us much of Brutus—in fact, all that is to be learned—during the first scenes of the play, and thus we know Brutus better than Cassius does. But, as we have noted, he tells us very little of Antony, and what he does tell proves inaccurate. All that we learn of Antony that proves to be accurate is what we learn from Cassius's repeated warnings, and so, since Cassius is our instructor, it appears we should recognize his awareness of Antony as superior to our own up until the time that Antony suddenly emerges, full-grown, just before his oration. The difficulty is that, though he warns repeatedly, Cassius also yields so easily to Brutus on the critical issues respecting Antony as to appear unsure of his grounds; in any event he can hardly have known as much of Antony's capabilities in advance as we ourselves shortly afterwards come to know, for, had he known, he would surely have offered more than a token resistance to Brutus's fatal decisions. Since we are in the position of dramatic spectators while he is an active participant, probably we take Cassius's warnings more seriously than he himself does; paradoxically, thus, our awareness of Antony may be superior to his even while he is the source of our misgivings. Late in the play, when all has been lost, Cassius is bold enough to say, in effect, 'I told you so' to Brutus: 'Now, Brutus, thank yourself; / This tongue had not offended so to-day / If Cassius might have rul'd.' But the fact is that he gave in to Brutus twice on vital issues, and did so without a struggle. What is more, and worse, just after he has been so unkind as to say 'I told you so' to Brutus, he contemptuously repeats the same old characterization of Antony

that Caesar had given: Octavius, he insists, is 'A peevish schoolboy'
who is 'Join'd with a masker and a reveller!' So steeped is he in his
own sourness that he never does come to recognize, or to acknowledge, that this masker is a Titan.

Cassius's vision is imperfect in two additional scenes, but so is our
own. All told, in *Julius Caesar* Shakespeare limits our vision on four
occasions: first, when Cassius whets Brutus against Caesar and we
are not advised until Cassius's later soliloquy that his motives are
dishonest; second, when Brutus grants Antony permission to speak
in Caesar's funeral and we learn only later how faulty was his
judgement; third, much later, when Brutus and Cassius have their
bitter quarrel and we are left until the end to learn the true cause of
Brutus's strange behaviour; and, fourth, later still, when we do not
learn until after Cassius's suicide that Brutus has won, not lost, his
battle. In all of Shakespeare's other plays together there are hardly
as many as four scenes in which key secrets are withheld from us
until the end of the scene to which knowledge of the secret is vital.
In *The Comedy of Errors* we are denied knowledge of Aemilia's
survival until the end, and in *The Winter's Tale* the dramatist cold-
bloodedly deceives us about Hermione's 'death', giving us the same
false report that is given the participants. Nowhere else are we
deliberately deceived about matters of first importance. Those about
which we are either deceived or simply allowed to remain ignorant
in *Julius Caesar* are of lesser magnitude, but the fact that they involve
four occasions indicates a deviation from the dramatist's usual way
of deceiving participants at will but keeping us advised in advance.

The first two occasions, both of which concern the inadequate or
misleading information we are given about Antony, we have already
noted. The third and fourth, to which we now turn, do approach
outright deception, though we are not given, as in *The Winter's Tale*,
directly false information. In the first of these the dramatist
astonishes us over a period of more than a hundred lines of the
Brutus–Cassius quarrel scene in IV. iii by the incredible alteration
that appears to have taken place in Brutus's character since we last
saw him. No longer the gentle Brutus whose very mildness has
involved him in these dire straits, he is here inexplicably quarrelsome, rude, contemptuous, even savage. Though we recognize that
he has cause to be angry with Cassius, we are quite unprepared, as
is Cassius, for the ferocity of his verbal onslaught which quickly
reduces Cassius from initial shock and feeble defence to whimpering

and tears: 'Strike, as thou didst at Caesar; for, I know, / When thou didst hate him worst, thou lov'dst him better / Than ever thou lov'dst Cassius.' When the Poet comes to the door to urge the disputants to 'Love, and be friends', Brutus shocks us by silencing him with a snarl: 'Get you hence, sirrah; saucy fellow, hence!' And he continues nastily: 'What should the wars do with these jigging fools?' The shaken Cassius voices our own disbelief: 'I did not think you could have been so angry.' It is only at this point, after he has sustained our astonishment for over a hundred lines, that Shakespeare abruptly reveals the cause of Brutus's uncharacteristic behaviour: 'Portia is dead.' This scene is unique in Shakespeare; elsewhere we sit comfortably while the dramatist exploits the unawareness of participants, but here he actually exploits our own. That the delayed revelation of Portia's death creates a sudden, sharp emotional effect when it comes is obvious; possibly, even, the dramatic experience is greater than if we, but not Cassius, had known in advance. But the point is not that the effect is more or less potent; the point just now is only that Shakespeare's handling of the scene represents a reversal of his regular method, and resembles most of all the earlier deviation by which he kept us from knowing Antony's strength until the time was ripe to give us a special thrill by revealing it.

The fourth deviation is similar though on a much reduced scale. In this final scene of the kind there is no sustained exploitation of our unawareness, but there is an unaccountable and clearly deliberate misleading of our awareness. As in the preceding scene, it is Cassius's unawareness that we are made to share. We are first advised that Brutus's forces have been defeated:

> O Cassius, Brutus gave the word too early;
> Who, having some advantage on Octavius,
> Took it too eagerly. His soldiers fell to spoil,
> Whilst we by Antony are all enclos'd.
>
> (v. iii. 5–8)

Cassius sends Titinius to investigate 'Whether yond troops are friend or enemy', then quickly sends Pindarus to observe and report Titinius's findings. Watching from afar, Pindarus reports that Titinius has been seized by enemies, whereupon Cassius has Pindarus stab him. Only thereafter are we told that Brutus had won, not lost, and that Titinius had alighted among friends, not foes. 'O hateful error,' says Messala, 'Why dost thou show to the apt thoughts

74

of men / The things that are not?' In the tragedy that just preceded
Julius Caesar, Romeo kills himself unaware that Juliet is not dead
but only sleeps; here, Cassius kills himself unaware that Brutus has
not lost, but won. But there the parallel ends, for of course we know
beforehand that Juliet only sleeps, whereas here we have been made
to share Cassius's belief that Brutus has lost.

CHAPTER IV

The Hero as Practiser: *Hamlet*

'... let me speak to th' yet unknowing world
How these things came about.'

THOUGH in notable ways *Hamlet* is the most abundantly Shake-spearian of Shakespeare's tragedies, its critical history has made it appear conspicuously un-Shakespearian because of apparent ambiguities that have occasioned a plethora of divergent theories about its hero and the action in which he is involved. 'What happens in *Hamlet*' is not merely the title of a particular book; it is also the abiding preoccupation of virtually all *Hamlet* critics, whatever may be their announced area of concern. Whether, as first appears, it is the dramatist's management of awarenesses that has occasioned such major problems of interpretation in this play is a question to be asked and will bear the main emphasis of this chapter.

To suggest that the handling of awarenesses here contrasts with his usual methods is by no means also to imply that in *Hamlet* Shakespeare slights or abandons his regular practice of creating and exploiting unequal degrees of awareness. On the contrary, the tragedy provides the richest field of all for a study of the subject, even though a few other plays show a higher proportion of scenes in which we hold advantage over participants. At one time or another we hold advantage over every person in the play, and in fifteen of the twenty scenes our vantage point stands above that of some persons. Once the Ghost has divulged its secret to Hamlet in the last scene of Act I, we hold tremendous advantage to the very end of the play over all participants but Hamlet and, during part of the action, Horatio. Awarenesses are stair-stepped in scene after scene, from the mere obliviousness of the unknowing world up to the not-quite-equal awarenesses of Hamlet and the King. As in the most complex romantic comedies, we hold multiple advantages in all the major scenes, and each advantage creates a dramatically exploitable gap.

In the number of secrets opened to us and closed to the participants, and in the multiplicity of practices perpetrated by various participants upon others, *Hamlet* is truly exceptional, rivalling such busy comedies as *Twelfth Night* and *The Merry Wives of Windsor*; among the tragedies only *King Lear*, with its trio of avid practisers—Edmund, Edgar, Kent—comes near it in this respect.

As in the most fully packed and bustling comedies, too, the action of *Hamlet* is propelled by a succession of practices, chiefly those perpetrated by Claudius (with Polonius sometimes his confederate) and by Hamlet. The King's total of distinct and countable practices, six in all, exceeds the number devised and executed by any other practiser, whether villain, prankster, or masquerading heroine, in all Shakespeare. These practices include the following: (1) the initially enveloping practice—the secret murder of King Hamlet—by which all Denmark is deceived; (2) the employment of Rosencrantz and Guildenstern in an effort to solve the mystery of Hamlet's 'madness'; (3) the use of Ophelia for the same purpose; (4) the use of the Queen for the same purpose; (5) the dispatching of Hamlet to England to be executed; (6) the use of Laertes in the 'fencing match' designed to kill Hamlet—this complex final practice itself being fitted out with multiple deadly prongs: Laertes is supposedly the better fencer and, besides, enters the match with the intent to kill, not play; Laertes' weapon is unbated; his weapon is anointed with poison; Hamlet is to be offered a deadly drink; all else failing, Laertes is to use a 'pass of practice'—a treacherous thrust. Against the King's formidable battery of practices, Hamlet opposes a mere three, the last of which, besides, is only gratuitous: (1) the 'antic disposition'; (2) the 'mousetrap'; (3) the forged letter ordering the deaths of Rosencrantz and Guildenstern. It is by means of this total of nine practices, together with the play and counterplay that they engender, that the action of *Hamlet* is propelled. Throughout this action our own vantage point remains fixed at the top; we are not, as in *Julius Caesar*, denied crucial information until the ends of scenes, but are regularly advised in advance of action, with the result that exploitation of the gaps between unequal awarenesses produces virtually incessant dramatic effects of one sort or another. But these passing effects are not, as they are in *Titus Andronicus*, the main functions served by the fact of unequal awarenesses; what is of more importance for the tragedy as a whole is that, as in *Romeo and Juliet* and *Julius Caesar*, and as in most tragedies that follow *Hamlet*, the unawareness of certain participants provides the channel through which action moves towards the catastrophe.

If the allowance of some degree of ambiguity in the management of awarenesses in *Hamlet* is one deviation from the dramatist's normal practice of making important matters unmistakable, one other deviation is of at least equal significance: *Hamlet* is Shakespeare's sole tragedy in which the highest level of awareness, next our own, is occupied by the hero. In the comedies the highest level is typically held by bright-eyed heroines: Julia, Portia, Rosalind, Viola, and Helena are sole proprietors of the topmost secrets in their respective actions; the perfect examples are Rosalind, over whose awareness the dramatist never permits even our own to gain an edge, and Helena, who for a considerable period during the climactic portions of the play is actually elevated to a station above our own. In the later comedies the heroines lose their high place: Isabella, Marina, Imogen, Perdita, and Miranda are, in terms of relative awareness, steps descending from the pinnacles of Rosalind and Helena. Even so, in these later comedies, partially excepting *Cymbeline*, where Iachimo holds a prize secret, the highest awareness is kept from the possession of villainy. On the other hand, heroes in the comedies fare badly in terms of awareness: Valentine, Bassanio, Claudio, Orlando, Orsino, and Bertram share a dim and generally mistaken vision of their situations; later heroes—Posthumus, Florizel, and Ferdinand— fare even worse. Among heroes of the comedies and romances only Duke Vincentio and Prospero, the super-heroes, stand high, over-peering everyone else, 'like power divine'. But whether given to bright-eyed heroines or to super-heroes, the highest awareness, in the comedies, regularly resides with those who are essentially good people; and that, no doubt, is why they *are* comedies, with happy endings.

But in the tragedies in general neither heroes nor heroines nor any manner of good people stand high on the scale of awareness. High vantage points in tragedy are typically occupied by villains—or, where there are no mortal villains, then by whatever malign forces there are that tend to direct or to support villainy. Heroes regularly occupy inferior positions on the scale, and heroines, with few exceptions, fare no better. Further, whereas in the comedies the key practisers are almost invariably the brilliant heroines who go a-masquerading, in the tragedies they are almost always the scheming villains, whose victims are the heroes and heroines. The tragic heroes are rarely practisers of any consequence; when they do attempt to practise on their fellows, as Brutus and Macbeth do, they prove inept and bungling. They are characteristically guileless and are commonly blinded from within or from without, or, as with Brutus, Othello, and Lear, from both within and without. At the top of the ladder of awareness in the romantic comedies typically stands a sharp Rosalind plying her pranks for the love of mockery; her counterpart in the tragedies is a diabolical Iago, playing a deadly game with the good or semi-good figures who stumble about at a lower level, unsuspecting that they are his puppets and his dupes.

Standing as he does with the heroines of the comedies and the villains of the tragedies, Hamlet is virtually the lone exception among the tragic heroes. (The very complex and difficult problem of Antony's awareness in *Antony and Cleopatra* is best left without comment at this point.) If as a man of honour Hamlet stands with Brutus, yet as a man of awareness he stands with Cassius. He belongs not among the succession of duped or blinded heroes, but in the long line of busy practisers: masquerading heroines, incorrigible pranksters, devious counsellors, machinating villains, old dukes of dark corners. He, and not the villain Claudius, is sole proprietor of the topmost secret in his play. Claudius's secret abuses all Denmark, but Hamlet's secret is that he knows the King's; thus Hamlet's secret draws a circle that shuts the King's in.

In point of actual time, of course, the King's is the prior practice and the prior secret, though we learn nothing of the fact until the final scene of Act I. During the first four scenes the dramatist opens no gaps between awarenesses, but prepares shrewdly for their eruption in the fifth scene. The creation of dramatically exploitable discrepancies thus comes relatively late in *Hamlet*: in *Romeo and Juliet*, thanks to the Prologue, a gulf separates our awareness from the participants at the moment the action begins; in *Othello* we learn at the forty-second line of the first scene that 'honest Iago' is not honest; in *Lear* our great advantage comes more gradually, but all of it comes during the opening scene as we learn, while the King does not, that the older daughters are false and the younger one true; in *Macbeth* both of the principal gaps are opened in the third scene. We hold advantage over Romeo, Othello, and Macbeth even before we meet them, and we begin to gain advantage over Lear in the midst of his preposterous first speech. In *Hamlet* the two great

secrets, the one enveloping the other, are divulged simultaneously
to us with the Ghost's words in the fifth scene:

> ... know, thou noble youth,
> The serpent that did sting thy father's life
> Now wears his crown.
>
> (I. v. 38–40)

The first secret, that Claudius is his brother's murderer, gives us an
advantage that we share with Hamlet and Claudius until the end of
the play—and beyond, for Horatio has still to inform 'th' yet unknowing world'. The second, that Hamlet knows the first secret, we
share, until later, when Horatio is made Hamlet's confidant, with
Hamlet alone. With Hamlet, therefore, we hold advantage not only
over the unknowing world but over Claudius himself.

At the end of Act I, then, as result of the Ghost's revelation, the
structure of awarenesses has three levels: at the top, ours and
Hamlet's; next, Claudius's; finally, all Denmark's. Except that
Hamlet later takes Horatio into his confidence, this disposition of
awarenesses holds until the end of the play so far as the two great
secrets are concerned. Within the fixed arrangement occur frequent
variations occasioned by incidental practices and counter-practices
carried on between the two main opponents; but these changes of
position are of brief duration: if, for example, Hamlet is momentarily unaware of a device invented by the King and Polonius, he
quickly smells it out and recovers his place at the top. Other participants, like Rosencrantz and Guildenstern, Ophelia, the Queen, and
Laertes, all of whom are at various times 'in' on plots against Hamlet
and accordingly hold momentary advantage over him, are nevertheless all the while ignorant of the two great enveloping secrets, hence
always belong with the unknowing world.

For both the great secrets are jealously guarded. The action of the
play after Act I takes the form of a deadly game between the
'mighty opposites', and, though many lesser figures occasionally
participate in it, the true nature of the struggle is so shrouded in
secrecy that it is both unseen and unsuspected by the general
participants; though evidences of the continuing tension are everywhere, none knows what is really taking place. Polonius, Ophelia,
Rosencrantz and Guildenstern, Laertes, and the Queen play their
respective roles without guessing what they are involved in, and
eventually all die ignorant of what it was that killed them. Indeed,
so closely guarded is Hamlet's secret that the King himself actively

competes in the conflict without glimpsing the truth of it and dies clinging to his own painful secret unaware that he has none. Only Hamlet, with us, sees the conflict in its true light from beginning to end.

In this battle of mighty opposites Hamlet maintains the edge not alone because of the Ghost's disclosure but because of the extraordinary qualities of his mind: Prospero aside, Hamlet is Shakespeare's most circumspect creation. But the political power belongs absolutely to Claudius, who is himself a shrewd dissembler, and the battle of the opposites is therefore a very close one; that is one of the reasons for the intense and continuous excitement that *Hamlet* generates. It is a far closer contest between hero and adversary than exists elsewhere in Shakespeare. In *Othello* there is no evenness in the contest between Iago and Othello, for Iago knows all and Othello nothing. Neither Romeo nor Macbeth is more than a puppet in the hands of Fate: Romeo acknowledges that he is 'fortune's fool', and Macbeth completely mistakes his own role in Fate's design. Lear is a feeble opponent for the villains banded against him, and Brutus is an innocent in the hands of either Cassius or Antony. The contest in *Hamlet* is unique because the opponents are so evenly matched and because—for just once—the hero has the edge in awareness and carries the fight to the villain.

The mask, or outward sign, of Hamlet's secret is his 'antic disposition'. The antic disposition is his chief practice, like Iago's pretence of honesty and Rosalind's masquerade as Ganymede. Yet differences also exist between Hamlet's and the other practices: no participant probes to find the cause of Rosalind's transformation, for no transformation is suspected, Orlando and others simply assuming that Ganymede is as she has always been; neither does anyone ever probe for what lies behind Iago's mask, the mask itself not being suspected. But Hamlet's drastic change is marked by all: 'Not th' exterior nor the inward man / Resembles that it was.' If because it invites probing Hamlet's antic disposition is not a perfect mask for his secret, yet it serves its purpose well throughout the play; for, as we shall observe, it is by this device that the hero pressures the King into taking the successively more desperate actions that ultimately expose and destroy him. The antic disposition sets a false trail that keeps pursuers off the true one, and entirely successfully, for neither the King nor any other person ever discovers Hamlet's secret. Through a succession of counter-practices the King seeks the cause of Hamlet's lunacy when in fact there is no lunacy; thus the play consists in large part of futile efforts to discover what we know to be non-existent, and it is the gap between Hamlet's and our knowledge, on the one side, and the King's ignorance of this fact that supplies the main exploitable condition during four acts.

Exploitation begins in the first scene of Act II. At the close of the preceding scene Hamlet had stated that he might 'think meet / To put an antic disposition on', and thus only the very brief Polonius-Reynaldo scene (in which Polonius devises a practice for Reynaldo to use in discovering how Laertes is behaving in Paris) stands between the statement and the next words that we hear about Hamlet; says Ophelia,

> My lord, as I was sewing in my chamber,
> Lord Hamlet, with his doublet all unbrac'd,
> No hat upon his head, his stockings foul'd,
> Ungart'red and down-gyved to his ankle,
> Pale as his shirt, his knees knocking each other . . .
>
> (II. i. 77–81)

It is Ophelia's shocked recital of her offstage encounter with Hamlet which first notifies us that he has indeed put his antic disposition on; and he has acted shrewdly in first advertising his lunacy before the dutiful girl, who, as he doubtless calculated she would, instantly carries word of it to her father. Polonius, in his turn—sycophantic busybody that he is and remains until his own nosiness kills him—will run straight to the King with some ingenious but erroneous theory about Hamlet's madness. And so he does: 'This is the very ecstasy of love,' he tells Ophelia, and continues, 'Come, go we to the King.' Hamlet's practice is well launched; he could not have tossed the bait to a surer fish.

From Hamlet's opening move follows everything to the end of the play; like Iago, he is the puppeteer and the King and others his puppets. Controlled, deceived, they frame theories to account for his 'lunacy': to Polonius and perhaps Ophelia the cause is frustrated love; to the Queen it is 'no other but the main, / His father's death and our o'erhasty marriage'; to Rosencrantz and Guildenstern it is a mishmash of thwarted ambition and the disease of melancholy; to the King it is as yet a nameless affliction and a cause for uneasiness. Hamlet feeds the several theories indifferently, having no preference which trail the pursuers may follow, so long as it is false. Excepting the passages that deal with the ambassadors' return from Norway

and the welcome for the players, the very long second scene of Act II
is occupied with the launching of counter-practices by the King and
Polonius designed to fix the cause of Hamlet's supposed madness.
Here, as nearly everywhere else, the main emphasis of exploitation
for dramatic effect is upon the gap between our awareness of Hamlet's
antic disposition and the Danish court's misunderstanding of it,
rather than on that between our knowledge of the King's guilt and
the court's ignorance of it. Though Polonius, Ophelia, Rosencrantz
and Guildenstern, and the Queen, in actively aiding the King's
practice, are ignorant that their efforts serve the ends of a murderer,
it is not their ignorance of Claudius's secret but their ignorance of
Hamlet's that the dramatist exploits. With two notable exceptions,
Shakespeare leaves the King's guilty secret untouched.

Further, the evidence of the text obliges us to recognize that in the
general search for the cause of Hamlet's 'madness', the King stands
on the same footing as his accomplices. 'What it should be,' he tells
his spies, 'More than his father's death, that thus hath put him / So
much from th' understanding of himself, / I cannot dream of.'
Though we know of the King's guilt and are occasionally reminded
how it torments him, we know also that he has no cause to suspect
Hamlet's knowledge of it; hence, in his command that his spies find
'Whether aught, to us unknown, afflicts him thus', he must be
understood to be no nearer the right trail than the others. When he
approves of Polonius's ignorantly ingenious device for finding
'Where truth is hid, though it were hid indeed / Within the centre',
we are obliged to recognize that he holds no advantage over Polonius
in either suspicion or knowledge of Hamlet's true secret. His sole
advantage over Polonius and the others is that he knows his own
guilt, and if the consciousness of this prompts him to be more
suspicious than others of Hamlet's behaviour, yet the fact gives us
no basis to suppose that he ever guesses, or comes close to guessing,
the truth. The King, indeed, plays out his part and dies ignorant
that Hamlet knew the details of his father's murder.

Just as there is no real ambiguity about the state of the King's
awareness, neither can there be reasonable doubt about the state of
Hamlet's awareness during the several attempts of the King to find
the cause of his lunacy. Though we hold a fleeting advantage over him
because we were present and he was absent when the King and
Gertrude instructed Rosencrantz and Guildenstern to sift him, yet
our advantage is too fleeting to permit exploitation for dramatic

effect. The formalities of greeting and fashionable bantering duly completed, Hamlet summarily inquires why his former schoolmates are visiting Denmark. Thrice phrased, each time more sharply, his question flushes out the answer that was presumably in his mind at first sight of the pair: 'My lord, we were sent for.' His suspicion thus confirmed, Hamlet instantly puts on his antic act, and here, for the first time, we learn the specific variety of lunacy that he has chosen to mimic: '. . . this majestical roof fretted with golden fire, why, it appears no other thing to me than a foul and pestilent congregation of vapours'. He continues for twenty lines, mimicking —for his immediate spies and eventually the King's consumption— the classic symptoms of melancholia. Before this speech we were once told that he would put on an antic disposition and were thereafter thrice shown that he had done so—first by Ophelia's account of his wild appearance when he confronted her, next by the King's reference to the abrupt transformation that had occurred in him, and, finally, by the exaggerated show staged for Polonius's benefit: '. . . Excellent well; you are a fishmonger.' Surely, for us, these reminders of his practice have been more than ample, even as Rosalind's would be if, playing Ganymede, she should suddenly turn full-face and remark, 'Remember, I am really Rosalind in disguise.' If, after such broad demonstrations besides Hamlet's explicitly announced intent to play mad, we should nevertheless mistake the practice and think him mad in fact, the fault is surely not Shakespeare's, but ours.

So in their attempt to sift Hamlet the spies and their employer get nowhere; the King's first practice thus fails utterly, whereas Hamlet's, his antic disposition, has served admirably by preoccupying and worrying the King and setting him on a false trail. Briefly in III. i Shakespeare raises our vantage point a degree above Hamlet's with respect to two matters. At the close of the soliloquy that ends Act II, the hero expresses a wish to have 'grounds / More relative' than the Ghost's word of the King's guilt. Until now, we have ourselves possessed exactly the same evidence, gained at the same time, that Hamlet possesses; but now, in a flash, we are provided privately with those very 'grounds / More relative' that Hamlet's second practice is designed to procure for him. Polonius, moralizing on the propriety of having Ophelia found reading a book, 'That show of such an exercise may colour / Your loneliness', remarks '. . . that with devotion's visage / And pious action we do sugar o'er / The

84

devil himself'. A general moralization of Polonius's standard variety, the remark is innocent enough—but it is not so for Claudius, who speaks aside,

> How smart a lash that speech doth give my conscience!
> The harlot's cheek, beautied with plast'ring art,
> Is not more ugly to the thing that helps it
> Than is my deed to my most painted word.
>
> (III. i. 50–3)

Just a moment before, we heard Hamlet devise his second practice designed to catch the conscience of the King: 'If he but blench, / I know my course.' Now, in our sight, the King has blenched already, giving us prior confirmation of the guilt that Hamlet's device will not discover until the end of Act III. But our brief advantage over Hamlet is relatively insignificant and is different in kind from that which Shakespeare more often gives us over tragic heroes. Usually our special light enables us to see that the hero is quite mistaken in his judgement and his choice of action, as Brutus is when he approves Antony's bid to speak, or as Othello is when he trusts Iago's honesty. We gain no such advantage over Hamlet; rather, our privately gained information only confirms the rightness of his conviction. It is the villain, not the hero, who is on the false trail.

As his next step on that trail the King, prompted by the chronically mistaken Polonius, prepares his second practice:

> ... we have closely sent for Hamlet hither,
> That he, as 'twere by accident, may here
> Affront Ophelia.
> Her father and myself, lawful espials,
> Will so bestow ourselves that, seeing unseen,
> We may of their encounter frankly judge.
>
> (Ibid. 29–34)

Since the trap is set in our hearing, we again gain the advantage of Hamlet; further, since the plans are known also to Ophelia and the Queen, they, too, hold brief advantage. Even so, the main structure of awarenesses remains as always, with Hamlet above and the others below; for it is only because they are ignorant of Hamlet's own secret that the King and Polonius waste their time in attempting to discover 'If't be th' affliction of his love or no / That thus he suffers from'. Their device, of which Hamlet is briefly unaware, is thus circumscribed by Hamlet's.

We are presumably to understand that Hamlet's 'To be or not to be', which immediately follows, is not overheard by the King and Polonius, even though they have just withdrawn for the purpose of eavesdropping, or by Ophelia, who is seated with her book in hand. The first words overheard by the King and Polonius, then, are those spoken when Hamlet notices Ophelia: 'Soft you now! / The fair Ophelia!' During the next forty lines, until his abrupt 'Where's your father?' Hamlet is evidently unaware of the King's trap. The surest evidence thereafter that he has recognized it is contained not so much in his own subsequent speeches as in those of Ophelia that evince her startled reaction to his changed manner. During the first portion of the scene Ophelia had reacted calmly to Hamlet's remarks, even playfully: 'My lord! ... What means your lordship? ... Could beauty, my lord, have better commerce than with honesty? ... Indeed, my lord, you made me believe so ... I was the more deceived.' But at the point of 'Where's your father?' when Hamlet recognizes a trap, he evidently puts on his antic act with great violence, for from that point on Ophelia's exclamations are drastically altered: 'O, help him, you sweet heavens! ... O heavenly powers, restore him! ... O, what a noble mind is here o'erthrown!' Indeed, so terrifying to her is Hamlet's sudden change that she is reduced to a state of shock within minutes, and, after the entrance of the eavesdroppers, remains prostrated, her condition being acknowledged only by Polonius's grossly insensitive remark: 'How now, Ophelia! / You need not tell us what Lord Hamlet said; / We heard it all.'

How Hamlet recognizes that he is being overheard, no text makes explicit by dialogue or stage direction; but it seems certain that Shakespeare intended no warning as obvious as a shadow, a bulge in the curtain, a momentarily protruding elbow, or a sneeze. To think that this particular hero needs so gross a clue is to underestimate the quality of his senses, his alertness that never winks behind the antic mask. Earlier, he had instantly and easily divined the reason for the presence of Rosencrantz and Guildenstern at court and had responded with a classic description of man and the universe as observed by the bilious eye of melancholia; so here, having sensed that a larger audience lurks behind the curtain, he abruptly accentuates the signs of lunacy and shocks Ophelia into the state in which her father finds her.

This scene of Hamlet's interview with Ophelia offers the most elaborate structure of awarenesses yet erected in the play. Four secrets are involved: the King's guilt, Hamlet's knowledge of it, Hamlet's feigned lunacy, and the eavesdropping practice itself. Of Hamlet's secrets the King, Polonius, and Ophelia remain ignorant; of the King's secret, Polonius and Ophelia are ignorant; of the eavesdropping Hamlet is briefly unaware but becomes aware while the perpetrators of the practice suppose him still ignorant, and acutely mad. Thus Hamlet not only quickly regains his usual topmost post with respect to the main situation but turns the tables and masters the immediate situation as well. As from the practice that involved his two spies, again the King learns nothing from the Ophelia interview, and Hamlet's secret remains as secure as before. For Ophelia the ordeal is a repetition of her earlier, offstage encounter with Hamlet—an encounter that was, to this, as a dumb show to a play. Wholly ignorant of the enveloping situation, Ophelia is callously exploited by her father and the King; at the same time, the dramatist exploits with poignant effect the gap between her ignorance and our awareness of the true state of Hamlet's health. Her speech—'O, what a noble mind is here o'erthrown!'—takes its place with many others in Shakespeare that are uttered by persons totally mistaken about their situations and that assault our emotions with simultaneous, conflicting demands. Such is Imogen's speech in *Cymbeline* (IV. ii. 291–332) when she wakes from a deathlike sleep to find stretched beside her the headless body of the despicable Cloten and, because the body is dressed in her husband's clothes, mistakes it for Posthumus. In that incident Imogen's own anguish demands sympathetic response, but our awareness that the truth is far better than she knows demands laughter at the very grotesqueness of the situation; yet, at the same time, our will to laugh is blocked by our awareness of the enveloping situation in the play, with its seemingly insurmountable barrier still dividing husband and wife. Similar is the scene in *Romeo and Juliet* when the Capulets pour out their grief for the 'dead' Juliet whom we know to be only asleep; but again we are prevented from rejoicing by our knowledge that Fate has doomed both lovers and is even now in process of destroying their lives. So it is in the case of Ophelia's distraught lament for the supposed downfall of Hamlet's princely mind: even while we know that the Prince is not at all 'blasted with ecstasy', but is in full possession of all his brilliant faculties, we know also that, as Hamlet himself has told us, he is engaged in a deadly game to which we can expect but one ending.

To the middle of Act III no change has occurred in the fixed structure of awarenesses: the world remains ignorant of both the King's and Hamlet's secrets; the King remains ignorant of Hamlet's; and we, with Hamlet, know all. But now the dramatist reveals that at some unmarked earlier time Hamlet has shared his secret with Horatio. Speaking to his confidant about the play that is to be performed before the King, he alludes to 'the circumstance / Which I have told thee of my father's death'. Horatio thus becomes the first of the unknowing world to learn of the King's guilt, and Hamlet's sharing of the secret represents a momentous step. Hitherto he had proceeded with utter secrecy; indeed, even before he himself had learned of the King's guilt, from the moment that Horatio informed him of the Ghost's visit, he urged strict secrecy: 'I pray you all,' he then said, 'If you have hitherto conceal'd this sight, / Let it be tenable in your silence still.' Thereafter Shakespeare lays conspicuous stress on Hamlet's preoccupation with secrecy. When the Ghost has revealed the King's guilt, Hamlet first masks his new knowledge with the 'wild and whirling words' remarked by Horatio —and, perhaps, the idea of putting on an antic disposition first suggests itself to him here. He then proceeds to an eerily solemn ritual of swearing his companions to silence: 'Never make known what you have seen to-night,' he tells them—not once but five times. 'Upon my sword', he thrice insists, 'Never to speak of this that you have seen. . . . Never to speak of this that you have heard . . . Never . . . with arms encumb'red thus, or this headshake . . . to note / That you know aught of me.' And Hamlet's urgent demands the Ghost echoes from below: 'Swear! . . . Swear! . . . Swear!' And yet, after they have sworn and all have started from the platform, Hamlet charges his companions again: 'And still your fingers on your lips, I pray. / The time is out of joint.' In all, after the Ghost vanishes, eighty of the hundred lines remaining in the scene are used to emphasize Hamlet's insistence on the 'dreadful secrecy' that afterwards shrouds the action of the tragedy. And with Hamlet's sense of the need for secrecy the Ghost has agreed from the start: it will not speak to Bernardo and Marcellus, it will not speak to Horatio, it will not speak even to Hamlet with the others present; and it raises its voice four times to back Hamlet's demand for his companions' silence.

What is more, Hamlet's demand is made even though his companions are perfectly ignorant of what secret, if any, the Ghost has divulged; they know only that the Ghost appeared. If Hamlet places such weight on the less important secret, we can but guess what significance he assigns to the safe-keeping of the great one. For it is by keeping this secret that he holds advantage over Claudius and the Danish court. He regards it as imperative that the gulf between his awareness and the world's be maintained until his moment for revenge comes. But thereafter the gulf must be bridged, or else, to the world, his bloody act will seem only another savage incident marring Denmark's unenviable reputation: 'They clepe us drunkards, and with swinish phrase / Soil our addition.' For Hamlet, who shows himself always to be as circumspect as honourable, the only right occasion for taking his revenge will be that which also makes possible the disclosure of the full truth to the world—the world that, until then, will not so much as have dreamed of Claudius's murder to gain his brother's crown and queen.

Further, he must leave behind him, should he not himself survive the crucial moment, one possessed of the truth, with evidence to prove his case. That is why the astutely cautious Prince opens a narrow crack and admits Horatio into his deeply hidden secret. It is his initial step towards bridging the gap between his knowledge and the unknowing world's ignorance, and Hamlet takes it carefully. No one else in his circle, only Horatio, would serve: Ophelia is manifestly unstable and is subservient to her father; Rosencrantz and Guildenstern, his former schoolmates, are now the King's spies; the Queen has proved herself untrustworthy by her abrupt marriage to her husband's brother; Polonius is a tattle-tale and a great fool. Of what Hamlet may have said at the time he first divulged the secret to Horatio, we know nothing; but during the moments that precede the playing of 'The Murder of Gonzago' before the King, his remarks to Horatio suggest the care that he has taken to make sure of his man. Shakespeare gives him twenty lines that he uses like hoops to bind Horatio to him before assigning him to observe Claudius during the play 'Even with the very comment of thy soul'. Hamlet's assessment of Horatio's character, to the extent that it is ever revealed to us, appears just but also ingratiating: '. . . thou hast been / As one, in suffering all, that suffers nothing, / A man that Fortune's buffets and rewards / Hath ta'en with equal thanks'. Hamlet ends the prelude to his request of Horatio with words that embarrass both men: 'Give me that man / That is not passion's slave, and I will wear him / In my heart's core, ay, in my heart of heart, /

As I do thee.' Though Hamlet's first words in the long eulogy are 'Nay, do not think I flatter', it is clear that he intends his emotional remarks to bind Horatio even more closely to him in affection and loyalty—for who, hearing his qualities thus praised, could find it in his heart ever to betray the speaker? In the last scene of the play Hamlet makes a similarly pointed bid for the friendship of Laertes, whose prior commitments, however, are so compelling that he does betray Hamlet.

In designing his drastic practice meant to catch the conscience of his father's murderer, Hamlet seeks indisputable proof of the King's guilt *not for himself alone, but for Horatio*: Horatio must witness the King's reaction, for it is his testimony that will be needed when the time comes to expose the truth to the world. The immediate 'mouse-trap' practice is of course a practice within a practice, for Hamlet's antic mask goes on and off repeatedly throughout the scene: 'I must be idle', he tells Horatio as a flourish announces the approach of the King and his court. During the scene the dramatist has the advantage of three exploitable gaps: those that divide our awareness from the participants' ignorance of the 'mousetrap', of the King's guilt, and of Hamlet's feigned lunacy. Hamlet exhibits his 'madness' in turn to Ophelia, Gertrude, Polonius, and Claudius; and repeatedly, as he does so, the dialogue shoots sparks across the several gaps. Ophelia speaks for all the unknowing world when, from the depths of the general obliviousness, having witnessed the Dumb Show, she inquires, 'What means this, my lord?' And the Queen, to Hamlet's sharply probing question, 'Madam, how like you this play?' replies from the same undisturbed depth of innocence with a comment of purely disinterested dramatic criticism: 'The lady protests too much, methinks.' Alternating with the flashes of unconscious irony are flashes of conscious irony, all of them emanating from Hamlet, so that the several gaps are illuminated from both sides. When the uneasy King questions, 'Have you heard the argument? Is there no offence in't?' Hamlet replies, 'No, no, they do but jest, poison in jest. No offence i' th' world.' But the main exploitation of the im-mediate gap is made by the 'Gonzago' play itself. To most of the spectators it is but a play; to the King it is a play that inexplicably recapitulates the murder of which he is guilty and of which he believes himself to have sole knowledge; to Hamlet and Horatio it is a play within a play, a mousetrap set to catch the King's conscience; and to ourselves it is a play within a play within a play, for it is within *Hamlet* that we watch Hamlet and Horatio watch the King, who watches the play. In all, it might be a deadly version of our complex experience in *A Midsummer Night's Dream*, when, within the frame of that play, we watch the royal couple as they watch Quince's actors perform 'Pyramus and Thisby'.

Like his enveloping practice, the antic disposition, Hamlet's second is entirely successful, for the King's guilt unkennels itself with such a degree of finality as could hardly be bettered—and it does so not only in Hamlet's view but in Horatio's, whose testimony must one day convince the world that Hamlet's cause was just and that the King was a murderer and a usurper:

> *Ham.* O good Horatio, I'll take the ghost's word for a
> thousand pound. Didst perceive?
> *Hor.* Very well, my lord.
> *Ham.* Upon the talk of the poisoning?
> *Hor.* I did very well note him.
>
> (III. ii. 297-301)

What is of greatest immediate significance to Hamlet is that his calculated campaign against the King has begun to produce results: his device shocks the King into a state of near-desperation. Yet we cannot suppose that the King, though startled by the nearness of the 'Gonzago' plot to his own act of murder, forms any idea either here or hereafter that Hamlet has somehow learned his secret. The King reacts to Hamlet's practice even as he had earlier reacted to Polonius's innocent remark 'that with devotion's visage / And pious action we do sugar o'er / The devil himself'; though he reacts more violently to Hamlet's practice, it is with no more suspicion that Hamlet has discovered the truth than that Polonius has discovered it. To think otherwise is to suppose that Shakespeare has effected a very drastic readjustment in the basic structure of awarenesses without first notifying us explicitly of the fact—and it is a principle of his method that he does not make such changes without making the fact unmistakable, as is demonstrable in plays from *Henry VI* to *The Tempest*. Had he wished us to know that the 'mousetrap' tips Hamlet's hand to the King, he would have given the King an 'aside' just prior to his abrupt 'Give me some light. Away!' outburst, or a soliloquy at the opening of the following scene just before his conversation with Rosencrantz and Guildenstern begins. Further, had the dramatist wished us to understand that Hamlet recognizes the fact that the King has caught on, he would have given the hero lines in his subsequent conversation with Horatio to advise us; but no line in this dialogue conveys so much as a hint that Hamlet questions the safety of his secret. On the contrary, his mood here is one of elation. Finally, somewhat later, in his conversation with Gertrude, he makes it directly clear that he has no thought of the King's having discovered his secret:

> Not this, by no means, that I bid you do:
> Let the bloat king tempt you again to bed,
> Pinch wanton on your cheek, call you his mouse,
> And let him, for a pair of reechy kisses,
> Or paddling in your neck with his damn'd fingers,
> Make you to ravel all this matter out,
> That I essentially am not in madness,
> But mad in craft.
>
> (III. iv. 181–8)

Indeed, Hamlet's course hereafter until the end proceeds on the certainty that the king knows nothing of his secret. After his return from the enforced sea voyage, he knows that the King will soon have word from England that Rosencrantz and Guildenstern are dead and will thus learn that Hamlet knows of the orders by which he was himself to have been executed: 'It will be short; the interim is mine.' But his concern even at that very late point in the play is only that this particular secret will have been discovered, not that the King has found or will find the main secret.

The King's words and conduct after the 'mousetrap' evince, like Hamlet's, that no change has occurred in their relative levels of awareness, for he quickly sends for his spies and orders them to continue prying into the cause of Hamlet's lunacy; further, he approves Polonius's plan to 'o'erhear the speech, of vantage' while Hamlet talks with his mother. It is true that when we first see the King after his frantic departure from the 'Gonzago' scene he is in the act of commissioning his spies to escort Hamlet to England in order to end the 'Hazard so dangerous as doth hourly grow / Out of his lunacies'. But this furnishes no evidence that the play has taught him Hamlet's secret, or even increased any suspicion that he may have had; for even before seeing the play he had stated to Polonius that Hamlet 'shall with speed to England / For the demand of our neglected tribute'. He said nothing at that time of having Hamlet killed in England, nor does he say anything to that effect directly after the play, but discloses this intent only after Hamlet has killed

Polonius. Whether he had planned from the first to have Hamlet slain in England, or decided suddenly after the death of Polonius, we are never informed. In any event, the evidence of the King's conduct after the play and the lack of explicit statement that the play has taught him Hamlet's secret are in accord: no change has occurred in the structure of awarenesses. No critic has yet argued that because Polonius pricked his conscience with an innocent remark, the King suspects Polonius of knowing about the elder Hamlet's murder; neither should it have been argued that because 'Gonzago' pricks his conscience, he suspects Hamlet of knowing about it. However this view may have originated, it cannot have resulted from an objective study of the text of *Hamlet*; rather, it appears to have grown from a Romantic and post-Romantic paranoia that found Hamlet oblivious, sick, helpless, over-peered, sneaked up on from behind, taken un-awares. But in fact we have no more cause to tremble for the safety of Hamlet than we have to fear for Prospero when Caliban's silly accomplices emerge from the filthy-mantled pool and make towards his cell intent on murder. For both Hamlet and Prospero are marvellously circumspect, and in command of their situations. Always, it is Hamlet who is the hunter, Claudius the hunted.

The King fears Hamlet not for his suspected knowledge but literally for the reason that he gives for sending him to England: majesty cannot endure 'Hazard so dangerous as doth hourly grow / Out of his lunacies'. Passively, without conviction, the badly shaken and frightened King accedes to Polonius's nomination of a new practice, this time using the Queen very much as Ophelia was used before. In the earlier incident Hamlet caught on midway through the interview, and turned the practice to his own advantage. This time he catches on to the fact of a trap more swiftly—indeed, even as it is being set. The King's spies enter to invite him to his mother's presence; he deliberately prolongs his conversation with them, and the delay inevitably brings in an impatient Polonius, who of course has a proprietary interest in the plot. To this point the sequence of events repeats that of an earlier occasion, when, after Rosencrantz and Guildenstern had advised him of the Players' approach, Hamlet correctly predicted that Polonius, entering, has come for the same purpose. But this time Polonius's urgency is clue enough: 'My lord, the Queen would speak with you, and presently.' Like one with pressing business who will agree to anything rather than suffer delay, Polonius—who is elsewhere always an intolerable quibbler—assents instantly to every proposition proposed by Hamlet: that a certain cloud is shaped like a camel, a weasel, a whale, or anything. His incautious eagerness is clue enough for Hamlet's quick apprehension: 'Then will I come to my mother by and by,' he says—and knows that he will walk into another trap.

In the boudoir scene Hamlet is never again as violent as in the first seconds. Shouting wildly, roughly seizing his mother, and forcing her to sit, 'Where you may see the inmost part of you', he so intimidates her that she betrays the plot by crying out: 'What wilt thou do? Thou wilt not murder me? / Help, help, ho!' As he had gambled they might, the Queen's shrieks flush out the eavesdropper: 'What, ho! help, help, help!' echoes Polonius from behind the arras. Hamlet strikes through the curtain, directly at the sound, neither knowing nor caring whether the echo that his tactic has produced has come from the King, Polonius, or some other. In view of his long-range strategy, to kill the King at this point would be undesirable; but the circumstances are such that he has no choice. Should it be the King behind the curtain, then Hamlet's revenge must fail of perfection, for there is not yet admissible evidence enough in hand to guarantee that the world will be made to believe the truth. In the immediately preceding scene Hamlet had passed up an easy chance to kill the King at prayer; the opportunity had come suddenly and unexpectedly, and he had there fought off temptation by stifling his impulse with a savage sort of rationalization: 'Up, sword, and know thou a more horrid hent, / When he is drunk asleep, or in his rage, / Or in th' incestuous pleasure of his bed, / At gaming, swearing, or about some act / That has no relish of salvation in't . . .' Thus justifying his delay with argument drastic enough to stay his hand under the pressures of the moment, he had dashed on to his mother's room, quickly flushed out the expected eavesdropper, and seemingly risked defeat of his best hopes by blindly striking at the unknown spy who might or might not be the King: 'I took thee for thy better,' he remarks upon finding that he has slain Polonius. And yet Hamlet cannot really have expected the body to be that of the King, whom he had left busily endeavouring to pray only a moment before; even in the instant of acting he must have known that the risk of its being the King was negligible. Of the quickness with which Hamlet's brains work in times of stress we shall have added proof on board the ship that would have carried him to his execution in England.

As for the state of the Queen's awareness, here as well as before and after, it holds a steady course at the same level as the unknowing world's. In the end the Queen will die ignorant of all that has occurred; she will not even be privileged to hear Horatio's documented account of events that began with the elder Hamlet's murder and ended with Prince Hamlet's death. She was herself, as the dramatist makes clear, the main prize of the initial murder, and Hamlet's biting words to her during their interview are partly just: 'Almost as bad, good mother, / As kill a king, and marry with his brother.' But she can only echo him with words that are empty of meaning for her: 'As kill a king!' She did not know earlier, does not now know, and will never learn that she did in effect 'kill a king'. She had been won, said the Ghost, to the 'shameful lust' of Claudius, who then murdered to win her: 'My crown, mine own ambition, and my queen', says the King in his futile effort at prayer, and 'those effects for which I did the murder' are here set in a climactic order: the Claudius whose soul is anguished for his crime would not have committed it only to gain the crown, especially since the crown thereafter must still be won through an election, when he had 'Popp'd in' between the election and Hamlet's hopes. The crown was a secondary consideration; Claudius killed his brother for Gertrude. Though he may earlier have suspected his mother of complicity in the murder— for he tested her as well as Claudius in the 'Gonzago' play—Hamlet is now aware of the limitations of her knowledge and her guilt: the daggers that he speaks to her in the boudoir scene are aimed not at finding whether she was Claudius's accomplice, or even whether she knew of Claudius's crime, but at awakening her conscience to prick and sting her because she had been won to the 'shameful lust' of Claudius. For Hamlet's concern for his mother's soul is second only to his desire for revenge: '. . . for love of grace, / Lay not that flattering unction to your soul, / That not your trespass but my madness speaks'. Unlike Claudius, who suffers for his crime, the Queen has plainly felt no remorse at all for the lust that drove her to her husband's brother's bed with unseemly haste.

For Hamlet, then, it is imperative that his mother's eyes be turned into her very soul so that she can recognize the 'black and grained spots' that have been present all the while but of which, until now, she has been blissfully unaware. In the boudoir scene he compels her to examine these spots; it is a major accomplishment, and it is the only expansion of Gertrude's narrow awareness that occurs in the entire play. The text does not support a view that the Queen learns, here or hereafter, of Hamlet's 'antic disposition', even though he tells her with terrible emphasis that she is not to allow the King's bedtime overtures to induce her 'to ravel all this matter out, / That I essentially am not in madness, / But mad in craft'. His wildness at the start of their interview, his fierce accusations that brand her guilty of crimes of which she knows nothing, his violent thrust through the curtain at the eavesdropper, his wild and whirling words and manner in the presence of the Ghost that she herself cannot see —all these manifestations only heighten her conviction of his lunacy: 'Alas, he's mad!' When, in the next scene, she reports of him to the King—'Mad as the seas and wind, when both contend / Which is the mightier'—she is not shrewdly inventing excuses for his killing of Polonius, however agreeable we would find the thought that, in accordance with her son's wishes, she dissembles in order to hide the truth of his feigned lunacy from her husband. Whatever are the special competencies that have made her so desirable to both her husbands, they are presumably more physical than mental, and the art of subtle dissembling is not among them. 'In his lawless fit' Hamlet slew Polonius, she tells the King, 'O'er whom his very madness, like some ore / Among a mineral of metals base, / Shows itself pure; he weeps for what is done.' In saying so much, she reports only what her limited faculties have enabled her to apprehend. She will never know anything more.

At the opening of Act IV, then, the fixed structure of awarenesses still stands as before: the King knows his own guilt but never guesses that Hamlet also knows it; Hamlet knows the King's secret and has now divulged it to Horatio; the rest of Denmark remains the unknowing world, ignorant of the meaning of the battle raging between the 'mighty opposites'. But now the character of the King's practices has changed. All his earlier devices, with Polonius as accomplice, were designed to find the cause of Hamlet's madness. All three—the spying by Rosencrantz and Guildenstern, the eavesdropping with Ophelia as bait, and the repetition of the same practice with the Queen as bait—failed utterly, and the last of them fell on the inventors' heads, killing Polonius and shocking the King into a state of true desperation. It is because of this desperation to which Hamlet's own practices have driven him that the King turns to devices of a different character: they are directed not at discovering Hamlet's secret, but at killing him.

At what precise time the King's purpose shifted, the dramatist leaves inexplicit. Much earlier, at the end of the Hamlet–Ophelia interview, the King said to Polonius,

> I have in quick determination
> Thus set it down: he shall with speed to England
> For the demand of our neglected tribute.
>
> (III. i. 176–8)

It was then that he first used the word 'danger', and concluded with a mildly ominous prescript worthy of Polonius himself: 'Madness in great ones must not unwatch'd go.' Next, directly after Hamlet's 'mousetrap' practice had startled him to the edge of desperation, he explicated his intent more fully, saying to his two spies,

> I your commission will forthwith dispatch,
> And he to England shall along with you.
> The terms of our estate may not endure
> Hazard so dangerous as doth hourly grow
> Out of his lunacies.
>
> (III. iii. 3–7)

Though our view of the large structure of awarenesses remains unimpaired, the dramatist does here allow it to cloud with respect to the immediate situation. We are left uninformed, during the considerable time between the King's first statement of intent to send Hamlet to England and his eventual revelation of his true purpose at the very end of IV. iii—'The present death of Hamlet'—that he intends to have Hamlet executed upon his arrival. Was this his intent at the time he first mentioned sending Hamlet 'For the demand of our neglected tribute'? Or was it not in his mind until after the fright given him by the 'mousetrap'? Or was it not in his mind until, on learning of Polonius's slaying, he tells the Queen that 'The sun no sooner shall the mountain touch, / But we will ship him hence'? Or was it not even yet in his mind when, at the opening of IV. ii, he tells his spies that 'This sudden sending him away must seem / Deliberate pause'? Four times in all, the dramatist makes the King allude to his plan of dispatching Hamlet to England without hinting at any purpose so drastic as immediate execution at his arrival. In leaving the matter ambiguous, Shakespeare clearly deviated from his more usual method of making all details of a major practice known to us in advance; but perhaps the deviation serves a useful purpose.

Whatever else, it serves to enhance our sense of Hamlet's uniqueness as a tragic hero gifted with a quality more regularly bestowed on villains—superior awareness. As we have noted, in each of the three traps previously set for him, Hamlet has performed brilliantly. He quickly recognized and exposed the suborned function of of Rosencrantz and Guildenstern, and tossed them only false bait to present to the King; just as quickly, he apprehended that his interview with Ophelia was being staged for the advantage of some eavesdropper; and even before he entered his mother's boudoir he had satisfied himself that some eavesdropper would again be in the wings. And now Shakespeare grants him the privilege of recognizing the true intent of the King's fourth practice *even before we recognize it.* For it is in fact Hamlet himself who first notifies us that the proposed voyage to England is to be more than an innocent airing meant to improve his health. 'I must to England,' he says to his mother, 'Alack,' she replies, 'I had forgot. 'Tis so concluded on.' And Hamlet now explicates what the King, who has repeatedly alluded to the voyage, left untold:

> There's letters sealed, and my two schoolfellows,
> Whom I will trust as I will adders fang'd,
> They bear the mandate. They must sweep my way,
> And marshal me to knavery.
>
> (III. iv. 202–5)

It is not until the end of IV. iii—three scenes later—that we hear the King confirm Hamlet's words: he has ordered England to effect 'The present death of Hamlet'.

Besides heightening Hamlet's stature as a circumspect hero, this demonstration of his acuteness partly resolves the problem that attends the King's early announcements of his intent to send Hamlet to England. For if Hamlet can assert in III. iv that his two schoolfellows 'bear the mandate' for his murder, it follows that, by this time at latest, the purpose of murder was in the King's mind. We have final confirmation of Hamlet's prior-reading of the King's mind in their exchange that just precedes Claudius's private statement to us of his intent:

> *King.* For England.
> *Ham.* For England?
> *King.*   Ay, Hamlet.
> *Ham.*   Good.
> *King.* So is it, if thou knew'st our purposes.
> *Ham.* I see a cherub that sees them.
>
> (IV. iii. 48–50)

Evidently, then, the plan to murder Hamlet in England was formed even before the slaying of Polonius, for Hamlet has read the King's mind accurately before the King knows that Polonius has been slain. But Shakespeare leaves us forever unsure whether the King had execution in mind as early as his first words about sending Hamlet to England. Nor do we ever learn how, other than by operation of his active sixth sense, Hamlet first learned of the King's true intent, and tells us of it before the King does.

By means of unrelenting pressure generated by his practices, Hamlet has forced the King upon a desperate remedy—murder. It is the King's position that is less enviable at this point, as, in fact, it has been less enviable since, on the battlements, Hamlet learnt of the earlier murder. Throughout Act IV Hamlet's vantage point remains higher than that of any other tragic hero, and, on the opposing side, Claudius holds a lower one than that given to any other tragic villain, being ignorant that Hamlet has long known his secret, that Gertrude has become a limited ally of her son, and that Hamlet, just now being hustled off to his death, is in fact aware of the murderous purpose, and on his guard. Such is the King's state as Act IV begins, and shortly afterwards he is caught in yet another fold of ignorance when we learn privately of Hamlet's escape from the death-trap and his safe return to Denmark. We learn of the hero's safety at the end of IV. vi, and during some thirty-five lines as the next scene opens the dramatist exploits to the full the widest gulf yet opened between the King's awareness and ours. Having just heard the reassuring words of Hamlet's letter to Horatio, we hear the King, reeking with confidence, boast thus to Laertes:

> You must not think
> That we are made of stuff so flat and dull
> That we can let our beard be shook with danger
> And think it pastime. You shortly shall hear more.
> (IV. vii. 30–3)

On this cue a messenger enters with letters—not, indeed, as the King expects, from Hamlet's executioners, but from the Prince himself, announcing his safe return and, with ominous irony, begging Claudius's leave 'to see your kingly eyes' on the morrow.

By now Hamlet should have been dead in England. The cryptic message to the contrary administers to the King the most severe shock in the quick succession of shocks that began with the 'Gonzago' incident. If Hamlet's war of nerves had shaken him earlier, driven him to desperation, it has now reduced him to incoherence: 'From

Hamlet! . . . What should this mean? Are all the rest come back? . . .
"naked"! / And in a postscript here, he says, "alone". . . . If it be so,
Laertes,— / As how should it be so? How otherwise?—' It is in
this frantic state to which Hamlet's aggressive campaign has forced
him that the King takes to himself a new ally and victim, Laertes,
and devises his fifth and last practice—one with which he quite
overreaches himself, for it provides Hamlet with exactly the right
circumstance for revenge.

Between this time, when the King devises his final practice, and
the time it is set in motion, Shakespeare places the graveyard scene,
with its elaborate structure of unequal awarenesses and abundant
materials for exploitation. It can hardly have been without calculated
purpose that the dramatist brings on the grave-diggers just at this
moment, directly after the agonized interview that closes the preceding scene, in which are exhibited in extreme form the emotions
of the King and Laertes—the King gnawed by conscience and
driven into a frenzy of hatred and fear of Hamlet, Laertes so wildly
incensed that he would seize the chance to cut Hamlet's throat 'i' th'
church'; and in which, at the very end, we witness an extraordinary
outpouring of grief and bitterness upon news of Ophelia's drowning.
From this depiction of the violent passions that now infect the court
we go abruptly into the company of the First and Second Clowns,
who represent not merely the unknowing but the indifferent world
that goes on about its usual affairs regardless of all others; Shakespeare never quite lets us forget that such a world is always present,
surrounding that in which his main figures undergo their special
agonies. If Hecuba was nothing to the First Player, neither Ophelia
nor, indeed, a King's murder or a Prince's revenge is of any moment
to the First Clown: 'The crowner hath sat on her, and finds it
Christian burial.' Casually digging the grave of Ophelia, the Clowns
quibble and jest for more than sixty lines with no more sense of the
Danish court's past, present, and continuing anguish, with which
our awareness is laden, than has the Porter, for example, who answers
the knocking at Macbeth's gate, of the appalling events that have
just occurred within.

Intruding here, Hamlet himself stands briefly unaware. We
deduced from the First Clown's 'Is she to be buried in Christian
burial that wilfully seeks her own salvation?' that the grave is
Ophelia's—a fact that remains unknown to Hamlet during two
hundred lines of half-serious, half-whimsical philosophizing,

bantering, and jesting with Horatio and the diggers. Briefly, at a point or two, the dialogue directly touches and thereby exploits Hamlet's unawareness:

Ham. Whose grave's this, sir?
1. *Clo.* Mine, sir. (Sings)
Ham. I think it be thine indeed, for thou liest in't.

(v. i. 127–32)

And again,

Ham. What man dost thou dig it for?
1. *Clo.* For no man, sir.
Ham. What woman, then?
1. *Clo.* For none, neither.
Ham. Who is to be buried in't?
1. *Clo.* One that was a woman, sir; but, rest her soul, she's dead.

(Ibid. 141–7)

'How absolute the knave is!' cries Hamlet, and goes on jesting with the Clown, half in merriment, half in vexation. At these points the effects resemble those in the comedies when contradictory demands are placed on the emotions, as in *Twelfth Night*, for example, when Count Orsino remains unaware that his 'Cesario' is in reality a girl who has fallen in love with him, and when Olivia, by falling in love with this same 'Cesario', breaks her vow to weep seven years for a dead brother. Here the basic situations are comic in nature, and so demand laughter, but at the same time the plights of the participants and the poignant poetic tones in which Shakespeare has cast the dialogue create contrary demands—for compassion, even for some mild private tears. But though they are reminiscent of such situations in the comedies, these comparable moments in *Hamlet* are in fact the converse of them: in the comedies the basic situation remains comic while incidental effects occasioned by the poignantly expressed sentiments of the participants are sobering at the least; but of course in *Hamlet* (and other tragedies as well) the basic situation remains grim while the incidental effects created by jesting dialogue are comic. In the immediate instance, it is, after all, the grave of poor drowned Ophelia over which Hamlet and the First Clown exchange unknowing and unfeeling retorts. The effects are somewhat different in the exchange that immediately follows:

1. *Clo.* It was the very day that young Hamlet was born; he that was mad, and sent into England.
Ham. Ay, marry, why was he sent into England?

1. *Clo.* Why, because 'a was mad. He shall recover his wits there; or, if he do not, it's no great matter there.
    Ham. Why?
    1. *Clo.* 'Twill not be seen in him there; there the men are as mad as he.
                                                                    (Ibid. 160–70)

Here the incidental effects are exclusively comic, rising from the surface of the triple-layered structure of awarenesses. The joke on England—''Twill not be seen in him there'—is one that Shakespeare's own audiences presumably relished, over and above the business of the play proper, as a private one in which even the speakers of the dialogue, being Danish, cannot themselves fully participate. But Hamlet is in position to relish the allusion to himself —'he that was mad, and sent into England'—for he perceives that the Clown, in the process of springing his joke about England, is unaware of the identity of the person whom he is addressing. At the bottom of the structure, the First Clown, an outstanding exemplar of the unknowing world, relishes his own wit in ignorance of his ignorance: he knows neither Hamlet nor, except by rumour, what is happening at the court. But his references to Hamlet's madness and his being sent off to England sharply prod our own remembrance of the deadly game being played at court and perhaps, after all, inject an element of grimness into an interlude that is otherwise strictly comic.

If our awareness of the struggle in which Hamlet and the King are engaged should nevertheless have been dulled by the nearly two hundred and fifty lines of the grave-diggers' interlude, it is abruptly revived by the entrance of the Court for Ophelia's funeral. For another dozen lines Hamlet remains unaware whose grave it is, until Laertes speaks the word 'sister', whereupon he leaps into the grave and challenges a brother's love with brave conceits of his own. But though his unawareness of the relatively tangential matter of Ophelia's death ends here, his unawareness of a central matter of terrible import first becomes exploitable here. At the end of Act IV, we heard the King and Laertes devise an elaborate practice aimed at Hamlet's murder in the course of what the unknowing world will perceive as a routine fencing match. Of this plot, when he leaps into the grave and strives to out-roar Laertes, Hamlet cannot possibly have a suspicion. Our own awareness—which scarcely needs reminding—is nevertheless sharply prodded by the King's private admonition to Laertes:

Strengthen your patience in our last night's speech;
We'll put the matter to the present push.

(v. i. 317–18)

Hamlet, at this point, has just left the scene; but the force of the King's reminder drastically affects our view of the scene between Hamlet and Horatio that immediately follows.

In this prelude to the final scene of the tragedy, we are first told the manner of Hamlet's escape on the seas—an escape wherein he evinced the same alertness of mind and quickness of decision that he had shown in all the earlier attempts of the King to trap him:

> Up from my cabin,
> My sea-gown scarf'd about me, in the dark
> Grop'd I to find out them; had my desire;
> Finger'd their packet; and in fine withdrew
> To mine own room again, making so bold,
> My fears forgetting manners, to unseal
> Their grand commission . . .
>
> (v. ii. 12–18)

'Being thus be-netted round with villanies,—' he tells Horatio, 'Ere I could make a prologue to my brains, / They had begun the play.' Moving swiftly, he rewrote the King's order so that it required not his own execution but the deaths of his two companions. The King's fourth, and, to this point, his deadliest practice Hamlet thus circumvented by the sheer alertness of his wits—whereas his own counterpractice, his third and also deadliest, will inevitably succeed: 'So Guildenstern and Rosencrantz go to't,' comments Horatio mildly. But even as he tells Horatio the facts of his escape and return to Denmark, Hamlet stands in imminent danger of the multi-pronged instrument of death newly devised by the King and Laertes. But though he cannot possibly know all the details of the King's final and most complex plot against him, we are left in no doubt that he is aware, even now, that he will shortly face a new attempt on his life. Horatio remarks that the King will soon learn from England 'What is the issue of the business there'—namely, that Rosencrantz and Guildenstern have been executed; the King is by no means so dull but that he will speedily realize that Hamlet had discovered and altered the order for his own death. At this point Hamlet's reply to Horatio is heavy with significance: 'It will be short; the interim is mine.'

'. . . the interim is mine': the line suggests that he trusts to have

103

an indefinite but brief period, during which his life will be comparatively safe, to plot his own next move in the deadly game. We know, of course, that the King has already designed a death-trap which is even now ready to be sprung: 'We'll put the matter to the present push.' Though Claudius remains ignorant that Hamlet knows of the plot to have him killed, yet the hero's own unawareness that the King has already mounted a new plot to kill him is clearly the more perilous of the two states of unawareness, for he is not to have 'the interim' at all, but only moments until the new device is sprung. 'I am very sorry', Hamlet tells Horatio, 'That to Laertes I forgot myself'—and the generous reference to the King's chief instrument for his next try at murder sharply illuminates the gap between our knowledge and the hero's. It accords perfectly with Shakespeare's typical method that, just after Hamlet has mentioned the 'interim' and Laertes, Osric enters with the King's invitation to a 'fencing match'.

As we have noted, the King's previous devices all have failed: his two spies discovered nothing in their effort to sift Hamlet; the use of Ophelia as bait ended in mere frustration for the King and brought on Ophelia's first signs of mental collapse; the use of the Queen as bait killed Polonius and brought a furious Laertes home; and the attempt to murder Hamlet in England resulted instead in the execution of the two spies. In contrast, Hamlet's own practices have prospered: his 'antic disposition', his 'Murder of Gonzago', and his brilliantly executed return to Denmark from the sea voyage have relentlessly pressured the King, agitating his mind and driving him to try increasingly desperate measures.

Hamlet's calculated campaign, in short, has forced the King at last to overreach himself by risking devices likely to explode in his own face and expose him to the yet unknowing world—or, at the least, to supply evidence for a credible case to be made against him. The King's fifth and last practice, designed under pressure, in sweaty haste, at word of Hamlet's strange return to Denmark, is the product of a mind pushed beyond its rational limits. This time the King will take no chances with failure, and it is because his device is so excessive in its precautions that it betrays the deviser and hoists him on his own petar. Hamlet's letter announcing his return uses terms shrewdly calculated to incite the King to some rash act: 'High and mighty,' Hamlet begins; he begs leave 'to see your kingly eyes'—insolent salutations both, to which are added ominous allusions to having been 'set naked on your kingdom' and to an intent to 'recount the occasions of my sudden and more strange return'. Though Hamlet stops just short of directly stating that he had discovered the secret order for his murder in England, he comes near enough that the King's brain is set whirling. If the earlier 'Gonzago' incident affrighted him, this one brings him to the edge of panic. And it is in this state that he devises the final, multi-pronged instrument for Hamlet's death.

We should now count the fatal prongs: (1) Laertes is, by repute, the better fencer; (2) Laertes will be fighting with an intent to kill; (3) Laertes' weapon will have no button at its tip; (4) Laertes' tip will be anointed with a deadly poison; (5) Hamlet, dry and heated from the match, will be offered a poisoned drink; (6) Laertes, all else failing, will strike Hamlet 'in a pass of practice'—i.e., use a treacherous stroke while Hamlet is off guard. So elaborate is the device that the odds against the possibility of Hamlet's surviving are obviously overwhelming. The King is himself aware that his last try may expose him, and is hesitant especially about Laertes' modest contribution to the device—the poison-tipped weapon: 'If this should fail, / And that our drift look through our bad performance. / 'Twere better not assay'd.' Even in the midst of panic, the essential egotism that rules Claudius remains dominant; it is only to ensure himself against the chance that the skills and advantages of Laertes may fail that he includes the 'chalice for the nonce'. In the last analysis, thus, he does not fully trust his final accomplice; but at the same time, because of the same egotism, he totally misjudges the character of his intended victim: Hamlet, he assures Laertes, is 'remiss, / Most generous and free from all contriving'.

As we have seen, up to the moment that the final trap is sprung Hamlet's record in detecting and dealing with practices has been perfect. The King has failed four times, Hamlet none. Except that it is multi-headed, the King's final device should have proved the easiest of all for the hero to identify in time to save himself. He knows that the King has tried four times to trap him and that the fourth attempt was on his life. He has sent an insolent letter to the King, the wording of which implies his intent to goad his enemy into making a new and rash attempt. All circumstances considered, then, Hamlet would be a dull and muddy-mettled rascal indeed if he were to suppose that, when the King invites him to take part in a fencing match just at this time, the purpose is innocent sport. Our own knowledge of the King's plot is of course secure, so that the moment Osric, with many flourishes, enters, we know his reason. Hamlet cannot, like us, know instantly; but he begins at once to sift the sycophantic messenger, even as, on earlier occasions, he sifted Horatio and Marcellus about their encounter with the Ghost, the pair of spies about the purpose of their visit, Ophelia on the where-abouts of her snooping father, and Polonius about the shape of a cloud, when that unfortunate came to summon him to be spied on in the Queen's closet: 'They fool me to the top of my bent.' For Hamlet the King's invitation is thus an old story with a familiar pattern. At what precise point he comes to suspect the hidden purpose of Osric's embassage, it is difficult to ascertain; perhaps it is as early as Osric's second line: 'Sweet lord, if your lordship were at leisure, I should impart a thing to you from his Majesty.' In any event he has surely caught on by the time Osric has spelled out the terms of the match:

Osr. The King, sir, hath laid that in a dozen passes between you and him, he shall not exceed you three hits; he hath laid on twelve for nine; and that would come to immediate trial, if your lordship would vouchsafe the answer.

Ham. How if I answer no?

(v. ii. 172–7)

He does not bother to pick Osric's mind as he had earlier picked the minds of Rosencrantz and Guildenstern to make them confess they were sent for; presumably he surmises, and correctly, that Osric is innocent of the sinister purpose behind the invitation.

Hamlet requires no confirmation of his suspicions after the with-drawal of Osric; but if he should require any, it is furnished by the arrival of another Lord, sent hard on the heels of Osric by the King: 'He sends to know if your pleasure hold to play with Laertes, or that you will take longer time.' When Polonius had followed similarly hard on the heels of the two spies to bid him come to his mother's closet, Hamlet quickly saw the truth: he was being drawn into a trap. Here the pattern is repeated: 'I will come by and by,' he had told Polonius; and here he tells the Lord, 'I am constant to my purposes; they follow the King's pleasure.'

Even so, perhaps the dramatist does not quite trust us to be as astute as Hamlet and to recognize that the hero has sensed a new death-trap; true to his habit of making assurance doubly sure, he makes Hamlet say to Horatio, 'But thou wouldst not think how ill all's here about my heart', to which Horatio replies. 'If your mind dislike anything, obey it. I will forestall their repair hither, and say you are not fit.' Hamlet's final speech on the matter leaves no room for doubt that he apprehends the immediate danger:

> There's a special providence in the fall of a sparrow. If it be now, 'tis not to come; if it be not to come, it will be now; if it be not now, yet it will come; the readiness is all. Since no man has aught of what he leaves, what is't that he leaves betimes? Let be.
>
> (Ibid. 230–5)

Hamlet's 'prophetic soul' had made him distrust the King even before the Ghost told him of the murder; 'I see a cherub that sees them' he said to the King of the hidden motives behind the voyage to England; lying aboard ship, he experienced 'a kind of fighting' in his heart that would not let him rest—and he arose, searched his escorts' packets, found the order for his death, and forged new orders for their deaths. Hamlet is himself aware of the native habit of alertness that, when danger is near, sets his brains to work: 'Ere I could make a prologue to my brains,' he tells Horatio of the events on board ship, 'they had begun the play.' Though Shakespeare has occasionally exposed Hamlet in moments when he is unaware of some hidden aspect of a situation, this hero is never unaware in these moments *because* he is Hamlet, as Brutus, Othello, and Lear are unaware because of a special blindness; on the contrary, when Hamlet is unaware he is so *despite the fact* that he is Hamlet: the details hidden from him in certain situations are such that he could not possibly detect them, however circumspect he is, unless he were very Prospero with Ariel at his side.

He knows, then, that he is summoned to a new death-trap; moreover, not only is he certain of the death-trap, but he expects, this time, to die in it. He accepts the match not because he wishes to die but because he believes the occasion will also provide the right circumstances for his revenge: 'The interim is mine.' He cannot possibly determine in advance the precise quarter from which the blow that kills him will be struck, nor can he know exactly when the moment will come for his own blow at the King. But it will come, and when it does he will strike without needing to make a prologue to his brains.

At the entrance of the King and his court, then, Hamlet holds the highest vantage point even as he has held it throughout the deadly game. Our own advantage over him is thin but crucial: we know all the spikes on the King's latest instrument—the unbated weapon, the anointed tip, the 'chalice for the nonce', the 'pass of practice'. On the level next below Hamlet's is the King, whose areas of unawareness exceed his areas of awareness. He knows what the court does not know, that he is a murderer, and he knows all the deadly points of the device that he has prepared; but he does not know that Hamlet knows him for a murderer, or that Hamlet discovered the order for his death in England, or that Hamlet suspects the immediate trap and intends to convert it to his own purpose. To the King—whose lot was never less enviable than now—Hamlet is 'remiss, / Most generous and free from all contriving'. Next below the King is Laertes, who knows the details of the immediate practice but who is otherwise ignorant of the entire context in which the present moment is set, and so belongs still with the Queen, Osric, and the various Lords and Attendants who make up the unknowing world. But Laertes' ignorance is more wretched than that of the others, for while they merely stand by as innocent onlookers, he, in his ignorance, has placed himself in the service of a murderer and made himself the key tool in the King's deadliest practice against Hamlet. Whatever are otherwise his merits, Laertes certainly does not represent the perfect avenger, the man Hamlet should have been in order to perform his own assigned task more creditably. Laertes has judged and acted rashly ever since he returned from France. At first he was eager to strike down the King himself and any others who bore a semblance of guilt in the death of Polonius. But he is like a furious child whose head can be turned quite around by clever blandishments; too quickly, truly like a child in the King's hands, he succumbed to the witchcraft of the King's wit, switched his rage from the King to Hamlet, and grew so inflamed, indeed, that he shouted his readiness to cut Hamlet's throat in the church. He assented without hesitation to the King's order that Hamlet be struck down 'in a pass of practice', and even went the King one better: 'I bought an unction of a mountebank'.

Laertes, in short, a man of well advertised honour, is so far diverted from the path of honour by his own passion and the King's persuasive skill that he is eager to commit an act of foul dishonour even in the course of a fencing match, an occasion that traditionally epitomizes the very idea of honour. It is this one facet of the many-faceted situation that Hamlet cannot anticipate: that Laertes would be capable of stooping to dishonour on such an occasion, even to the extent of using a poisoned tip and a treacherous stroke. For Hamlet is himself a man of uncompromised and uncompromisable honour. Though he speaks of himself as 'indifferent honest', yet in his most probing soliloquy he leaves no doubt about the purity of his concept of honour: 'Whether 'tis nobler in the mind', he proposes, 'to suffer / The slings and arrows', or to oppose them and so die. His question is not whether one should or should not take the way of honour, as if that were a matter for debate; rather, his question asks only which *is* the way of honour, it being quite understood that that is the way to be taken once it can be found. What he admires in Fortinbras is his unquestioning readiness 'to find quarrel in a straw / When honour's at the stake': but he would surely never applaud Fortinbras's neglect to ascertain whether honour is truly at stake before acting. Quietly, unostentatiously, Hamlet stands at the very head of Shakespeare's representatives of the ideal of honour as a principle of thought and action. His honour is not, like Hotspur's, a trophy to be won and hung up boldly for the world to admire; neither is it, like Brutus's, a slogan to be emblazoned across the breast and stroked each time one speaks: '. . . let the gods so speed me as I love / The name of honour more than I fear death'. Hamlet's honour, as we shall shortly see, is at once the reason for his refusal to take early, precipitate action against Claudius and the fault that makes him, at last, vulnerable to Laertes' treachery.

Though honour permeates Hamlet's being, it is not, like Brutus's, a generally blinding honour. Brutus's blinds him even to men like Cassius and Antony, whose practices would be apparent to anyone whose vision was not impaired around the full circle. But Hamlet's narrow segment of blindness extends no wider than the sliver of a single degree—just wide enough to block his perception of the fact that a man noted for honour, like Laertes, during a fencing match, could so far fall from honour as to be capable of treachery. Up to the instant of Laertes' treacherous thrust, Hamlet's history, as we have seen, has been one of sharp-eyed and cautious vigilance: his soul distrusted Claudius from the beginning; he was sceptical when first informed of the Ghost; he tested even the Ghost's word by exposing Claudius's conscience; he distrusted his old school friends; a sixth sense alerted him to the presence of eavesdroppers on the occasions of his interviews with Ophelia and Gertrude; a 'cherub' alerted him to the King's intent in sending him to England; a 'kind of fighting' in his heart prompted him to search his escorts' packet. And on being invited to win the King's wager in the match with Laertes, he distrusted the situation with perhaps less strain on his extrasensory faculties than had been required on the earlier occasions.

But he is helpless to fix the precise quarter from which Claudius's anticipated dart will fly. Before the match commences, responding to his mother's injunction to 'use some gentle entertainment to Laertes before you fall to play' with a remark pregnant with meaning, 'She well advises me', he takes Laertes' hand and addresses him seriously, at great length, very much as, just before the 'Gonzago' play, he had addressed Horatio before asking him to observe Claudius 'Even with the very comment of thy soul'. On the former occasion his prefatory remarks, embarrassing to both—'Something too much of this'—were calculated to bind Horatio's allegiance with hoops of steel. Just so, his long and probing speech to Laertes is designed to reassure himself that, wherever it may come from, the death blow will not come from this quarter: 'I've done you wrong, / But pardon't as you are a gentleman.' And after many more ingratiating words, he concludes: '. . . I have shot mine arrow o'er the house / And hurt my brother'. Laertes replies disarmingly, 'I do receive your offer'd love like love, / And will not wrong it.' 'I do embrace it freely, / And will this brother's wager frankly play,' Hamlet concludes; he has allayed whatever feelings of unrest had troubled him: Laertes is a man of honour.

But though his own honour prevents him from suspecting Laertes, he remains alert to danger elsewhere: twice he refuses the cup, first proffered by the King—'Give him the cup'—and next by the Queen, who is herself preparing to partake:

> *Queen.* The Queen carouses to thy fortune, Hamlet.
> *Ham.* Good madam!
> *King.*   Gertrude, do not drink.
> *Queen.* I will, my lord; I pray you, pardon me.
> *King.* (Aside) It is the poison'd cup; it is too late.
> *Ham.* I dare not drink yet, madam; by and by.
>   (v. ii. 300-4)

Perhaps Hamlet's 'Good madam!' is only a pleasant acknowledgement of the Queen's good wishes; or perhaps, all things considered, it is much more—an exclamation voiced in fear of the drink. And, again, perhaps his 'I dare not drink yet' is only the expression of an athlete's awareness that to drink during strenuous activity is inadvisable; but, all circumstances again considered, perhaps it expresses suspicion of the drink. In any event, Hamlet twice refuses the poisoned cup.

But he does not avoid the 'pass of practice' that the King had urged Laertes to use if the progress of the match should require it. And the progress of the match has indeed shown the need for treachery: in the first two bouts Hamlet has scored two hits and Laertes none; further, Hamlet has scorned the proffered drink. The time has come:

> *Laer.* My lord, I'll hit him now.
> *King.* I do not think't.
> *Laer.* (Aside) And yet 'tis almost 'gainst my conscience.
>   (Ibid. 306–7)

That Shakespeare means us to understand that Laertes does in fact use a treacherous thrust, probably while Hamlet's head is turned between bouts to hear Osric's latest verdict of 'Nothing, neither way', there can be no more question than if we had an explicit stage direction in the dramatist's own hand. Having lost in successive bouts, Laertes would surely lack confidence to boast summarily that 'I'll hit him now', as though it were an easy matter; and surely there would be no occasion for his adding that to do so is '''gainst my conscience' unless he meant to take unfair advantage.

Neither can there be reasonable doubt that, at the touch of the unbated point, Hamlet's quick wit apprehends the situation: it *is* Laertes, after all, who is the King's instrument; the blow that he had anticipated, though from what quarter he could not know, has been struck. Instantly he drops all pretence of supposing the match to be a friendly game; with the same alacrity with which he had struck through the arras and killed Polonius and with which he had boarded the pirate boat, he wrests the deadly weapon from Laertes and mortally wounds him with it. Hamlet's own wound is superficial, and he might yet survive but for the poisoned tip; of the fact that the tip is poisoned he remains ignorant through the space of about fifteen lines, during the busy action in which he wounds Laertes, sees the Queen fall while gasping that the drink is poisoned, and hears the cries of the King and all the Danish Court. He shows no surprise when Laertes divulges that the match has been a 'foul practice' devised by the King, for he had suspected as much from the instant Osric arrived with the King's invitation; similarly, the Queen's dying words, 'The drink, the drink! I am poison'd', only confirm the suspicion that was aroused when the King had first offered him the cup. His sole expression of surprise comes when Laertes reveals that the tip of the foil was not only unbated—as its touch had warned him—but poisoned as well: 'The point envenom'd too!' Thus his only lapse from omniscience during the entire match was his failure to understand that Laertes, a man of honour, could stoop to treachery; and that lone failure derived from his own sense of honour as a thing incorruptible.

It is just this idea of honour that dominates the closing moments of the action. Though it has been the driving force behind Hamlet's behaviour from the time he accepted the Ghost's command, it has remained rather implicit than explicit—for Hamlet does not, like Hotspur and Brutus, reaffirm his devotion to it with every breath. But in the end the fact is made explicit: honour has been his abiding concern, the determiner of his procedure in carrying out his task of revenge. It is therefore that, while Laertes lies dying and blurting out the ugly facts of the situation as he knows them, Hamlet makes no move against the King though his hand holds the means—'The treacherous instrument is in thy hand, / Unbated and envenom'd'—and though the King's defenceless body is within reach; for these are only convenient but subordinate circumstances. What matters to Hamlet is that the perfect moment for revenge shall have arrived before he strikes—the moment that will make possible the bridging of the great gulf that still separates his knowledge from the unknowing world's ignorance, the moment towards which he had long and relentlessly pushed the King. Hence he does not strike in the next instant after wounding Laertes, or in the next instant after Gertrude has fallen, crying that she is poisoned. Weapon in hand, he stands while Laertes gasps out eight lines disclosing his own treachery, informing Hamlet that he, too, is slain, revealing that the foil is envenomed, and affirming that the Queen is indeed poisoned. Hamlet waits resolutely through these revelations until at last Laertes cries out the magic words in the hearing of all the Lords and Attendants of the Danish Court: 'The King, the King's to blame.'

It is the first notice of the King's guilt that the world has heard, and it comes not from Hamlet but from another. On the instant that Laertes gasps out the truth, and with all the palpable proofs now lying about him, Hamlet takes his revenge.

But though it is now bridgeable, the gap is not yet bridged. For when Hamlet strikes the King, the whole Court raises the cry of 'Treason! treason!' and the stricken King still has the bravado to shout, 'O, yet defend me, friends.' Hamlet had once charged himself with thinking too precisely on 'the event'—the final outcome, the killing of the King and its consequences. Revenge not thought on precisely enough and accomplished imperfectly would have meant his own immediate and dishonourable death, with none left behind to establish the truth, or even to guess at it. For Hamlet that kind of dismal 'event', that aftermath, was unacceptable. Hence, long ago, he had begun equipping Horatio with evidence to be used at need, and in the end the body of evidence is at hand, complete. Horatio had seen the Ghost, knew that Hamlet talked with it, and must have surmised, despite Hamlet's protestations to the contrary, that it had imparted some momentous secret; further, Hamlet soon divulged the Ghost's secret to him, and next, perhaps less from his own need for 'grounds more relative' than from his desire to supply Horatio with convincing evidence, staged the 'Gonzago' play while Horatio witnessed the King's violent reaction; thereafter Hamlet pointedly furnished him with details of the King's attempt to have the Prince's head struck off in England—and, to climax all, actually handed him the King's own written order for the murder: 'Here's the commission; read it at more leisure.' This damning document presumably still remains in Horatio's possession at the end, or within his reach; and, finally, Horatio, along with all the Court, has seen the Queen die asserting that the cup was poisoned and has heard Laertes' dying charge, 'The King, the King's to blame.' In all, Hamlet's calculated procedure, his precise—but not too precise—thinking on 'the event', joined with the actual occurrences of the final moments, has supplied Horatio with evidence enough, and to spare. Hamlet's honour will stand untainted.

Even so, but for one violent dying effort, all Hamlet's careful campaign might yet go for nought, for he has no sooner given Horatio his final instructions than this indispensable survivor seizes the poisoned cup and, crying 'Here's yet some liquor left', puts it to his lips. Stalwart, stable, stolid, solid, implacable, unshakable, loyal, but, at this moment, not too brilliant, Horatio has somehow missed the point of it all. Nearly dead from Laertes' unction, Hamlet must wrest the cup away by force to prevent his man from killing himself and burying the truth, together with Hamlet's honour, forever.

How profoundly Hamlet values the publication of the truth is rendered peculiarly conspicuous by the fact that in each of the last three speeches that he makes in the play, a single idea is repeated:

Horatio, I am dead;
Thou liv'st. Report me and my cause aright
To the unsatisfied.

(v. ii. 349–51)

O good Horatio, what a wounded name,
Things standing thus unknown, shall live behind me!
If thou didst ever hold me in thy heart,
Absent thee from felicity a while
And in this harsh world draw thy breath in pain
To tell my story.

(Ibid. 355–60)

So tell him, with the occurrents, more and less,
Which have solicited—The rest is silence.

(Ibid. 377–8)

'. . . draw thy breath in pain / To tell my story': Hamlet's conduct of the battle between 'mighty opposites' has enabled Horatio to speak with proofs to all those who yet 'look pale and tremble' and 'are but mutes or audience' to the slaughter that has marked the end of the ordeal.

Polonius, Ophelia, Rosencrantz and Guildenstern, Gertrude, and Laertes—all have died ignorant of the centre of the vortex, the old King's murder. Not Hamlet's practices but Claudius's are responsible for their deaths: Polonius died while eavesdropping in the service of a sovereign of whose past trespasses and immediate intent he was wholly ignorant; Ophelia was a casualty of her father's misguided service to the King; Rosencrantz and Guildenstern, unwitting pawns in a war of which they knew nothing, died while aiding in the King's first attempt on Hamlet's life; Gertrude died impaled on one prong of the King's last, multi-pronged device meant to kill Hamlet; Laertes died on another prong of the same device.

The end, then, is Hamlet's triumph, not his failure. Had he acted earlier, his deed would have appeared to the world as a mere bloody assassination occurring in a land already notorious and for the reputation of which he early expressed concern: 'They clepe us drunkards, and with swinish phrase / Soil our addition.' No man could have done more, and no man of uncompromisable honour could have done as much and also saved his own life. Brutus dies in dubious battle, because he is Brutus; Othello, the doer of a monstrous, unwitting wrong, dies because he is Othello. Hamlet dies with unsullied honour, and dies not because he is Hamlet but in spite of the fact.

CHAPTER V

The Villain as Practiser: *Othello*

'Fall'n in the practice of a cursed slave.'

WITH its brilliant hero usurping the role of master-practiser that is elsewhere played by bright-eyed heroines, scheming villains, and old dukes of dark corners, *Hamlet* is unique not only among Shakespeare's tragedies but among all his plays. (Only *Henry V*, in this respect, stands comparison.) *Othello*, coming next in the tragic succession, reflects an abrupt and complete reversal, with the practiser's role restored to the villain and the hero consigned to a pit of darkness.

Like *Romeo and Juliet*, *Othello* is a tragedy of unawareness. In the earlier play, all participants stand alike ignorant of the design exposed to us by the Prologue, and Fate uses their ignorance as the avenue to its own end. In *Othello* the participants stand equally ignorant of Iago's malign nature and intent, and Iago uses their ignorance as means to enmesh and destroy them all. In both plays, thus, unawareness is more than merely a dramatically useful condition the exploitation of which yields various spectacular but incidental effects (though indeed both plays exhibit such effects in profusion); it is the open road to catastrophe.

Among the tragedies, too, *Othello* stands nearest to *Romeo and Juliet* in the proportion of scenes during which we hold significant advantage over participants, and in the number of participants over whom we hold advantage. In the earlier play the Prologue gives us advantage over all participants even before action commences, and we hold it steadily thereafter to the end—and beyond, since in *Romeo and Juliet* the gulf between our knowledge and the participants' ignorance is never fully closed: there is no Horatio left behind to tell Verona 'How these things came about'. No Prologue advises us in *Othello*, but the great central gap begins to open with Iago's first words to Roderigo about his hatred of Othello, widens abruptly with his blunt statement that he follows the Moor only 'to serve my turn upon him', and attains its permanent breadth with the speech that follows:

> In following him, I follow but myself;
> Heaven is my judge, not I for love and duty,
> But seeming so, for my peculiar end;
> For when my outward action doth demonstrate
> The native act and figure of my heart
> In compliment extern, 'tis not long after
> But I will wear my heart upon my sleeve
> For daws to peck at. I am not what I am.
>
> (i. i. 58–65)

In its significance for what follows, the speech rivals the Prologue of *Romeo and Juliet*, for it tells us unmistakably that Iago is not what he seems to others in his world. During the course of subsequent action, and always as result of specific practices contrived by Iago, numerous incidental discrepancies arise between the participants' awareness and ours; but all are subordinate to and enveloped by the main dramatic condition of the tragedy: that we know what Iago is, and the participants do not—not even the dupe Roderigo, though he hears the villain's initial statement as plainly as we do.

The main gap between us and the unknowing worlds of Venice and Cyprus remains open throughout fourteen of the fifteen scenes of *Othello*—from about the middle of i. i to the middle of v. ii—though a momentary first-and-last glimpse across the breach is given Roderigo at v. i. 62, when, to his great surprise, he is stabbed by 'honest' Iago: 'O damn'd Iago! O inhuman dog!' During this long period we share our awareness with none but Iago: not only Othello, but Desdemona, Cassio, Roderigo, the Venetian and Cyprian officials, and Iago's most intimate associate, his wife—all are oblivious of the false mind behind the 'honest' mask and of the many specific machinations by which they are victimized. The basic condition of the tragedy, thus, is static, framing the action from the first scene to the last. But within this frame the incidents that make up the action are propelled by the swift succession of Iago's practices— even as the action of *Hamlet* is propelled by the practices and counter-practices of the 'mighty opposites'. Yet just here appears also a noteworthy contrast between the two tragedies: in *Hamlet* the opponents are indeed mighty, and the action exhibits a continuous battle between their rival practices; the tensions of *Hamlet* are therefore taut, for neither opponent is hopelessly outclassed,

Claudius having highest authority and Hamlet highest awareness. But in *Othello* no comparable tension is possible: Iago's is the sole awareness, and Othello must take his place with Brutus as the high-minded hero who is simply no match for the force that opposes him. Equal in their dedication to the ideas of honour and justice, Hamlet and Othello are opposites in their awareness of themselves and their worlds. Hamlet sees himself, his task, and his situation clearly and wholly; in this respect he is blood-brother to the bright heroines of the romantic comedies and runner-up to the god-figures of the late plays, Duke Vincentio and Prospero. Othello, on the other hand, never learns until the end who is his adversary or even that he has one.

The action of *Othello*, then, like that of *Hamlet*, is propelled by a succession of what may be called 'operational' practices within the enveloping one. Here the equivalent to Hamlet's mask of lunacy is Iago's mask of honesty. Unlike Hamlet, Iago does not put his mask on after the action commences; he has presumably worn it from birth onwards—or at least from the start of his military career. Neither does he, like Hamlet, frequently remove and then replace his mask as different occasions allow or require; Iago's mask has grown to his face and is irremovable. Even when, with Roderigo, a very limited confidant, he puts on his most disarming show of candour, insisting that he follows the Moor only to serve his turn upon him, the mask remains tightly in place, for he is even then urgently bent on deceiving Roderigo in order to keep his purse available. This mask of honesty by which all Venice and Cyprus are deceived creates and holds open the main gap between the awarenesses of participants and audience, and it is the steady exploitation of it alone that produces the virtually incessant flashes of irony and other incidental effects that accompany the action of the tragedy from start to finish. And not only is this exploitation incessant; typically it is also the least subtle, most outrageously bold exploitation of its kind in all Shakespeare, and to find a rival in this respect we must look at such strokes as those in *Richard III*, when, for example, the innocent and doomed York chatters amiably with his uncle:

> *York.* I pray you, uncle, give me this dagger.
> *Glou.* My dagger, little cousin? With all my heart.
>
> (*RIII*, III. i. 110–11)

. . . and when Clarence, also doomed by Gloucester, begs a boon of his appointed assassins:

If you are hir'd for meed, go back again,
And I will send you to my brother Gloucester,
Who shall reward you better for my life
Than Edward will for tidings of my death.

(Ibid. I. iv. 234–7)

So Desdemona, in her hour of direst need:

Alas, Iago,
What shall I do to win my lord again?
Good friend, go to him . . .

(IV. ii. 148–50)

In all, Iago is called 'honest' no fewer than sixteen times, in addition to the several occasions when he advertises his own 'honesty':

O wretched fool,
That lov'st to make thine honesty a vice!
O monstrous world! Take note, take note, O world,
To be direct and honest is not safe.

(III. iii. 375–8)

Iago's profession of honesty is a kind of grotesque travesty of Brutus's insistent protestations of honour. Othello, Desdemona, Cassio, all call him 'honest' in such a manner as to single him out especially, beyond other men, for his possession of the virtue of honesty: 'This fellow's of exceeding honesty', says Othello—precisely at the moment of crediting the villain's defamation of Desdemona. Othello debates with himself, and with Iago, the honesty of others, of Desdemona and Cassio: 'I do not think but Desdemona's honest.' But Iago's own honesty is never called in question: 'My friend, thy husband, honest, honest Iago'. To Othello, Iago's is the absolute by which others' honesty is to be measured. Iago is also repeatedly identified by other salutary expressions: he is 'good' Iago, 'kind' Iago, 'bold' Iago. But it is chiefly by the insistent drumming of 'honest' that the dramatist carries on a continuous exploitation of the great gap between the participants' and our awareness. Shrewdly, it is during the roughly four hundred lines that represent Othello's transformation from the serene and noble Moor 'whom our full Senate / Call all in all sufficient' to the howling, horn-mad husband—'I'll tear her all to pieces' and 'Damn her, lewd minx! O, damn her! damn her!'—that Shakespeare most persistently hammers at the word 'honest' and so exploits his fertile gap for effect. But of course the repetition of 'honest' neither begins

nor ends with this climactic, terrible scene: 'Good night, honest
Iago,' says Cassio gratefully, just after the villain has diverted him
from his assigned military duty, got him drunk, and witnessed him
sacked by Othello; and again, just after Iago has opened the door for
Cassio's innocent interviews with Desdemona that lead straight
on to catastrophe: 'I never knew / A Florentine more kind and
honest.'

Once we have seen what Iago is, in contrast to what his victims
think him, the usual effects that arise from exploitation of discrepant
awarenesses follow almost automatically: Iago need but appear, and
our minds grow busy with thoughts of the difference between the
appearance and the reality; indeed, he need not even be present in
order to set our minds going, for everything that other participants
engage in in our sight, whether the villain is visible or not, is
conducted in his shadow and prods remembrance of the universal
misjudgement of his character. No other figure in Shakespeare's
plays so monopolizes both the drama on the stage and the drama in
our mind's eye as does Iago, for we cannot watch a moment of the
action, overhear a segment of dialogue, without instantly associating
it with Iago and his devices. Once Shakespeare has established the
initial dichotomy—and he wastes no time, but establishes it early in
the opening scene—the effects of its exploitation might continue of
themselves even though the dramatist had neglected to punctuate
dialogue with reminders like 'O, that's an honest fellow.' But of
course it is rarely Shakespeare's way to trust an exploitable situation
to exploit itself; hence in *Othello*, where the stakes are uncommonly
high, the potentialities of effect unusually rich, we find an extraordinary number of occasions on which our minds are sharply
prodded. If we do not really need to be told so often that Iago's
victims continue to take him for the soul of honesty, yet the repetition, especially at those moments when an unsuspecting victim
beseeches the villain to extricate him from a predicament into which
the villain's own practice has cast him, no doubt intensifies the
dramatic experience; each repetition of 'honest' at such times is like
a new and potent injection.

So much for the enveloping condition, the gap between the
participants' and our perception of Iago; but of course the action and
effects of the play do not derive from this condition alone. As in
Hamlet, they move on a succession of practices, and the first of these
is the practice on Roderigo, to which allusion is made in the first

speech of the play; says Roderigo, 'I take it much unkindly / That thou, Iago, who hast had my purse / As if the strings were thine, shouldst know of this.' In short order we learn the basic facts: Iago has long practised on Roderigo by feeding his hopes of achieving Desdemona, his own reward for this supposed service being ready and continuing access to Roderigo's purse; but now Desdemona has wed the Moor, and the dupe, recognizing that the chase is hopeless, will give it up—and shut his purse.

Measured on the scale of later monstrous events—the sacking of Cassio, the maddening of the Moor, the murder of Desdemona, Othello's own death—Iago's practice on Roderigo in order to use his purse appears relatively innocuous—hardly worse than Sir Toby's similar practice on Sir Andrew in *Twelfth Night*. But Shakespeare is not in the habit of starting a play by introducing trivial issues, or issues extraneous to the great central action soon to be developed; rather, he starts head on, preparing that action: so in *Romeo and Juliet*, as we have noted, when the servants' baiting of one another leads directly into the main business and ultimately to the pile of young bodies in the Capulet tomb; so in *Hamlet*, where in the very first line the *relieving*, not the on-duty sentry demands 'Who's there?' and is pointedly corrected, 'Nay, answer me!'—thus serving notice without delay that something is rotten in the state of Denmark; so in *King Lear*, where the first thirty lines raise the issues of the division of the kingdom and Edmund's bastardy, and drive wedges into both plots of the tragedy; but, indeed, so it is in all the plays, comedies and histories as well as tragedies, where the first scene regularly ushers us into the main action.

It cannot be regarded as an idle thing, then, that Shakespeare begins *Othello* where he does: with Iago's bait (Desdemona) lost and his prize fish (Roderigo) just taking off for other waters. In this circumstance the villain has no choice but to move suddenly and drastically. Seizing on Roderigo's convenient cue—and throughout what follows Iago will always be quick to seize on cues—'Thou told'st me thou didst hold him in thy hate', he assures his victim that such is indeed the case, that he does hate Othello with a passion, for has not the Moor unjustly passed him over in favour of the incompetent Cassio? Seeming to fan his own fury as he speaks, he quite overwhelms Roderigo's puny gust of anger that touched off this venomous tirade: he hates Othello, he has only contempt for Cassio, he hates the military service where 'Preferment goes by letter and affection, / And not by old gradation'. To cap his show of outrage, he confides that his only motive for following the Moor is 'to serve my turn upon him'.

At this point the dupe gives him another cue: 'What a full fortune does the thick-lips owe, / If he can carry't thus.' And again Iago snaps up the cue, this time with action, not mere words: if Roderigo has doubts that Othello can 'carry't thus', then these are doubts to build on, and quickly. 'Call up her father,' he cries, 'Rouse him. Make after him, poison his delight, / Proclaim him in the streets.' There is vigorous life yet in the same old practice by which he has kept Roderigo's purse open: if the dupe can be made to suppose that Brabantio will divorce Othello and Desdemona, then his hope of winning Desdemona for himself can be restored after all, and sustained until the next crisis shall arrive. Nor, certainly, does his own conviction that neither Brabantio nor the State will dare to divorce Othello deter the villain even for a moment. Once Brabantio has been brutally aroused and, in Iago's foulest language, advised of Desdemona's elopement, the villain, who never lacks boldness, even confides his thought to his dupe: 'I do know, the state, / However this may gall him with some check / Cannot with safety cast him.'

Many times in the past, no doubt, Iago's lucrative practice on Roderigo has been endangered and the villain has had to invent fresh devices in order to sustain his gull's flagging hopes. The present device works admirably, and at the end of Act I Iago's practice has survived another crisis: 'I'll sell all my land', asserts the dupe; and later we hear again how costly have been his vain hopes of having Desdemona: 'I have wasted myself out of my means. The jewels you have had from me to deliver Desdemona would half have corrupted a votarist.' Viewed in this perspective, Iago's initial practice on Othello, in arousing Brabantio, inflaming him, and directing him to the Sagittary, is a direct implementation of his continuing practice on Roderigo. Once we have recognized this fact, we are obliged to review with a cautious eye the opening dialogue in which Iago vehemently asserted his hatred of Cassio, his hatred of Othello, his insistence that he follows the Moor only 'to serve my turn upon him'. Reconsidering the circumstances in which these protestations are made—Iago's dire need of sudden, drastic speech and action if he is to keep his fish now that Desdemona has married—we are obliged to question the truth of everything spoken by the villain in our hearing during the opening scenes. Need we believe that he does in fact hate Othello and Cassio when we know that everything he says was prompted by the occasion, calculated to revive Roderigo's hopes of having Desdemona? Should we not suppose that, had the fair Venetian married some noble young Venetian instead of Othello, he would yet have devised a set of arguments to convince Roderigo that he himself had long hated this particular gentleman for sundry injurious acts done by him, that the marriage of the couple was vulnerable, and that Roderigo would shortly enjoy Desdemona after all?

Whatever monstrous growths develop later, and however it all ends, this is how Shakespeare chose to initiate the tragedy of *Othello*: with Iago's passion to keep his dupe's purse at hand. Whatever later motives come to supersede the original one, the motive that initiates the tragedy is the purse. The source of the campaign against Othello is the practice on Roderigo. Othello, Desdemona, and Cassio are, at the outset and for long thereafter, mere pawns, unfortunate incidentals. In the end they are innocent victims of a deadly game the origin of which they never learn: 'Thus do I ever make my fool my purse.'

The initial practice on Othello is a direct extension of the practice on Roderigo. After arousing Brabantio, Iago thus justifies his own immediate departure: 'It seems not meet, nor wholesome to my place, / To be produc'd—as, if I stay, I shall— / Against the Moor.' He has already begun juggling balls in the air; and he goes on: 'I must show out a flag and sign of love, / Which is indeed but sign.' In the next scene, having aroused Othello, he warns him of Brabantio's approach—omitting to mention, of course, that it was he himself who aroused and inflamed him, then directed his way to Othello's lodging. When Brabantio's party arrives and fighting appears imminent, Iago boldly singles out Roderigo: 'You, Roderigo! come, sir, I am for you'—a show of valour that Othello and Roderigo each would take as evincing good will on his side. The incident ends abruptly when Othello and Brabantio agree to carry their differences to the Duke in council. Iago has contrived to demonstrate loyalty alike to Senator, dupe, and Moor.

The villain and his practices remain briefly in the background during Othello's confrontation with state officials; for about three hundred lines, while the marriage is approved by the Duke and the newly-weds are ordered to prepare for Cyprus, Iago is silently present, and there is no overt exploitation of the gap between the participants' and our awareness of what his mind holds. It is the first and last such interval in the play; afterwards there will be no measurable period during which the basic discrepancy is exploited neither explicitly by spoken lines nor silently through the strength of our abiding consciousness of Iago's sinister intent. It is Othello who ends the interval when, replying to the Duke's order that he leave behind some officer to bear the council's commission, he names Iago: 'A man he is of honesty and trust.' And then again, as he departs: 'Honest Iago, / My Desdemona must I leave to thee.'

At the moment of Othello's departure for Cyprus, Iago has no active practice in operation against the Moor; our awareness holds only his earlier statement, 'I follow him to serve my turn upon him', made in the heat of his need to convince Roderigo that he would yet have Desdemona. The first episode, by which he sought to appease Roderigo by arousing Brabantio and directing him to the couple's lodging, was ended by the Duke's sanction of the marriage and the commissioning of Othello for Cyprus. Here matters might simply rest forever: whatever true grudge against the Moor that Iago may have—if, indeed, his profession of undying hatred was anything more than a convenient sop for Roderigo—might remain quiescent. That the villain himself had considered the whole affair ended is apparent when, after Othello's exit, Roderigo accosts him:

> *Rod.* Iago,—
> *Iago.* What say'st thou, noble heart?
> *Rod.* What will I do, think'st thou?
> *Iago.* Why, go to bed and sleep.
>
> (I. iii. 302–5)

Here it is made clear that Roderigo, not Iago, revives the threat to Othello. The dupe is again as desperate as when he learned that Desdemona had married—for now she will join her husband on Cyprus and be irretrievably lost: 'I will incontinently drown myself.' To keep the purse at hand, Iago is again obliged to make drastic promises and devise drastic practices.

His immediate response to the new challenge is to unleash the force of his persuasive skill: Desdemona, he assures the dupe, will quickly tire of the Moor, and the Moor of her; it is but a matter of time, for 'If sanctimony and a frail vow betwixt an erring barbarian and a super-subtle Venetian be not too hard for my wits and all the tribe of hell, thou shalt enjoy her.' During the course of his immensely plausible harangue, he boldly instructs Roderigo to 'Put money in thy purse'—not once but eight several times, so that the repetition of this solemn injunction becomes a cord to bind the entire speech together. And at the end the dupe's hopes flourish again: 'I'll go sell all my land.'

It is thus again the purse that stimulates Iago's brain to devise a plot against the Moor. Of the roughly three hundred and fifty lines in the play given Iago in the forms of soliloquy, 'aside', and confidence to Roderigo, nearly one hundred appear here, at the end of Act I, when the villain's lucrative practice on his dupe meets and survives a second challenge. Like the opening lines of all Shakespeare's plays, the closing lines of the first act are regularly weighted with significance for subsequent action; they often stand like the thesis statement of an expository piece. Here the closing lines confirm the emphasis of the opening lines and leave no reasonable doubt that at this point, at least, Othello and Desdemona, their marriage and their very lives, are but incidentals in the operation of Iago's practice on Roderigo. It is not, of course, that Roderigo himself, aside from his purse, is of the least consequence either way to the villain: 'For I mine own gain'd knowledge would profane / If I would time expend with such a snipe'. The being of Roderigo matters only for his purse, even as the being of Desdemona matters only as bait. Again at the close of the act as at the start, Iago assures his dupe of his hatred for Othello: 'I have told thee before, and I re-tell thee again and again, I hate the Moor.' And as before, the utterance is caused by a crisis in the continuing practice. It chances to be Othello's wife for whom Roderigo lusts; since that is so, the continued prosperity of the practice necessarily involves her, and, since she is married to Othello, it must involve him as well. If the object of Roderigo's lust were another than Othello's wife, there would be no need for Iago to comb his brain in search of reasons to hate Othello.

In the soliloquy that ends Act I, Iago says yet again, 'I hate the Moor'—and because the line is spoken in soliloquy rather than in confidence to Roderigo, we should ordinarily take it for the truth; but here the placement of the line compromises its literal truth, for it immediately follows four lines that concern the practice on Roderigo —the practice that inspired the entire soliloquy:

> Thus do I ever make my fool my purse;
> For I mine own gain'd knowledge should profane

If I would time expend with such a snipe
But for my sport and profit.

(I. iii. 389–92)

Then, with no transition, Iago continues: 'I hate the Moor'. His preoccupation with the problem of Roderigo prompts the reassertion of his hatred for Othello even as it had prompted it just before, when he had soothed Roderigo with 'I have told thee often, and I re-tell thee again and again, I hate the Moor.' In the next breath, still without transition, he adds, 'And it is thought abroad that 'twixt my sheets / He has done my office'. The causal connection that we should have expected is absent: not 'for' he has done my office, but only 'and' he has done it. Earlier, knowing that words alone, however persuasive, could not forever appease his dupe, Iago acted by arousing and inflaming Brabantio and directing him to Othello's lodging; now, having again soothed Roderigo with words, he knows that further productive action must be undertaken if access to the purse is still to be his. His earlier action in bringing Brabantio to confront Othello had injured the Moor not at all, as, indeed, he knew it could not, being aware that the state needed him against the Turks. But the next action, if it is to content Roderigo for long, must be more drastic and must actually hurt the relations of Othello and his bride. It is precisely at this point that even Iago's amoral mind appears to need some kind of self-justification—hence 'And it is thought abroad that 'twixt my sheets / He has done my office'. His 'motive-hunting', a form of rationalization, thus commences just at the outset of what will grow into a deadly campaign against Othello and Desdemona.

'Thus do I ever make my fool my purse': the practice on Roderigo has now made necessary some kind of monstrous practice on Othello, the shape of which is not yet clear in Iago's mind; but it will somehow involve Cassio as means—and for any incidental injury done to Cassio the villain's groping mind finds quick and easy justification: 'To get his place and to plume up my will / In double knavery'. But 'How, how?' he asks himself: 'Let's see:— / After some time, to abuse Othello's ear / That he is too familiar with his wife.' The soliloquy ends on a note of frightful resolution but uncertain means: only, somehow, Cassio is to serve as the means to Othello, and Othello will serve as means to the continued enjoyment of Roderigo's purse. All three who are sure to be most hurt— Cassio, Othello, Desdemona—are thus still, at the end of Act I, in

Iago's yet-bleary perspective, mere incidentals. The Moor and his wife are at this point in a position not unlike that of Romeo and Juliet: Fate chooses to use the lovers' deaths as means of ending their parents' feud; and Iago chooses to sacrifice Othello and Desdemona, should it come to that, in order to keep his gull living in hope. Yet Shakespeare gives his young lovers at least inklings of disaster: 'Is she a Capulet? / O dear account!' cries Romeo just after their first meeting, 'my life is my foe's debt'; and Juliet, 'My only love sprung from my only hate! / Too early seen unknown, and known too late!' But the older newly-weds, embarking for Cyprus, have no hint of danger: 'So please your Grace, my ancient,' Othello tells the Duke; 'A man he is of honesty and trust. / To his conveyance I assign my wife.' It is appropriate to note also, just at this point, how drastically Shakespeare's method of exploiting the main gap between the participants' and our awareness differs in the two tragedies. *Romeo and Juliet* is loaded with such utterances as those above—forebodings, ill-divinings, hints of doom; but in *Othello* the exploitative phrases are exactly the contrary—expressions of confidence, trust, perfect security: 'Honest Iago'.

With the same sense of well-being that marked their parting in Venice, Othello and Desdemona greet each other on Cyprus: 'If after every tempest come such calms, / May the winds blow till they have waken'd death!' And, joyously kissing his wife, Othello concludes: 'And this, and this, the greatest discords be / That e'er our hearts shall make'. These blissful utterances we hear just after hearing Iago, who has observed Cassio holding Desdemona's hand: 'He takes her by the palm; ay, well said, whisper. With as little a web as this will I ensnare as great a fly as Cassio.' And now, watching the newly-weds kiss, he speaks aside again: 'O, you are well tun'd now! / But I'll set down the pegs that make this music, / As honest as I am.' As yet the couple are totally oblivious of the 'monstrous birth' that remains at the moment only embryonic in Iago's brain, and Shakespeare continues to exploit their unawareness by exhibiting their bliss. Unlike the young lovers trapped in Fate's web, they have no misgivings.

The Fate of *Romeo and Juliet*, as we learn from the Prologue's categorical declaration, is in any event more clear-sighted about its objective than Iago, who began with a modest, relatively innocuous purpose—to keep Roderigo's purse open. But by the end of II. i his purposes appear to have multiplied: to keep the purse, to get

Cassio's place, and to get even with Othello 'wife for wife'. These aims would be clear enough if each were distinct and separate, but all are intertwined and their relationship complicated because each is at once an end in itself and a means to another end. Thus, in enlisting Roderigo's aid in his first direct practice on Cassio, Iago argues that Desdemona loves Cassio and is likely, once she has tired of Othello, to choose Cassio rather than Roderigo; hence, in the dupe's perspective, the practice on Cassio is to serve two purposes, one for himself, the other for Iago. But, all the while, Roderigo is himself the victim of Iago's lucrative practice, ignorant that the villain's overriding purpose is continued access to the purse; unwittingly, thus, in agreeing to participate in the practice on Cassio, Roderigo enlists in the practice on himself.

Even if the tangle of means and ends stopped there, it would yet be fairly simple: a guller with two gulls, using one against the other in a double practice designed to serve only his own two purposes—the purse and Cassio's office. But it does not stop there. Earlier, in the soliloquy that ends Act I, Iago groped for some means of abusing Othello's ear with the charge that Cassio is too familiar with Desdemona; now, in the soliloquy that ends II. i, he gropes for a way to relate his newly conceived device for disgracing Cassio to what at this point appears a new and unrelated purpose, that of destroying Othello's peace. ''Tis here, but yet confus'd', he mutters —meaning, perhaps, that he does not himself clearly distinguish what are means and what are ends. To this point in the play, Iago has been our sole informant about the plots that are in process of development—a corrupt informant, too, whose 'candid' remarks, even in soliloquy, must be sifted and discounted if we are to separate the false from the true. But from this point on we begin to gain an advantage over our informant himself, for the dramatist has enabled us to see in his villainous scheming some signs of a disordered intellect.

In the first place, it appears that Iago has begun to believe some of the very lies that, out of necessities occasioned by his practice on Roderigo, he had invented to appease his gull. Over and over—who can say how many times over an indefinite period of time?—he has told Roderigo that he hates the Moor and follows him 'but to serve my turn upon him'. The fiction spoken again and again, argued passionately for the sake of persuading Roderigo, has worked its way so deeply into his own consciousness as to seem like truth. 'I hate the Moor; / And it is thought abroad that 'twixt my sheets / He has done my office', he said in the soliloquy that ends Act I, echoing just what he had been telling Roderigo—and then adding, rather tentatively, a rationalization: Othello has cuckolded him with Emilia. But when he next speaks in soliloquy, at the end of II. i, his tentative addition is no longer limited to 'it is thought abroad', but has grown to 'I do suspect the lusty Moor / Hath leap'd into my seat'. A similar evolution has occurred in his notion of the Cassio–Desdemona relationship. First, seeing Cassio greet Desdemona on the Cyprus dock, he catches a glimpse of an opportunity: 'He takes her by the palm; ay, well said, whisper. With as little a web as this will I ensnare as great a fly as Cassio.' Next, in groping for a way to splice together his old practice on Roderigo and a new one on Cassio, he enlarges the initial thought into an elaborate fabrication for the purpose of persuading Roderigo to serve in the new practice: Desdemona loves Cassio; she cannot long be satisfied with Othello, but must have change; she will then inevitably turn to Cassio; for already 'They met so near with their lips that their breaths embrac'd together'. Always a master of the plausible, Iago uses nearly seventy lines to make sheer fantasy, born of nothing, look like fact; and at the end Roderigo is persuaded: 'I will do this.' But so passionately has Iago invented in order to persuade Roderigo that his fantasy has now become fact in his own mind also, for his soliloquy that immediately follows begins, 'That Cassio loves her, I do well believe't; / That she loves him, 'tis apt and of great credit.'

But indeed this entire soliloquy that ends II. i and marks the conception of the gratuitous campaign that will destroy Othello, Desdemona, Roderigo, Emilia, and Iago himself is an irresponsible and disordered tissue of falsehood and half-truth, impossibility and possibility, means and ends, all confused, entangled, crossing and re-crossing one another. Incoherent and illogical, improvised in the same moment it is spewed forth, the soliloquy makes a jumble of order and is the product of a hyperactive and lunatic mind: Cassio loves Desdemona and probably she loves him; the Moor, though I cannot bear him, has a noble nature and will prove a good husband; I love Desdemona too, partly out of lust but partly because I cannot rest until I have debauched Othello's wife as he has mine—or at least until I have driven him mad with jealousy; I'll use the fool Roderigo to compromise Cassio and abuse him to Othello—for I suspect Cassio also with my wife—and thus I will win the Moor's love and gratitude for driving him mad. Can Iago truly believe that Desdemona is corrupt, or corruptible? If so, does he entertain the idea of debauching her himself, or mean to leave that pleasure to Cassio—or to Roderigo? Does he intend to destroy Cassio in order to have his place—or to take revenge for having been cuckolded? Does he truly believe that Cassio has cuckolded him? Can he mean both to destroy the Moor, whether by actually debauching Desdemona or by making Othello believe her to be false, and, at the same time, to win, in this way, Othello's thanks, love, and reward? Can he possibly hope *both* to destroy Othello and to win Cassio's office under him? ''Tis here, but yet confus'd', the soliloquy concludes; and indeed Iago does not clearly apprehend either what he wants or why he wants it with respect to Othello, Desdemona, or Cassio; at this point he appears only to want to hurt everyone around him as badly as possible, and resembles nothing so much as a frustrated baby, red with rage, that hits and kicks in every direction.

Iago has seemingly persuaded himself that the lies invented to keep Roderigo's purse are true, and out of purposes and cross-purposes has flung together a monstrous engine that is ready to move at the opening of II. iii. The structure of awarenesses now has become more complex: on the bottom-most level are Othello and Desdemona, who, wholly ignorant that their bliss has an enemy, retire to enjoy the fruits of their marriage, leaving the cares of the watch on Cyprus to Cassio and to 'honest' Iago; hardly above their level is Cassio, target of a practice by which Iago means to get his place, take revenge for imagined familiarity with Emilia, sustain Roderigo's hopes of having Desdemona, and prepare the ground for a practice on the Moor; on the level next above is Roderigo, who knowingly participates in the practice on Cassio, thinking it designed to serve his own purpose, but who has no sense of the uses to which he is being put by Iago for *his* purposes, no sense of the true place he occupies in the villain's haphazard machinery, no sense of the contempt in which Iago holds him; situated high above these inferior levels is Iago, sole builder of the whole structure, in whose perspective all those below are merely puppets and pawns. But the dramatist has raised our own level above Iago's, even though the villain remains our only informant among the participants; for we now have insight into his madness and the madness of his apparatus.

Our awareness has been fully equipped for the practices that are set in motion in II. iii. At the end of Act I, Iago had first attacked the problem of getting at Cassio: 'Cassio's a proper man: let me see now: / To get his place and to plume up my will / In double knavery—How, how?' Next, in II. i, he instructed Roderigo to 'find some occasion to anger Cassio'. Finally, as the action resumes, Iago reveals that he has 'fluster'd with flowing cups' three Cypriots who are to help with the watch, and 'Now, 'mongst this flock of drunkards / Am I to put our Cassio in some action / That may offend the isle.' Noteworthy in these preparations are two characteristics of Iago as practiser that illustrate his method: first, his habit of fixing upon the particular propensity of his victim that will best serve his turn; second, his habit of leaving the details of a practice to be improvised at need. Thus Roderigo's deluded craving for Desdemona was the basis of the practice by which the purse is kept available, but all details of the actual operation were devised on the spot—as when Brabantio was aroused at night and directed to Othello's lodging, and now again as the device for disgracing Cassio is set. Cassio's propensity for blind anger when drunk obviously invites the practice, the details of which, however, are left to improvisation. Iago does not devise an exact procedure for Roderigo to follow in angering Cassio, but rather names wide-ranging possibilities, concluding 'or from what other course you please, which the time shall more favourably minister'. Iago always plays the game, so to speak, by ear, making no effort to control events so tightly that their precise course is predictable. The supreme egotism of the villain is nowhere better exemplified than by his confidence that he will be capable of dealing with any and all contingencies once his practice has been activated; and, indeed, once the whirlwind that he has generated has attained a swirling frenzy, he is here, there, everywhere, converting every chance to his special use. Thus in the present scene, having set the stage by getting Cassio drunk, he seizes on Montano—whose presence at the precise moment could not have been foreseen—and loudly deplores the fact of Cassio's one abiding frailty, in this way preparing Montano's reception of whatever may next occur. When Cassio staggers out—as, again, Iago could not have foreseen—he hastily dispatches Roderigo after him; then, when Cassio returns, pursuing and striking Roderigo, Montano, earlier alerted by Iago's remarks on the lieutenant's weakness and his warning that 'the trust Othello puts him in, / On some odd time of his infirmity, / Will shake this island', intervenes on cue to halt the brawl before it spreads: 'I pray you, sir, hold your hand.' And now the villain comes abruptly alive with energy, rushing Roderigo forth to 'cry a mutiny', while he himself creates din and confusion all about:

> —God's will, gentlemen;—
> Help, ho!—Lieutenant,—sir,—Montano,—sir;—
> Help, masters! Here's a goodly watch indeed!
> (Bell rings.)
> Who's that which rings the bell?—Diablo, ho!
> The town will rise. Fie, fie, Lieutenant, hold!
> (II. iii. 157–61)

Thus, by his own noise, he capitalizes on the situation, swelling what might have been a minor, local incident into a general alarm that arouses Othello from his bed and gets Cassio summarily cashiered. His incitement and spur-of-the-moment management of this first incident in the campaign against his victims are quite of a kind with his conception and direction of the entire campaign; seemingly reckless of consequence, he sets a practice going and then stands ready to convert to his own use whatever may fly out of it. Not once does he acknowledge a thought that what he starts might one day get out of hand; but by this point the thought should certainly have crossed our own minds.

Having, with one stroke, so completely discredited Cassio that Othello summarily sacks him—'Never more be officer of mine'— Iago could now logically break off his practice, for he has eliminated his rival, the incumbent lieutenant, and at the same time has given Roderigo at least an earnest of success in his expectations of having Desdemona. He can now expect to get Cassio's place, and his very egotism would assure him that, with facile argument and relatively easy manipulation, he can continue to use Roderigo's purse. But logic has no standing in his irrational brain, and he promptly initiates a new practice, one that involves Cassio only as a tool and is directed ultimately at Othello and his wife:

> ... whiles this honest fool
> Plies Desdemona to repair his fortune
> And she for him pleads strongly to the Moor,
> I'll pour this pestilence into his ear,
> That she repeals him for her body's lust;
> And by how much she strives to do him good,
> She shall undo her credit with the Moor.
> So will I turn her virtue into pitch,
> And out of her own goodness make the net
> That shall enmesh them all.
> (Ibid. 359–68)

131

Thus far Iago's practices have at least run parallel: he used Roderigo in ruining Cassio, and he used Cassio's ruin to keep Roderigo hopeful of Desdemona. Both practices were undertaken as means to his private ends: Roderigo's money and Cassio's place. But the turn now taken, at the end of Act II, is a perverse and irrational one, for it appears designed not to further either of the former purposes but to run counter to them. If the newly announced practice should succeed in destroying Othello and Desdemona—as 'enmesh them all' implies —then there will no longer be a Desdemona for Roderigo to aspire to, and no longer an Othello under whom to serve.

Why, then, when his past practices have won him what he sought, does Iago plunge into new ones that threaten to lose what he has gained? But it is futile to attempt a rational explanation of a decision that is itself irrational. The muddled condition of Iago's intellect has been noted earlier, and at the ene of Act II this condition has plainly deteriorated. The term 'decision' is itself improper, for in fact the villain reaches no decision, but merely strikes out anew when an opportunity presents itself. Elsewhere, Shakespeare's major figures, heroes and villains alike, tend to debate with themselves, whether to do or not to do, and each reaches the decision, for better or worse, that is consistent with his particular nature. But Iago never debates with himself about what is better or worse to do; he is single-mindedly concerned with *how* to do, and his soliloquies avoid philosophical considerations: 'So will I turn her virtue into pitch, / And out of her own goodness make the net / That shall enmesh them all.'

The new practice begins with no delay. Having been cashiered, Cassio laments his misfortune, and Iago, casually, with no seeming thought about the matter, offers him assistance in regaining his place: 'I'll tell you what you shall do. Our general's wife is now the general'—and so on. Cassio overflows with gratitude, agrees that he will 'beseech the virtuous Desdemona to undertake for me', and ends the interview with the usual testimonial to Iago's character: 'Good-night, honest Iago.' Then follows the soliloquy that is remarkable for its delineation of what is to be done and how it is to be done, and for its omission of any word on why the villain will proceed with practices that run counter to his own interest. The new practices can afford him nothing of personal gain; indeed, he runs the risk that Cassio will succeed too quickly by using Desdemona as intermediary and thus will recover his lieutenancy before Iago is able to advance his deadlier practice designed to 'enmesh them all'. Iago knows well that Desdemona can readily persuade her husband to restore Cassio's place: 'His soul is so enfetter'd to her love / That she may make, unmake, do what she list.' And his assessment proves accurate, for on the occasion of Desdemona's first conversation with her husband on Cassio's behalf, when she has hardly begun to entreat, Othello's opposition crumbles: 'Prithee, no more; let him come when he will, / I will deny thee nothing.' After Cassio's fall, the lieutenancy was as good as Iago's own; it is, then, irrational of him to teach Cassio the way to win his place again.

Obviously—for so the villain himself directly tells us—he does so as means to the ultimate end, to 'enmesh them all'. It is this objective that, to his twisted mind, warrants the risk of losing what in effect he has just gained. But Iago never once tells us what he hopes to gain by destroying Othello and Desdemona. He has said that he hates the Moor and will be even with him wife for wife; he has also said that he will 'Make the Moor thank me, love me, and reward me' for driving him into madness. But his declarations of hatred for Othello were made in the heat of persuading Roderigo to continue handing out money and jewels; there is no more reason for us to believe that they are literally true than for us to believe that the other assurances he gives Roderigo are literally true. And even if they were literally true, they do not mesh with his intent to 'Make the Moor thank me, love me, and reward me'.

The deadly practices new-hatched in Iago's brain are put to work at the start of Act III, and, once started, grow like monsters. Emilia, at her husband's command, enables Cassio to interview Desdemona. These three, of course, are entirely ignorant of their true roles in the villain's drama, and Othello, ignorant that anything at all is afoot, is preoccupied with the defences on Cyprus: 'This fortification, gentlemen, shall we see't?' Though single in its final aim, Iago's practice is a two-staged operation that requires, first, his moving Cassio and Desdemona to act in ways that will allow suspicion to be cast upon them, and, second, his inflaming the Moor with suspicion. Once started, the first stage proceeds of its own volition, being exactly fitted both to Cassio's love for Othello and his zeal to recover his lost place, and to the sheer goodness of 'th' inclining Desdemona'. Says she,

> My lord shall never rest;
> I'll watch him tame, and talk him out of patience;

His bed shall seem a school, his board a shrift;
I'll intermingle everything he does
With Cassio's suit.

(III. iii. 22–6)

Warped mass of lies though his brain is, Iago spoke truthfully in advising Cassio that Desdemona's intervention was the way to regain his place; so well does she play her role that Othello's resistance falters at once—'I will deny thee nothing'—and Iago is obliged to work with lightning speed in impregnating Othello with suspicion lest the entire scheme be aborted by the abrupt success of its preliminary stage: Cassio would regain his place, and the master practiser would have defeated himself, the shrewdly engineered device by which he had caused Cassio to be sacked having gone for nothing. The two stages of the practice are in fact in a curious kind of conflict with each other, for the more urgently those engaged with the first stage strive to achieve their goal, the more unlikely it is that the architect will have time to set the second stage in motion.

But Iago, working with fiendish efficiency and racing against time, manages to inflame Othello's mind before Desdemona, despite her zeal, can recover Cassio's place for him. The more insistently she pleads for the lieutenant, the readier is her husband to suspect her honesty. For Iago has gone to work at once, and the brilliance of his assault on Othello's peace of mind is attested by the suddenness with which the Moor is transformed from a doting husband of absolute faith to a savage of no faith at all. At the time Othello enters with Iago and recognizes Cassio just parting from Desdemona—when the villain's abrupt 'Ha! I like not that' launches the acute second stage of the practice—he would find unimaginable the mere possibility that he could ever be brought to suspect his wife of infidelity. Just one hundred lines later, he is convinced that Iago's mind locks in some dreadful secret and demands that it be revealed: '. . . speak to me as to thy thinkings, / As thou dost ruminate, and give thy worst of thoughts / The worst of words'. Forty lines later he bellows, 'O misery!' Another seventy, and he asks himself, 'Why did I marry?' Hardly three hundred lines after Iago's 'Ha! I like not that', his peace is lost forever: 'Not poppy, nor mandragora, / Nor all the drowsy syrups of the world / Shall ever medicine thee to that sweet sleep / Which thou ow'dst yesterday.' And after another hundred lines the savage has become a beast: 'I'll tear her all to pieces.' By the end of the scene he has sworn an oath 'by yond marble heaven', has ordered Iago to kill Cassio, and withdraws 'To furnish me with some swift means of death / For the fair devil'—but not before he apprises Iago of what he has won: 'Now art thou my lieutenant.'

That Iago succeeds so quickly in converting Othello's faith to jealousy is surely attributable as much to his own cunning as to any special trait in the Moor that renders him peculiarly susceptible. The key element in the success of the practice is Iago's well-established reputation for honesty. Given the platform of credibility from which his practice is launched, he has only to maintain his usual image throughout the innuendoes that he serves Othello, and the Moor's mind does the rest. It is the secret of the villain's genius that he leaves the conviction of Cassio and Desdemona to the Moor himself. He begins not by questioning either's honesty, but by grandly rejecting the possibility of Cassio's dishonesty: 'I cannot think it, / That he would steal away so guilty-like, / Seeing your coming.' It is just so that the rest of the game proceeds, and Othello, not Iago, first raises the question of Cassio's honesty: 'Is he not honest?' So, too, it is Othello who first shifts the question of honesty to Desdemona, whose name Iago shrewdly leaves out of the conversation until Othello introduces it. Proceeding by degrees, the villain then turns to the possibility of dishonesty among Venetian wives in general: 'Their best conscience / Is not to leave't undone, but keep't unknown', to which generalization he then quickly joins a pointed remark that Othello must acknowledge to be plain fact: 'She did deceive her father, marrying you.' When Othello lamely and somewhat negatively asserts Desdemona's honesty—'I do not think but Desdemona's honest'—Iago takes it upon himself to argue vehemently for Cassio's worth, for Desdemona's virtue, and against Othello's reaching a hasty conclusion: 'My lord, I would I might entreat your honour / To scan this thing no farther; leave it to time.' But by now Othello has swallowed the poison and is lost beyond recall; while Iago is briefly away, his mind tortures itself, and he is not speaking figuratively when he tells Desdemona, 'I have a pain upon my forehead here.' Then, while he is himself briefly out of our sight, the poison continues to work, permeating his system, so that when he returns to find Iago alone he has so lost control of his faculties as to be no longer recognizable as the noble Moor 'whom our full Senate / Call all in all sufficient'. Now his condition has become ripe for Iago's first direct assault instead of the succession of mock denials and restrained innuendoes that preceded; now Iago can unload his well-stocked store of foul accusations and fouler physical images without going too far: 'Would you, the supervisor, grossly gape on— / Behold her topp'd?' And again: '. . . as prime as goats, as hot as monkeys, / As salt as wolves in pride'. These gross images he tops at last with a fictitious account of having lain with Cassio one night, who said in his sleep, 'Sweet Desdemona, / Let us be wary, let us hide our loves' and then would 'lay his leg / Over my thigh, and sigh, and kiss'. It is here that what remains of the Moor's civilized veneer peels away: 'I'll tear her all to pieces.'

It is noteworthy that Othello's frightful threat here precedes any use by Iago of Desdemona's handkerchief, though that item has already come into his possession, and his quick brain, after its fashion of seizing upon accidental windfalls and converting them to his purpose, has earlier conceived how to employ it: 'I will in Cassio's lodging lose this napkin / And let him find it. Trifles light as air / Are to the jealous confirmations strong / As proofs of holy writ.' Still, it is not until after Othello's terrible threat that the villain mentions the handkerchief to him and thus introduces the final thrust of his practice: '. . . such a handkerchief— / I am sure it was your wife's— did I to-day / See Cassio wipe his beard with.' Immediately after hearing this piece of false information, and before Iago has staged the interview during which Othello witnesses the exchange of the handkerchief between Bianca's hand and Cassio's, the Moor takes a 'sacred vow' never to relent until 'a capable and wide revenge' shall have swallowed both Cassio and Desdemona.

Thomas Rymer to the contrary notwithstanding, it is made clear by the dramatist's ordering of these utterances and events that *Othello* cannot accurately be described, and so belittled, as 'the tragedy of the handkerchief', as if its catastrophic outcome depended upon Iago's use of this trivial item. By means of loaded innuendoes and other devices of persuasion, Iago, as we have noted, has already maddened the Moor before the handkerchief is even mentioned; we have heard his savage bellow, 'I'll tear her all to pieces', and must recognize that for him there can be no turning back. On the other hand, it would not truly diminish the force of the tragedy even if Iago had in fact built his entire practice around the handkerchief; the villain himself advises well when he tells us in confidence that 'Trifles light as air / Are to the jealous confirmations strong / As proofs of holy writ'. Were it indeed only 'the tragedy of the hand-kerchief', *Othello* need lose nothing of its plausibility, nothing of its numbing terror, nothing of its pity. But in fact the very great importance of the handkerchief does not lie in its being the indispensable item of evidence that persuades Othello and so destroys both himself and his wife; rather, the handkerchief is of the first importance because it supplies the indispensable item of evidence that ultimately destroys Iago. But with this fact we shall be concerned later.

By the opening of Act IV Iago has drawn all the other major persons of the play into practices upon one another and upon themselves even while in reality they serve only the villain's purposes. The dupe Roderigo, whose purse began it all, has been twice used—first in arousing Brabantio and leading him to Othello's lodgings, second in getting Cassio sacked; Emilia has served first by bringing Cassio and Desdemona together, next by passing the handkerchief on to her husband; Cassio, merely by soliciting Desdemona's aid in regaining his place, has made himself a key element in Iago's campaign against Othello; Desdemona, by meeting with Cassio and soliciting her husband on his behalf, has placed herself, her husband, and Cassio all in peril. Further, the villain has enlisted the Moor himself in the campaign by urging him to deny Cassio for a time so that 'You shall by that perceive him and his means. / Note if your lady strain his entertainment / With any strong or vehement importunity.' Finally, at the close of III. ii, Iago and his chief victim agree to a practice on the very lives of Cassio and Desdemona: Iago is to see that Cassio is dispatched 'Within these three days', and Othello will 'furnish me with some swift means of death / For the fair devil'. Thus, at the opening of Act IV, the villain has set his stage with active subordinates, all busily forwarding his private purposes towards their own destruction.

For the master practiser himself one of the busiest periods of the campaign begins now. Othello's frenzy has increased with each passing moment; even the villain is startled by the speed with which his poison has worked when the Moor, muttering incoherent phrases supplied in part by Iago in past interviews and partly by his own super-heated imagination, falls in a trance. Cassio's entrance at this moment, certainly unforeseen by Iago, he quickly converts to his use even as, from the start, he has capitalized on accidents as they occurred. Quickly dismissing Cassio lest Othello awaken and see him, he adds, 'When he is gone, / I would on great occasion speak with you'—and thus makes certain of Cassio's return at the proper moment. He next advises the revived but still groggy Moor to withdraw 'And mark the fleers, the gibes, and notable scorns' that will show in Cassio's manner as he is questioned about his relations with Desdemona. Here, then, the Moor himself turns practiser, engaging in eavesdropping, one of the more ubiquitous forms of practising in the comedies and the tragedies alike; Othello thus joins a motley parade of eavesdroppers—Claudius, Polonius, Duke Vincentio, a whole host of back-curtain witnesses in *Much Ado* and *Twelfth Night*, Julia of *The Two Gentlemen of Verona*, Oberon, Puck, and so on and on, even to Prospero himself, 'o'erlooking' while his child of wonder 'changes eyes' with an entranced prince. The practising eavesdropper sometimes discovers what makes him sad, like Julia when she sees her Proteus false; or what makes him roar with laughter, like Maria and her accomplices observing Malvolio's vain antics; or what makes him reverse his own image, like Benedick and Beatrice; or what makes him furious, like Polixenes watching his princely son wooing a shepherd's daughter; or what delights his soul, like Prospero seeing Miranda and Ferdinand fall in love. Usually the scene that is witnessed, whatever the effect it produces in the eavesdropper, represents the situation accurately. The exceptions, where the eavesdropper observes a false scene staged to deceive him, are, all things considered, surprisingly few in Shakespeare. *Much Ado* is unique in its inclusion of three such false scenes—those by which, successively, Benedick and Beatrice are fooled into believing that each dotes on the other, and that by which Don John convinces Claudio that his Hero is false, though in fact this scene is only reported to us.

Iago's staged scene, then, with Othello appropriately 'encaved' to observe, takes its place with the rarer type of eavesdropping incident. The scene both attests to the villain's skill in improvisation and evinces the heavy debt his campaign owes to happy accidents. Quickly and adroitly, he manipulates the conversation so that while he and Cassio jest about Cassio's whore, Othello—whose mind Iago has previously prepared to misconceive everything—readily assumes that they are discussing Cassio's relations with Desdemona. So cleverly does Iago work that he induces even Cassio, dispirited as he is, to laugh aloud:

> *Iago.* She gives it out that you shall marry her. Do you intend it?
> *Cas.* Ha, ha, ha!
> *Oth.* Do ye triumph, Roman? Do you triumph?
>
> (IV. i. 118–21)

The interview is wonderfully brief and efficient and stands as another masterpiece of Iago's tactical manœuvring. Even so, its crowning moment depends upon sheer accident—the timely entrance of Bianca, which the villain could not have foreseen and the usefulness of which he nearly destroys with his startled 'Before me! look, where she comes.' Since the preceding conversation has, to Othello's hearing, concerned Desdemona, Iago's sudden 'she' should also refer to her, not Bianca, and Iago's ejaculation mars the perfection of his performance; further, it is Cassio, not Iago, who unwittingly saves the situation for the villain by crying ''Tis such another fitchew! Marry, a perfum'd one.' To Othello, the 'such another' can mean only 'such another as Desdemona'—and thus all is well for Iago. In any event, Bianca's timely appearance proves a boon to him, for Othello witnesses her angry return of the handker-chief and Cassio's hurried pursuit of her. Moreover, the impromptu interlude enables the villain to speak a line that further inflames Othello's brain: 'She gave it him, and he hath given it his whore.'

Perhaps the staged incident with Cassio and the fortuitous arrival of Bianca serve to confirm the Moor's earlier decision to kill his wife; but they were not needed to help him reach this decision: we heard him cry, 'I'll tear her all to pieces' before the handkerchief was mentioned by Iago, and we heard him vow 'by yond marble heaven' never to turn back, but 'To furnish me with some swift means of death / For the fair devil' before he eavesdropped and saw the hand-kerchief returned by Cassio's whore. At the end of the two episodes, the one planned, the other accidental, he cries, 'I'll chop her into messes'—a threat that does not much transcend the savagery of the earlier 'I'll tear her all to pieces'. Possibly the staged scene hastens Othello's misguided revenge: 'Get me some poison, Iago; this night.' But the most obvious function of the handkerchief episode is twofold: for the dramatist it provides a towering dramatic climax at IV. i, just where the climax often occurs in the comedies and the tragedies alike; and for Iago it provides a veritable orgasmic triumph of his practiser's art. At the end of the episode, pouring additional venom into his victim's ear, he is positively euphoric: 'Do it not with poison; strangle her in her bed, even the bed she hath contaminated.' Cries Othello, 'Good, good'; and Iago, 'And for Cassio, let me be his undertaker.' Incredibly, both practiser and victim share a moment of sheer joy in contemplating the swift elimination of Desdemona and Cassio.

Iago's practices do not cease with the episode of the handkerchief; very soon afterwards he will re-enlist Roderigo in the shabbiest of all his practices, an ambush of Cassio as result of which the dupe actually supposes that he will come to enjoy Desdemona 'the next night following'. In the interim occurs an extended and uninter- rupted exploitation of the accumulted discrepancies that Iago's practices have created not only between the participants' and our awareness but also between the awarenesses of participant and participant. This interval extends from IV. i, the end of the hand- kerchief scene, to IV. ii. 171, when Roderigo enters. Throughout this period the villain practises on all his victims, but the emphasis of the scenes is not so much upon the forward thrust of his practising as on the obverse, the grievous ignorance of the victims, Othello, Desdemona, and, by virtue of a culpability that she would rather have died than knowingly have incurred, Emilia. Lodovico, too, just arriving from Venice and therefore oblivious of the state of affairs in Othello's household, contributes his own ignorance to the abundance of exploitable matter: 'How does Lieutenant Cassio?' he asks politely, and by so doing touches off the first flashes:

> Des.   I would do much
>   T'atone them, for the love I bear to Cassio.
> Oth. Fire and brimstone!
> Des.   My Lord?
> Oth.   Are you wise?
> Des. What, is he angry?
> Lod.   May be the letter mov'd him.
>   (IV. i. 243–6)

Coming from outside, as from another world—like Sebastian stepping obliviously into the chaos that his sister's masquerade has created in Illyria—with orders deputing Cassio to the governorship of Cyprus, Lodovico stands thunderstruck while Othello strikes his wife, harshly dismisses her with 'Out of my sight!' and brutally characterizes her: 'Sir, she can turn, and turn, and yet go on, / And turn again; and she can weep, sir, weep; / And she's obedient, as you say, obedient, / Very obedient.' In his amazement, Lodovico turns repeatedly for clarification to the one plausible-seeming member of the household, Iago: 'Alas, alas!' the villain laments, 'It is not honesty in me to speak / What I have seen and known. You shall observe him, / And his own courses will denote him so / That I may save my speech.'

Exploitation of the accumulated gaps reaches a new peak of intensity in the next scene, which opens ominously enough with Othello's efforts to pry into Desdemona's imagined secrets through Emilia. Though she had earlier questioned Desdemona about Othello's outbursts—'Is not this man jealous?'—Emilia has not guessed until now the nature of his suspicions, hence has shared Desdemona's level of complete ignorance. But if she now rises barely above that level on learning that the Moor suspects his wife with Cassio, she yet remains oblivious of her own husband's guilt: 'If any wretch have put this in your head, / Let Heaven requite it with the serpent's curse!' Though the remark sets off a flash of irony for us because we know who singly bears the guilt, it shows Emilia also to be a shade above Othello's level of awareness, for the thought that anyone at all has abused his credulity has never even occurred to him—nor does he now give it any hearing. Iago's practices have set the Moor in an ambiguous position on the structure of awareness: he stands above Desdemona in that he suspects her with Cassio while she is ignorant even that she is suspected, and above her in knowing that he intends her immediate murder while she will remain ignorant of his intent until the moment is upon her; but at the same time, because all his suspicions are entirely false, his awareness of the truth falls below hers, for of course she knows her own innocence even while she is unaware, as yet, that her husband doubts it. So pure, indeed, is her innocence, that even after the Moor roars at her, dismisses Emilia as the keeper of a brothel and she herself the whore, and tells her directly 'thou art as false as hell', she yet imagines that something quite apart from their private relationship has disturbed him: 'If haply you my father do suspect / An instrument of this your calling back, / Lay not your blame on me.' The depths into which Iago's practices have precipitated husband and wife are here bottomless: three times the Moor calls her 'whore' and 'strumpet' before her senses, deep in shock, can recognize a need even to deny his accusations; and when she does deny them, so insulated from truth has madness made him that her words cannot penetrate his senses.

At the end of the interview Desdemona calls in the very author of all her woes for advice and consolation, and the villain was never more honest-seeming than now, when his succession of bland questions and expressions of amazement and concern might almost suspend our own disbelief in his innocence and the sincerity of his compassion: 'Beshrew him for't!' he cries on learning how abusive the Moor has been: 'How comes this trick upon him?' In the end, Desdemona casts off what of pride remains and directly begs: 'Alas, Iago, / What shall I do to win my lord again? / Good friend, go to him.' And Iago obliges with the gentlest of reassurances: 'Go in, and weep not; all things shall be well.'

This exchange, when the shattered victim begs her destroyer to intercede for her and is assured that all will be well, is one of the dramatist's most unblushingly bold exploitations of the multiple gaps that Iago's practices have created; yet even it is no bolder than the speeches of Emilia that just preceded: 'I will be hang'd', cries the villain's wife to the villain and his stricken victim, 'if some eternal villain, / Some busy and insinuating rogue, / Some cogging, cozening slave, to get some office, / Have not devis'd this slander.' Replies Iago: 'Fie, there is no such man; it is impossible.' But neither Emilia nor the dramatist is yet done: 'The Moor's abus'd by some most villanous knave, / Some base notorious knave, some scurvy fellow.' That *Othello* survives such moments of shameless exploitation without degenerating into melodrama is partly attributable to the flawless meshing together of its poetry and its dramaturgy, but to nothing singly so much as to the absoluteness of its villain's plausibility. It is one of the great achievements of the Shakespearian magic that even while we perceive with perfect clarity what a monster Iago is, we can also believe that all his intimates, all his world, would call him 'honest Iago' and turn first to him in their moments of acute distress as do Roderigo, Cassio, Othello, and Desdemona.

For, indeed, there still is Roderigo. The dupe of the initial practice from which grew all the later deadly campaign reappears near the end of Act IV, having been absent from sight since the end of Act II, just after the cashiering of Cassio. On that occasion he approached Iago to complain that his means were wasted and that he would forthwith return to Venice 'with no money at all' and nothing but experience for his pains. His own head then being busy with the formulation of designs to 'enmesh them all', Iago summarily dismissed Roderigo with a platitude or two about patience, not foreseeing, at the moment, any further use for a gull whose purse he had already emptied. Thereafter, during Act III and most of Act IV, Iago's energies were consumed with the practices that, at the end of Act IV, have swept Cassio, Othello, and Desdemona to the edge of the precipice, and the sudden reappearance of his long-term but abandoned victim first strikes him as a mere annoyance: 'How now, Roderigo!' The gull has come to chide and even to threaten: if he cannot recover his investment, he will 'seek satisfaction' of Iago. Perhaps we should recognize here that for the first time Iago's crooked dealings have begun to overtake and embarrass him; but again the master improviser is equal to the unexpected turn and quickly converts it to his own advantage. If there is no longer reason to delude his dupe in order to enjoy his purse, yet there is cause to do so in order to shut him up and, at the same time, accomplish the immediate chore of dispatching Cassio. To entice Roderigo into this, his last of many practices, he needs only to dangle before him again the same familiar bait: 'If thou the next night following enjoy not Desdemona, take me from this world with treachery and devise engines for my life.' Though Othello has been ordered home to Venice, Iago instantly squelches Roderigo's hopes of following Desdemona there by assuring him that the Moor 'goes into Mauritania and taketh with him the fair Desdemona' and that the way to hold both on Cyprus is to kill the intended deputy, Cassio. 'I will hear further reasons for this,' declares Roderigo; but his anger and his will have been subdued already.

At this point Iago's means and ends have come full circle. At the start he used Cassio, Othello, and Desdemona strictly as means to Roderigo's purse; at the intermediate stage he used all his puppets as both means and ends: Roderigo as means of getting Cassio cashiered, but also Cassio's cashiering as means of reviving Roderigo's hopes of having Desdemona, thereby continuing the availability of his purse; but now, finally, with the purse drained, his intent is to use Roderigo strictly as means against Cassio and, ultimately, the Moor and his wife.

At the beginning of Act V, it remains for the villain only to execute the final practice already designed. Like its predecessors, this practice is loosely structured, with leeway for the master to improvise at need: 'Now, whether he kill Cassio, / Or Cassio him, or each do kill the other, / Every way makes my gain.' He has good cause to wish Roderigo dead: 'Live Roderigo, / He calls me to a restitution large / Of gold and jewels that I bobb'd from him / As gifts to Desdemona.' And obvious cause to wish the same for Cassio: 'He hath a daily beauty in his life / That makes me ugly; and, besides, the Moor / May unfold me to him.' Aside from the initial motive that precipitated the whole campaign—Roderigo's purse—these are Iago's most rational, convincing expressions of motive; but, more importantly, they mark also the point at which he himself recognizes both the enormity of the monster that he has brought forth and the need to obliterate all traces of his personal involvement in case his practices go wrong. For the fact is that things have begun to get out of hand, and this final practice—though the villain is quick enough to patch over the trouble—goes badly awry from the start. Roderigo bungles his job: instead of killing Cassio, he is himself wounded by Cassio; then Iago, from behind, thrusts at Cassio but only wounds him. Thus the practice fails dismally, leaving alive two victims as threats to the villain's security. One, Roderigo, he summarily kills, using his old swiftness and cunning to take advantage of the wild confusion and din—to which, of course, he himself contributes most.

Despite its inauspicious beginning, Iago thus manages to salvage something from the wreckage of his practice: he silences Roderigo permanently, at the same time loudly placing the blame on him for the wound that he himself gave Cassio; he obviously impresses the important Venetian officials, Lodovico and Gratiano, with his bravery and efficiency in routing—as they are led to suppose—sundry 'bloody thieves' and murderers who have had a hand in the fracas; and of course he ingratiates himself with those about him by his solicitude for the comfort of Cassio: 'O, a chair, a chair!' Finally, and quite gratuitously, he seeks to lay the whole blame for the bloody interlude on the innocent whore Bianca: 'Gentlemen all, I do suspect this trash / To be a party in this injury.' And, after Cassio has been borne out of earshot, he piously slanders him as well: 'This is the fruits of whoring.' Yet the very fervour with which he works to salvage what he can from the miscarriage of his initial purpose—the death of Cassio—suggests more than a degree of desperation; his final remark of the scene evinces his anxiety: 'This is the night / That either makes me or fordoes me quite.'

The final scene of the play indeed 'fordoes' him, for in the process of setting Othello and Desdemona at the edge of the precipice he has set himself there as well, and when they topple, so does he. The early portion of this scene, when Othello smothers his wife, exhibits a painful parallel with the corresponding portion of the final scene in *Romeo and Juliet*, where Romeo, erroneously believing Juliet dead, kills himself; here Othello, erroneously believing his wife false, kills her. Othello's error is worse by a turn of the screw:

Romeo kills only himself, but the Moor kills another, a total innocent. So deep is Romeo's unawareness at the critical moment that he misjudges the signs of Juliet's imminent revival—her crimson lips and cheeks; and so obsessed is Othello with the notion of his wife's falsity that he heeds neither his own true instinct—'Oh, balmy breath, that dost almost persuade / Justice to break her sword!'— nor her repeated protestations of innocence and pleas for mercy. The Moor rejects her pleas that Cassio be sent for and the truth demanded of him; he does so on the grounds that, as he erroneously thinks, Cassio is dead; but here his unawareness is unimportant, for so resistant to truth has his madness made him that he would refuse to summon Cassio even though he were aware that Iago's trap has failed. Indeed, even if Cassio should intervene without a summons and reveal the truth, it is clear that the Moor would not take his word against Iago's any more than, earlier, he took Emilia's angry denials of illicit meetings between Cassio and Desdemona. With Othello, the villain's practices have succeeded absolutely.

In her own deep ignorance of what her husband is and has done, Emilia could not save her mistress; but it is she and she alone who brings to light the one item of evidence that 'fordoes' Iago. Yet she would be helpless to recognize the truth and reveal it but for Iago's own fatal error in adding the superfluous 'proof' of the handkerchief to convince an already-convinced and maddened Othello of his wife's falsity. Indeed, but for the handkerchief, no argument could cause Emilia to accept the fact of her husband's guilt. When Othello tells her 'Thy husband knew it all', referring to Desdemona's meetings with Cassio, she hears him with the same incredulity with which she has heard the Moor's accusations against his wife. 'My husband!' she cries out three times, not in disbelief of Iago but in disbelief and scorn of Othello: 'My husband say that she was false!' When Iago enters, she continues to vent her rage not upon him but on Othello: 'Disprove this villain, if thou be'st a man,' she screams at her husband. 'He says thou told'st him that his wife was false. / I know thou didst not, thou'rt not such a villain.'

In the presence of the Moor, Iago's options all run out, and he has no choice but to admit that he had spoken of Desdemona's falsity— but insists that he had told no more nor less than truth: 'I told him what I thought, and told no more / Than what he found himself was apt and true.' Even so, his persuasive genius might extricate him, for Othello has not been shaken by Emilia's charges but continues to

uphold Iago: 'Nay, stare not, masters; it is true, indeed,' and, again,
'O, she was foul!' and yet again,

> 'Tis pitiful; but yet Iago knows
> That she with Cassio hath the act of shame
> A thousand times committed. Cassio confess'd it;
> And she did gratify his amorous works
> With that recognizance and pledge of love
> Which I first gave her.
>
> (v. ii. 210–15)

Between them, Iago and his chief victim would easily convince
Lodovico and Gratiano despite Emilia's shrill, shrewish voice and
would send them back to Venice with news of Desdemona's deserved
execution by her wronged husband. As for the wounded Cassio—
who yet believes in 'honest' Iago—he would be silenced abruptly and
permanently by some ingenious new device, and then the villain
would have it all.

Indeed, given hours in which to improvise, the plausible Iago
would convince even Emilia that Desdemona was guilty. But once
Emilia has heard Othello utter a single key sentence—'I saw it in his
hand; / It was a handkerchief . . .'—her blurred vision comes clear,
and then it requires but one piercing sentence to penetrate the Moor's
insulated mind: '. . . that handkerchief thou speak'st of / I found by
fortune and did give my husband.'

Like Claudius, who loaded the odds by one too many in his final,
multi-pronged practice on Hamlet, Iago, a masterful but, after all, a
flawed practiser, overreached and destroyed himself by adding the
weight of a handkerchief to a practice that had already succeeded.

CHAPTER VI

Practice as Diversion: *King Lear*

'. . . my practices ride easy.'

In *Hamlet* and *Othello*, to cite the nearest chronological examples, the force of the dramatic machinery of practice and counter-practice is centripetal, tending steadily inward upon a single, central matter of continuing concern from first to last. In *Hamlet*, hero and villain remain locked in a death struggle, with each inventing practices or counter-practices designed to gain advantage; with the trivial exception of Polonius's device for discovering how Laertes is behaving himself in Paris, *Hamlet* includes no practices not directly related to this struggle. The practices of *Othello* hew even more rigorously to the centre: Iago is the sole practiser (though he contrives to enlist all his unwitting victims in sub-practices against themselves and one another), and all his machinations ultimately serve that which destroys the Moor and his wife. In neither play does the machinery of practices divert our concentration from the centre; on the contrary, its force is calculated to prevent attention from wandering afield.

The machinery of practices in *King Lear*, on the other hand, does not converge upon a single magnetic centre. All the major, durable practices of the play—and there are many—tend outward rather than inward; never steadily related to the central action, they are more diversionary than integral. In *Hamlet* and *Othello* the major practices are actually more than closely related to the main action: they *are* the action. In *King Lear* they are, for the greater part, a kind of dramatic dressing. In all the earlier tragedies great central secrets created mainly as result of practices provide the basic condition for the continuing byplay between the participants' awareness and our own. In *Romeo and Juliet* the secret of Fate's purpose, known to us and to none other, envelops a second great secret, that of the lovers' marriage, and from this fixed arrangement arise the dramatic effects of the tragedy. In *Hamlet*, where the deep secret of the King's guilt is soon enveloped by the secret of Hamlet's knowledge of it, no peripheral or extraneous dramatic machinery is needed to provide the material for action and attendant spectacular effects. And so also in *Othello*, where Iago's parade of practices aimed towards the destruction of Othello and Desdemona, and the abiding secret known to us alone of what Iago *is*, prove ample in themselves to pack the play with action and effect. In *King Lear* alone the dramatist finds it useful to surround the inner core of his tragedy with an elaborate machinery of practices that are not germane.

His source, *King Leir*, offered him neither a ready-made awareness–unawareness gap nor the possibility of creating one within the frame of the essential story—at least not one that would be both central and durable, lasting throughout the tragedy as in *Hamlet* and *Othello*. In the source, the old king's error in assessing the characters of his three daughters is quickly perceived by himself, so that the play comes immediately to centre not on his unawareness of his error but on his suffering once he has seen the truth. In his own version Shakespeare could not both retain the original emphasis and at the same time maintain an exploitable gap with his protagonist continuing throughout five acts to misjudge his daughters while the audience perceives their characters clearly. He must either develop a tragedy without an exploitable main gap, and with little opportunity to create minor ones, or else surround the story of Lear's travail with an elaborate added structure that offered abundant opportunity to develop exploitable secondary gaps to compensate for the absence of a central one. Obviously he took the latter choice, using bits for the story of Gloucester and his sons from Sidney's narrative of 'the Paphlagonian unkinde King, and his kind sonne'.

It would be presumptuous in the extreme to suggest that at the date of *King Lear* Shakespeare had developed an absolute dependence upon dramatic situations that provided him with one or more durable and richly exploitable discrepancies in the awarenesses of audience and participants; it is nevertheless true that from the very beginning he had fixed on a favourite kind of dramatic situation involving a serviceable awareness–unawareness gap that he could work across, back and forth, touching off showers of varied effects at will. From the earliest plays his preference was for at least one main gap of long duration and an abundance of incidental secondary ones. In the comedies the mainstay was the heroine's masquerade as a young man, with her true identity often kept from all but ourselves until the final moments; then, too, the main gap was augmented by an assortment of minor ones created by subordinate practisers like Maria, Sir Toby, and Feste in *Twelfth Night*. The abundance of practices and practice-created gaps in *Hamlet* and *Othello* approximates the abundance of the same in the mature comedies. But the abundance exhibited in *King Lear* surpasses that found in any of the other plays, whether comedies or tragedies.

All told, the active practisers of *King Lear* number five—Regan, Goneril, Kent, Edmund, and Edgar; and, among them, these five perpetrate no fewer than twenty practices, counting those that are brief and incidental as well as those of longer duration. No other tragedy approximates such a total of practices, and among the comedies those with the greatest number—*Cymbeline* with thirteen and *The Merry Wives of Windsor* with eleven—also fall notably short, even though these plays, especially when acted, appear literally to be of practices all compact. Inevitably, too, in *King Lear*, the large number of practices creates a large number of gaps between awarenesses: twenty-two of the twenty-six scenes involve persons who are unaware of some aspect or aspects of situations, and at one time or another we hold advantage over every participant except Cordelia. Since, in the sheer number of practisers, practices, and practice-created discrepancies, *King Lear* out-tops all other plays of Shakespeare, we might expect it to stand also as the play in which the machinery of unequal awarenesses contributes most heavily to its cumulative power and its ultimate effect as tragedy; but in fact, if we take the core of the tragedy to be the suffering and redemption of Lear, this vast and complex machinery contributes only negligibly.

To say so much is not to imply that it contributes negligibly also to the effects of *King Lear* as a whole. On the contrary, the gaps opened by the devices of Edgar alone are fantastically productive, yielding a succession of spectacular effects throughout the play. But because almost all of the machinery of practices and exploitable gaps is confined to the Gloucester plot or, at most, to the points where this plot touches the Lear plot, it does not render the kind of service to the tragedy of Lear that the similar machinery in *Hamlet* and *Othello* renders those tragedies. Whereas the practices devised by Hamlet and Claudius steadily fix our awareness upon the main issues of their play, as do Iago's also in *Othello*, the devices of the most prolific practisers of *King Lear*, Edgar and Edmund, tend rather to divert our minds from the story of Lear. Perhaps the gaps created by the practices of these two compensate and even over-compensate by their very abundance for the absence of a great central gap like that in *Hamlet* or *Othello*, and indubitably they add immeasurably to the excitement of the total play; but because they are diversionary they cannot begin to generate the degree of centripetal force that derives from the central discrepancies in *Hamlet* and *Othello*.

Though such a central gap is indeed opened in the first scene of *King Lear*, when we first recognize and Lear does not that Regan and Goneril are false and Cordelia true, it is already half closed by the end of Act I, when Lear perceives the falsity of Goneril, and totally closed before the end of Act II, when he discovers that the Fool prophesied correctly: Regan tastes as like Goneril as a crab does to a crab. Thereafter Shakespeare never develops a suitable substitute for the lost central discrepancy, but busily exploits multiple gaps resulting from a welter of practices devised by Edgar, Edmund, and Kent.

Probably we cannot fix the exact point at which the dramatist means us to recognize our advantage over Lear in the opening scene, but it is surely very early. Goneril's patently puffed and sycophantic protestation of love, quickly followed by Cordelia's melancholy aside, 'What shall Cordelia speak? Love and be silent', presumably opens a crack between our awareness and Lear's. Regan's equally inflated profession, followed also by a sharply pointed aside from Cordelia, can leave us in no doubt: the older sisters are false, and their vows are cruel hoaxes; Cordelia is loving and true. When, then, it comes Cordelia's turn to profess her love, our senses have been well advised, and we can relish a brief exploitation of the gap between us and Lear:

> *Cor.* Nothing, my lord.
> *Lear.* Nothing!
> *Cor.* Nothing.
> *Lear.* Nothing will come of nothing. Speak again.
> (I. i. 89–92)

Direct exploitation of the initial gap continues without interruption until Lear's angry exit near the end of the scene. A key figure in this exploitation is Kent, whose bold speeches, each louder and more insistent, beat furiously at the truth that we have already seen and that, presumably, all participants but blind Lear see as clearly as we. Though Kent's angry speeches are not needed to advise our awareness of Lear's error, they serve a vital function by delineating the features and dimensions of the monumental egotism, vanity, and plain obstinacy by which Lear is blinded, his mind impenetrably insulated from the truth. It is only by perceiving him so that we can accept his failure to recognize what we, along with Kent, Cordelia, and the rest, have recognized so easily. So palpably false are the older sisters' protestations that they barely qualify as practices in the usual sense: an Iago would hoot at their transparency. The hidden intents of Shakespeare's subtlest practisers are typically revealed to us through speeches not heard by their victims; but Lear hears all that we hear and yet mistakes it utterly, and continues to hold his stubborn opinion despite Kent's efforts to drive the truth into his senses. Perhaps the nearest analogy occurs in *The Winter's Tale* when Camillo and others vainly seek to remove the baseless misapprehension that has suddenly afflicted Leontes' mind. Lear, like Leontes, is more self-deceived than deceived from without, more blind than blinded.

Yet perhaps we should not disallow the possibility that Lear sees more truly than he is willing to acknowledge; blinded as he assuredly is by vanity and stubbornness, he is even more wilful than blind and would be quite capable, if it came to that, of literally cutting off his nose to spite his face. Perhaps he neither wholly believes Goneril and Regan nor wholly doubts Cordelia; but the hypocritical sisters gave him exactly what he sought—fulsome protestations of love— whereas Cordelia, her father's daughter in more ways than one, stubbornly refused to do so.

But whether from native or merely wilful blindness, Lear is in any event grossly at fault in his decision, and Shakespeare takes his usual elaborate measures to make the fact unmistakable. Kent's outspoken, risky challenge of his king's will might easily be enough in itself to fix our own view; but when Kent has gone, France presents a quiet but emphatic defence of Cordelia, and Cordelia herself both evinces her own inflexible honesty and with pointed remarks removes any lingering question we might yet entertain about her sister's characters. Sharply noting that she lacks the 'glib and oily art' of her sisters, the art 'To speak and purpose not', she coldly adds that 'I know you what you are', and takes her leave with a warning that in Shakespeare unfailingly signals the direction of subsequent scenes: 'Time shall unfold what plighted cunning hides.' And as though we might even yet mistake the situation after so conspicuous a signpost, the dramatist appends a brief conversation between the wicked sisters: 'Sister,' says Goneril, 'it is not little I have to say of what most nearly appertains to us both.' Their initial practice on their father has not ended with their lies about loving him; it has only begun, and its implementation, to which they allude with the same candour with which Iago announces his next practice, is yet to come. Says Goneril, '. . . let's hit together; if our father carry authority with such disposition as he bears, this last surrender of his will but offend us'. And to Regan's 'We shall further think of it', Goneril ominously replies. 'We must do something, and i' th' heat.'

The central gap between our awareness and Lear's that is so promisingly opened in the first scene begins to close in the next scene in which Lear appears, leaving the core of the drama to concern his coming to awareness and his subsequent agony of mind and spirit. Though sufficient for the anonymous author of *King Leir*, these resources could never satisfy Shakespeare, who, anticipating the loss of his central discrepancy, acts 'i' th' heat' to set new practices going, open new gaps, begin exploiting multiple situations. The first scene has no sooner closed with warning that the sisters' practice will continue than the second opens with notice of a device being readied by the chief professional practiser of the play, Edmund, whose machinations will not only create several highly serviceable discrepancies directly involving himself and many others but will also prompt Edgar to launch a practice of his own that, in turn, will create the most enduringly productive discrepancy of all.

Such justification as Edmund requires for launching his career as practiser he finds in Nature, which he perceives as the antithesis to the mere 'curiosity' of civilized law, or 'custom'. Nature has equipped him with attributes as fine as those of his legitimate brother; hence, he rationalizes, by Nature's law he should be entitled to the same benefits. Edmund, however, aspires not only to an equal share, but to all the shares, including those of both his father and his brother. He launches his practice as another entrepreneur might launch a ship, with a wish for it to prosper: 'Now, gods, stand up for bastards.' The initial act of this momentous practice of which the effects are to extend to the last moments of the tragedy amounts only to stuffing, with 'terrible dispatch', a previously forged letter into his pocket:

Glou. Why so earnestly seek you to put up that letter?
Edm. I know no news, my lord.
Glou. What paper were you reading?

Edm. Nothing, my lord.
Glou. No? What needed, then, that terrible dispatch of it into your pocket?

(I. ii. 28–33)

With this bold device that might have been appropriated directly from Bacon's 'Of Cunning', Shakespeare initiates the long train of practices calculated to remedy the basic defect of the *King Leir* plot, the lack of a durable central discrepancy between awarenesses. Considered only quantitatively, the device thrives almost alarmingly, spawning a rash of subsequent practices, creating numerous gaps, supplying the exploitable condition for eleven scenes, and finally wearing out only in the last moments of Act V when Edgar identifies himself to the fallen Edmund, initiator of the whole succession, whose comment then is apt: 'The wheel is come full circle'.

Once sprung, the initial device immediately catches both its intended victims, Gloucester and Edgar, and thus at one stroke opens the ignorance of each to exploitation. With a few seemingly reluctant words Edmund convinces Gloucester that Edgar pursues his life; then, when he enters for the first time in the play, Edgar already stands unaware, knowing nothing of Edmund's prior conversation with their father. When Edmund then, with a few more words, persuades him to go armed because Gloucester seeks his life, Edgar merely trades one condition of unawareness for a worse: formerly only ignorant, now he is actively abused. 'Some villain hath done me wrong', he tells the villain who has done it; and Edmund ends the interview with a gloating summary of what his quick practices have achieved:

> A credulous father and a brother noble,
> Whose nature is so far from doing harms
> That he suspects none; on whose foolish honesty
> My practices ride easy.

(I. ii. 195–8)

With the villain's highly auspicious double practice barely set going, it is necessary for the dramatist to abandon it briefly in order to advance the Lear plot—more precisely, the practice hinted in the conversation of Goneril and Regan at the end of the first scene. 'Put on what weary negligence you please,' orders Goneril, drawing Oswald into her stealthy practice on Lear, 'And let his knights have colder looks among you.' She will 'breed from hence occasions', so that the issue of her father's position will be quickly brought 'to question'.

Further, she will advise Regan 'To hold my very course'. Formerly victim of his own bad judgement, Lear is now to be victim of a cruel and active practice; when we next see him, he will occupy a truly pitiable condition of unawareness that seems the more pitiable because of his continuing haughtiness.

But again, before showing us Lear abused by the new practice, the dramatist shifts his ground to introduce yet another device that, like Edmund's, is to prove of long duration, extending from the fourth scene of the play until the end. But this practice, unlike Edmund's, is benign. Like a heroine of comedy donning disguise with the ultimate design of catching a hero, loyal Kent alters his identity in order to serve and protect the erring master who banished him. From this point on, in each scene that shows Kent and Lear together, we hold advantage over Lear—and thus the dramatist gains use of another exploitable gap. The character of Kent, together with his disguise and secret attendance upon Lear, was entirely Shakespeare's own invention, and the fact of this major addition underscores his sense of a need to furnish the main plot with a useful, lasting discrepancy. But though Kent's practice does indeed supply such a gap, the fact is that it proves only marginally effective. Whereas Othello's ignorance of Iago's dishonesty is always a matter of first importance to both us and Othello, Lear's unawareness of Kent's identity turns out to be a fact of tangential importance to either us or Lear. If we find it somewhat comforting, during the long ordeal, to know that the mad king's devoted servant, unrecognized, is at hand and watchful, yet Lear's own anguish would not be eased or his ultimate fate altered if he were immediately to discover his follower's identity. Hence little is at stake, and Kent's identity gradually becomes irrelevant: while Lear is in his purgatory it matters little whether Kent is Kent or merely Caius. That Shakespeare himself recognized the true irrelevance of Kent's masquerade is suggested by the fact that—though he introduces it early and stretches it out to the very end—he scarcely bothers to exploit the gap that it creates, and when he does do so it is only incidentally and in very minor (though expositorily necessary) scenes—for example, those that show Kent's meetings with the Gentleman. In these the Gentleman is kept ignorant of Kent's identity, while Kent—serving an obvious exploitative purpose—alludes mysteriously to it: 'I am much more / Than my outwall', and, again, 'When I am known aright, you shall not grieve / Lending me this acquaintance.' But the mystery is purely gratuitous:

it is surely a matter of the most trivial interest whether the Gentleman does or does not know Kent from Caius.

Even in the disguised Kent's first meeting with Lear after the banishment, Shakespeare appropriately leaves the King's unawareness unexploited, for here the point of crucial interest is elsewhere: the scene marks the beginning of Lear's coming to awareness, his realization that he has acted foolishly. Oswald's entrance reminds us at once of Goneril's instruction at the end of the previous scene: 'Put on what weary negligence you please, / You and your fellows; I'd have it come to question.' Now Oswald slouches in, and slouches out, the scrap of dialogue marking the precise start of Lear's long ordeal:

> *Lear.* You, you, sirrah, where's my daughter?
> *Osw.* So please you,— (Exit.)
>
> (I. iv. 48–9)

'My Lord,' Lear's attendant gratuitously offers, 'I know not what the matter is; but, to my judgement, your Highness is not entertain'd with that ceremonious affection as you were wont.' And Lear, as yet but vaguely troubled, replies, 'I have perceiv'd a most faint neglect of late.' At this heart-stabbing instant—for us, not yet for Lear—we cannot greatly care whether Kent is Kent or only Caius; nor, if he knew, would Lear.

In *Hamlet* and *Othello* the dénouement is reserved until the end of the tragedy. Then, in *Hamlet*, the unknowing world learns what we and Hamlet have known from I. v—that Claudius is a murderous usurper; and then, in *Othello*, the Moor learns what we have known from I. i—that 'honest' Iago is a 'Precious villain'. But Lear's private dénouement begins midway through Act I, when he just begins to see what we saw three scenes earlier. Oswald's first surly performance is soon followed by his return, when he is thrown out by angry Kent; and here the Fool begins to pick his single string: Lear's folly. Goneril's subsequent entrance and blunt command next makes unmistakable to Lear what the preliminaries have only suggested: 'Be then desir'd / by her, that else will take the thing she begs, / A little to disquantity your train.' After these blunt words Lear laments the 'small fault' for which he disinherited Cordelia, and a few lines later he beats at his head 'that let thy folly in'. Twice he asserts—is it to reassure himself?—that he has one daughter left: 'Who, I am sure, is kind and comfortable'. But in the brief final scene of Act I he offers no rejoinder to the Fool's remark that Regan 'will taste as like this as a crab does to a crab'. Desperately he clutches at one thin straw: that Regan will prove different from Goneril; but for all practical purposes our advantage over him has virtually ceased to exist at the end of Act I—and with its passing the dramatist loses the invaluable resource of such a great central gap as served him well throughout *Hamlet* and *Othello*.

To cover the loss, he opens Act II with quick, double practices by Edmund, first on Edgar, then on Gloucester. Here, with his speedy manœuvres, Edmund resembles Iago: confronting innocent Edgar— who manages to utter but one baffled sentence during the interview— with news that their father pursues his life, he stages a moment of mock sword-play, prompts Edgar to flee, pricks his own arm for blood, shouts for Gloucester, and gasps out a monstrous lie—that Edgar had sought to 'Persuade me to the murder of your lordship'. Deceived and shaken, Gloucester gratefully rewards the bastard's loyalty: '. . . of my land, / Loyal and natural boy, I'll work the means / To make thee capable'.

Gloucester's error in misjudging the characters of his sons too closely parallels Lear's in misjudging the characters of his daughters to be mistaken for accident: with this stroke and what immediately follows Shakespeare fuses the plots of Lear and Gloucester. No sooner have Cornwall and Regan reached Gloucester's castle than they are informed of Edmund's heroic feat, and Cornwall, impressed by the villain's 'virtue and obedience', claims him for his own—for 'Natures of such deep trust we shall much need'. It is of at least passing interest here that Cornwall, who soon proves to be even more vicious than Edmund, should be as easily deceived by Edmund's practice as are the 'credulous father' and the 'brother noble'; by the same token, it is curious that one so wicked as Cornwall should be attracted to Edmund because of his 'virtue and obedience'. In any event, by being thus deceived about Edmund's character, he gains the services of exactly the quality of man we might have expected him to prefer—an ambitious scoundrel like himself. Edmund's adoption by Cornwall serves to bind together the families of Lear and Gloucester and, simultaneously, the Lear and Gloucester plots. Further, the dramatist connects the families through Edgar:

> *Reg.* What, did my father's godson seek your life?
> He whom my father nam'd? Your Edgar?
> *Glou.* O, lady, lady, shame would have it hid!
> *Reg.* Was he not companion with the riotous knights

That tended on my father?
Glou. I know not, madam. 'Tis too bad, too bad.
Edm. Yes, madam, he was of that consort.

(II. i. 93–9)

This is in fact the key scene for interrelating the two plots, since not only do Edmund and Edgar serve this purpose but also the dramatist so manages events that the meeting of Cornwall and Regan with Gloucester and Edmund, and later with Albany and Goneril, is made to occur at Gloucester's castle, to which Lear himself soon comes. And, finally, it is in this scene that Gloucester's anguish, matching Lear's, has its beginning: 'O, madam, my old heart is crack'd, it's crack'd!' he laments to Regan—of all people—who will shortly drive her own father into the storm and madness.

The dramatist's own sharp practice in acting to consolidate his plots just here serves an obvious purpose: having lost his indispensable central gap at the end of Act I, and needing to make up for the loss in the best way possible by creating a rash of secondary gaps in the Gloucester plot, he worked to relate the two plots as intimately as possible, so that these secondary gaps would seem less peripheral. With the extent to which his stratagem actually succeeds we shall be concerned hereafter.

For the present it is necessary to shift our ground again, as Shakespeare does, in order to deal with a second practice executed by Kent. Already the proprietor of one secret, his disguised identity, Kent devises his new practice within the old, acting in the character of the servant Caius. The new practice is an interim measure, meant to produce quick and positive results. Reaching Gloucester's castle ahead of Lear, Kent–Caius, with coolly calculated purpose, conducts himself so outrageously that he is promptly set in the stocks. No explanation except deliberate intent can account for his behaviour, first to Oswald and then to Cornwall and Regan. Kent's own later explanation to Lear, that he had acted 'with more man than wit about me', is lame indeed, being not only inconsistent with the solid sense that he exhibits elsewhere in the play, but wholly at odds also with the evidence of the immediate scene, during which he clearly does have his wits about him constantly: he controls the entire episode from his first excessively insulting remark to Oswald, throughout his saucy, startlingly rude interview with Cornwall, and on to the very moment that he is set in the stocks. He directs the swiftly worsening situation not at all like a hothead who has been put beside himself with temper, but like a tactician who knows well what result he seeks. An alternative, but unacceptable, theory of his behaviour is that, striving to make his assumed role of plain, blunt servant entirely convincing, he overplays it and thus finds himself in trouble; but he has had considerable experience in the Caius role before the scene begins and has played it well enough to deceive Lear himself. It is therefore unlikely that he would now overplay it so violently as to deserve the penalty he gets.

Throughout the tragedy, Kent stands for clear perception and rational judgement; indeed, if any participant may be said to represent the point of view of the play as a whole—that which we ourselves are expected to hold—it is certainly Kent, and Kent alone. For, aside from Kent's, *King Lear* abounds in faulty and distorted visions. Starting with the prime case, Lear's monumental misjudgement of his daughters, it next displays the warped and vicious natures of Goneril and Regan and the overreaching egotism of Edmund. Gloucester and Edgar, though men of wholesome minds, are quickly and easily abused by Edmund, and afterwards they see nothing truly. The Fool's vision—unlike that of other professional jesters in Shakespeare—is narrowed from the first to focus on but one object, Lear's folly. Cornwall's quick embracement of Edmund impugns his judgement, and thereafter his cruelty shows his nature to be as warped as that of the wicked sisters. Albany, until near the end of the play, is a mere cipher who counts for nothing. And finally Cordelia: after her initial display of inherited obstinacy that implies a brittle rather than sound judgement, she vanishes until late in the play and never proves to be one from whom we might be expected to take our own bearings. Solid, honest, capable Kent is our only fixed star.

But, then, since the notion is unacceptable that such a man would so lose control of himself as to deserve being set in the stocks, leaving us with no choice but to understand his action as deliberate, how do we explain his wish to be thus publicly humiliated? Mindful of the charge that the knights who attend Lear are men 'so disorder'd, so debosh'd and bold, / That this our court, infected with their manners, / Shows like a riotous inn', he might rather be expected to restrain his impulses and deal gently even with the detested Oswald in order to avoid giving Goneril and Regan grounds for complaint. Instead, he not only forces a brawl with Oswald but braves and offers to fight Edmund on the spot, then grossly insults

Cornwall and Regan: 'I have seen better faces in my time / Than stands on any shoulder that I see / Before me at this instant.' These are gratuitous insults, totally out of place and crude in the immediate situation. By behaving so, Kent belies Lear's earlier insistence that 'My train are men of choice and rarest parts', and plays directly into the daughters' hands; for at the end of the episode of the stocks, it is Lear's repeated and furious demand, 'Who put my man i' th' stocks?' that reopens the vexed question of how many followers he should be permitted to retain. Once the question has been reopened, the daughters, vying with each other in their cold insensitivity, reduce the number to fifty, to half that; and then, cries Goneril, 'What need you five and twenty, ten, or five?' And finally Regan whoops, 'What need one?' It is upon witnessing this display of heartlessness that Lear vents his fury in curses and incoherent threats, and then stomps off into the storm. Whatever was Kent's intent in deliberately forcing public humiliation upon himself, one fact is clear: his act brings matters suddenly to a head and results in Lear's furious denunciation of his daughters and his raging departure.

Though Kent does not, like most practisers, take us into his confidence by advising us of the good that he hoped to accomplish by getting himself set in the stocks, we are obliged to assume that his deliberateness was pointed towards some worthy end. This assumption is soon corroborated by his manner of smug self-satisfaction once he has been locked into the device; when Gloucester expresses determination to have him released, Kent rejects the offer with unfeigned cheer:

> Pray, do not, sir. I have watch'd and travell'd hard;
> Some time I shall sleep out, the rest I'll whistle.
> A good man's fortune may grow out at heels.
> Give you good morrow!
>
> (II. ii. 162-5)

If this is resignation, it is smug and rosy resignation; Kent settles into the stocks like one who has won his goal against odds. Far from berating himself for his ungoverned temper, as we should have expected had he inadvertently brought punishment on himself, he appears to congratulate himself. His soliloquy after Gloucester has withdrawn, if it is not brimming with good spirits, is nevertheless relaxed and pleasant; he yawns in the process of reading Cordelia's letter, and in the next moment drops asleep on a cheery note: 'Fortune, good-night! Smile once more; turn thy wheel!'

Even as Kent would easily have foreseen, the very first sight that confronts Lear when he reaches the castle is that of his personal emissary ingloriously situated in the stocks. Lear had left Goneril's place in a rage and had rushed to Regan's in the ill-founded belief that he would be warmly welcomed; at Regan's castle he had been somewhat shaken to find his daughter gone from home even though he had dispatched his messenger to notify her of his coming. But, no doubt, much of his faith in Regan remains intact when he arrives at Gloucester's castle. Then, seeing Kent, he reacts with mingled disbelief and rage:

> They durst not do't;
> They could not, would not do't. 'Tis worse than murder
> To do upon respect such violent outrage.
>
> (II. iv. 22–4)

As if the sight were not shock enough for his master to sustain in an instant, Kent hastens to shake him further by telling of his earlier cold reception at Regan's house and of Regan and Cornwall's abrupt departure from home upon learning that Lear's arrival was imminent; they 'gave me cold looks', he adds. At this news Lear no longer entertains either expectation or hope, but knows the worst:

> O, how this mother swells up toward my heart!
> Hysterica passio, down, thou climbing sorrow,
> Thy element's below!—Where is this daughter?
>
> (Ibid. 56–8)

Grief shakes but does not overwhelm him, for grief itself is over-whelmed by a tougher rival: rage. Rage sustains him through sixty lines before Regan enters, then sustains him during the long and furious interview that follows with both daughters. Sometimes, but briefly, grief and self-pity join in getting the better of fury—

> Her eyes are fierce; but thine
> Do comfort and not burn. 'Tis not in thee
> To grudge my pleasures, to cut off my train,
> To bandy hasty words. . . . Thou better know'st
> The offices of nature, bond of childhood,
> Effects of courtesy, dues of gratitude.
>
> (Ibid. 175–84)

—but when Regan brutally cuts him off with 'Good sir, to th' purpose', the briefly repressed fury seizes control again, and he demands hotly, 'Who put my man i' th' stocks?' Twice more, in the course of the bitter interview, he interjects the same demand. In a final speech before his raging exit he fights off the grief that would make him go mad:

> ... touch me with noble anger,
> And let not women's weapons, water-drops,
> Stain my man's cheeks!
>
> (Ibid. 279–81)

Earlier, when he first discovered Goneril's ugly nature but still believed in Regan's goodness, his vanity sustained him; he was more offended than grieved, and what grief came was dissolved in curses, not tears. Kent, riding ahead and reaching Regan's castle before Lear, witnessed the effects there of Goneril's letter brought by Oswald: the sudden and brutally rude departure of Regan and Cornwall before Lear could arrive. If Kent had had any doubts hitherto of what Lear's reception from his second daughter would be, he could entertain them no longer. Lear would soon follow to Gloucester's castle, and there would encounter the torture of heartbreak.

Kent must do something, and 'i' th' heat'—bring things to an abrupt head, not let the agony of slow recognition gnaw at the old king, lingering out the inevitable moment of truth; against that moment Kent must kindle the fires of rage in Lear, insulate his heart with fury, clothe him in 'noble anger'. Kent's solution is to design the practice that we saw him execute when he conducted himself so abominably that he was promptly clapped into the stocks. Entering, his eye first falling on his servant thus publicly humiliated, Lear caught the full truth in an instant and reacted with the very rage that Kent had sought to kindle. When Gloucester returns to Regan and Cornwall shortly after Lear's departure, he reports that 'The King is in high rage'. And when we next see Lear, 'Contending with the fretful elements', the same fury still cloaks him; he rides out the storm in anger. Kent's drastic strategy had worked: mad is better than sad.

At this point we must again interrupt our concentration on the Lear plot and the practices related to it in order to take note of Edgar's initiation of a new practice, which quickly becomes the most spectacular, most exploitable, and most exploited of all; unfortunately for the tragedy as a whole, however, it is also, of all the practices in *King Lear*, the one with the loosest relation to the Lear plot. The very awkwardness of its timing—with Edgar stepping on to the stage while Kent sits asleep in the stocks and Lear's arrival is imminent—

illustrates the principal difficulty of running the Gloucester plot concurrently with the Lear plot. Edgar's brief soliloquy in which he first announces and then rehearses his role as 'poor Tom' intrudes into an otherwise closely integrated sequence of action; just when we have reached a critical moment in Lear's worsening situation, our attention is diverted to Edgar and his affairs. What is more, Edgar's performance in his first rehearsal of the lunatic's role is so startling, so bizarre, so intensely demanding of our attention that it momentarily erases all memory of Lear from our minds. Curiously, the lone function served by the intrusion of Edgar's practice at this particular moment is a trivial one: it lets us suppose that a period of time, perhaps a few hours, elapses while Kent sleeps and before Lear arrives; but in fact there is no real need to suppose that any time elapses here, for we could as conveniently imagine that Lear arrives within moments of Kent's falling asleep. The main effect of Edgar's intrusion is basically different from, say, that of the Porter in the gate scene of *Macbeth*; there, though the Porter pauses long in order to detail the effects of too much drink, our main concern remains with the murderers, for we know that whoever stands knocking at the gate brings business relevant to them and their deed. But Edgar's untimely announcement of his practice only butts the affairs of Gloucester's family irrelevantly into Lear's. That we need to be introduced to 'poor Tom' at some point prior to his meeting the Fool and Lear in the hovel is apparent; but we might have been introduced to him less awkwardly at the start of Act III, just before the entrance of Kent and the Gentleman—a mere expository scene the force of which could hardly have been diminished by any sort of intrusion. Edgar's practice, like all the practices of the Gloucester family, would have diverted attention from Lear at any point; yet it remains curious that Shakespeare chose to introduce it at what seems the worst possible time. But so much exploitable matter does this practice make available, so serviceable does it prove throughout the play, that the dramatist presumably thought only of getting it launched as quickly as possible.

It is during the great middle of the play, while Lear endures the storm and passes through raving madness into a state of physical exhaustion with mental and spiritual exaltation, that the compounded practices of Gloucester's family reap their harvest of effect. Paradoxically, it is also during this most intense period of Lear's ordeal that Shakespeare has greatest need of just such exploitable opportunities as they supply. As we have noted, once Lear learns, as he does in II. iv, that Regan is as wicked as Goneril, the exploitable gap of the Lear plot is lost. Thereafter, were there no Gloucester plot, dramatic interest would have to be sustained through two full acts, and more, by the continuous exhibition of Lear in his purgation. But even a Lear who rages more sublimely than any other dramatic character could not rage against the elements and his wicked daughters for two full acts without becoming unbearably tedious: two acts filled with repetitions of 'Blow, winds, and crack your cheeks! Rage! Blow!' would prove disastrous to the tragedy—even as Timon's continued scolding and howling, which we shall note hereafter, do to his sad piece. Perhaps we might continue fascinated almost indefinitely as we watch Othello's unawareness of his agonizing situation being exploited; but Lear, once he has discovered his error and thus closed the dramatist's exploitable gap, has little left to offer but direct raging against which we would soon rebel.

So Shakespeare would not and did not fill up Acts III and IV merely with raging. In III. ii, Lear has three raging passages amounting to thirty-two of the ninety-five lines in the scene. In III. iv, he rages through two speeches for thirty lines of the two hundred in the scene; he has short speeches also of three or four lines each, but he makes no more lengthy tirades in Act III. In the whole of Act IV he holds forth in three rages that total about forty-five lines; the remaining two long speeches in the final scene here, with Cordelia, do not count in this respect because they are subdued, marking the quiet after the storm. It is evident from these proportions that the dramatist held Lear's actual raging to something near an irreducible minimum; Acts III and IV would have been brief indeed had they consisted of it alone.

For the great bulk of these acts, Shakespeare goes to the exploitable matter made available by the practices of the play's three major practisers. First, though the contribution is relatively slight, there is Kent's masquerade. In III. i Kent and the Gentleman discuss Lear's perilous situation, and Kent reveals that French powers 'already have secret feet / In some of our best ports'. The Gentleman is instructed to show Kent's ring to Cordelia in Dover, 'And she will tell you who that fellow is / That yet you do not know'. Here the dramatist's intent to exploit the Gentleman's ignorance is obvious, but the gain is negligible: we cannot believe that the fact of Kent's identity is of much moment to the Gentleman, and therefore the exploitation of his ignorance of it provides no measurable thrill. But the fact that Shakespeare bothers at all to exploit such a trifle suggests how strongly he felt the lack, in this play, of his customary central, highly exploitable gap; hence he appears to salvage every exploitable scrap. One act later, in IV. iii, the same pattern is repeated: Kent again meets the Gentleman, and we learn that the King of France has returned home, that Cordelia reacted strongly to news of her father's treatment by her sisters, and that Lear is now in Dover but refuses to meet Cordelia. Though Kent, in the first interview, had said that at sight of his ring Cordelia would advise the Gentleman 'who that fellow is', it appears that she neglected to do so; says Kent,

> Some dear cause
> Will in concealment wrap me up a while;
> When I am known aright, you will not grieve
> Lending me this acquaintance.
>
> (IV. iii. 53–6)

Here, clearly, Shakespeare contrives to squeeze a second drop of effect from Kent's persistent but trivial secret. Nor has he yet quite finished: immediately after Lear and Cordelia have been reunited in IV. vii, Kent and the Gentleman again find themselves alone, and the old game is continued. Says the Gentleman, 'They say Edgar, his banish'd son, is with the Earl of Kent in Germany.' Replies Kent, still reluctant to relinquish proprietorship of his secret, 'Report is changeable.' But, tease the Gentleman as he will, for us, as in the two earlier exchanges between the pair, the dramatic excitement generated by this quick flash of exploitation cannot be much.

Though Lear's ignorance of Kent's identity obviously holds much greater potential for effect, it is in fact made to yield even less. Kent is at Lear's side in nearly every scene after his banishment, and though as a loyal servant he performs heroically for his master, the fact of his masquerade remains virtually useless to the dramatist, perhaps because, during the long course of the action, no appropriate moment occurs for exploitation of it—for Lear's travail is unceasing and so intense that the fact of his servant's true identity simply does not matter. At first beside himself with rage and self-pity and thereafter stricken with madness, Lear could not care less, during the great middle of the play, whether Caius is Kent or merely Caius; and so sharply concentrated upon Lear's ordeal are our own feelings that we are left with no margin for enjoying the fact that we know Kent while Lear does not. Perhaps, during the height of the

storm on the heath and during the mad 'trial' of the daughters in the
hovel, we derive a degree of comfort from knowing that Kent and
not Caius is at Lear's side; but it is more likely, so riveted is our
attention to Lear in these climactic moments, that we forget the very
presence of the servant, be he Kent or Caius.

Shakespeare never does let Kent divulge his identity to the
Gentleman; similarly, he keeps the secret from Lear, stretching it
out to the last moment of the king's life. If at the last moment the
revelation could produce a stunning dénouement like that, for
example, in *Measure for Measure*, when Lucio snatches off the friar's
hood and exposes the face of the mighty Duke, then we would know
why the dramatist bothered to stretch out a masquerade that yields
little effect along the way. But the ending of *King Lear* is so packed
with tremendous events that it allows little place for such a revelation as Kent's; indeed, the dramatist must make a place for it, an
awkward one. We have last seen Kent talking with the Gentleman
at the end of Act IV. He seems quite forgotten in v. iii, when Lear
and Cordelia are brought in as prisoners. Soon thereafter Edgar,
having challenged and wounded Edmund, describes the death of
Gloucester. 'But speak you on,' says Edmund; 'You look as you had
something more to say.' Upon this rather unlikely prompting, since
it comes from the villain lying mortally wounded on the ground,
Edgar speaks of his meeting with Kent just after Gloucester's
death: 'Whilst I was big in clamour came there in a man, / Who,
having seen me in my worst estate, / Shunn'd my abhorr'd society'—
and so he goes on at length to detail the occasion, without mention-
ing Kent by name. 'But who was this?' cries Albany. 'Kent, sir,'
replies Edgar, 'the banish'd Kent; who in disguise / Follow'd his
enemy king, and did him service / Improper for a slave.' Edgar's
emotional narrative is clearly designed to make the revelation of
Kent's identity moving; but of course it is not to Edgar, Albany, and
Edmund, but to Lear, that Kent's long, loyal service in disguise
should matter—and Lear is not present to hear the speech. Kent
himself now enters; the bodies of Regan and Goneril are brought in;
and finally Lear enters with dead Cordelia in his arms. These are
momentous events which together make perhaps the most potent
final scene in Shakespeare's tragedies. And it is an inauspicious time
for Kent to try to make himself known to Lear, in whose perspective,
as in our own, the revelation can hardly seem earth-shaking at the
moment. Yet the business must be somehow ended, and accordingly

Kent insists on Lear's attention, finally succeeds in being recognized, or partly so, but strives vainly to get the old king to understand that Kent and Caius were one and the same. It is surely the most awkward moment of Kent's long practice. With admirable tact Albany finally intervenes: '. . . vain is it / That we present us to him'. A less tactful Albany might well have said of Kent's persistence in forcing his identity upon the dying king what in fact he says when Edmund's death is reported: 'That's but a trifle here.'

Thus Kent's practice, though it is central in that it gives us advantage over the protagonist, yields only insignificant dramatic effect during the action and proves superfluous at the end; certainly Shakespeare never brought a major practice to a lamer conclusion. Conversely, Edgar's practice, though it creates a highly useful gap that yields many spectacular effects, is unfortunately only loosely related to Lear's story. We now return to trace this practice through to its great final moment.

Edgar's is one of only a few practices in Shakespeare that are undertaken because the practisers are themselves deceived. Edmund, with a double practice on father and brother, has made Edgar think that Gloucester seeks his life—as, indeed, he does, but only because he, too, is Edmund's victim. Thus Edgar assumes his lunatic's role ignorant of the truth, and ignorant he remains during the long period in which he deceives, with his disguise, all those he meets. As proprietor of the play's most exploitable and most exploited secret, he yet occupies the lowest rung on the ladder of the awarenesses, for even Gloucester, though he remains unaware until the end that his companion and benefactor is his son, gains the advantage of Edgar in IV. i when he learns that the bastard, not the legitimate heir, is his enemy:

> O my follies! then Edgar was abus'd.
> Kind gods, forgive me that, and prosper him.
>
> (IV. i. 91–2)

After this moment Edgar's scenes with Gloucester take on a special poignancy because each man is unaware of the one fact that, if known, would end his pain: Gloucester knows that Edgar is innocent but does not know that Edgar is by his side; Edgar knows, of course, his own identity and his innocence but remains ignorant that his father now loves him as before.

Edgar's involvement in Lear's story begins at the hovel when he is found and brought into the king's presence. The two scenes in Act

III in which he appears with Lear present the wildest extremes of his feigned lunacy, for he howls at fiends, whines, moralizes, lists the perils of a vagabond madman's life and the details of his diet, and, in short, exhibits impressive versatility as an actor. The incoherent fits and starts of his simulated madness present at once a contrast to and a welcome respite from the ravings of Lear that have dominated scenes ever since he first began to curse Goneril in I. iv. Lear is fascinated by 'poor Tom', insanely assumes that he, too, must have been reduced to his present state by his unkind daughters, first pities his nakedness and moralizes on the condition of 'unaccommodated man' who is 'no more but such a poor, bare, forked animal as thou art', then seeks to emulate him by tearing off his own 'lendings'. At the end of the scene the mad king has become so attached to his 'philosopher' that he refuses to part with him.

In III. vi, with Kent and the Fool, Edgar serves as a 'robed man of justice' in the 'trial' of Goneril and Regan, and here, as in the preceding scene, each member of the group is ignorant of certain aspects of the situation: Lear is unaware that 'Caius' is Kent and that 'poor Tom' is his godson Edgar; the Fool is ignorant of the same facts; Edgar is unaware of Kent's identity and remains unaware why his father sought his life; Kent is unaware of Edgar's identity; Gloucester, who moves in and out of these scenes, is unaware of both Kent's and Edgar's identities and still supposes that Edgar sought his life. As in the rich middle scenes of the mature comedies, where typically occurs the fullest exploitation of the accumulated unawarenesses, so here are gathered together the greatest number of participants ignorant of the greatest number of details that make up the total situation. But here also the resemblance ends. Whereas in the comedies the unawarenesses, individually and collectively, are key elements of the situation and accordingly occasion potent dramatic effects, in these scenes of *King Lear* most of the unawarenesses are irrelevant to the main purpose, which is to represent what has happened, or is happening, to Lear's mind. So far as the dramatic effects of the scenes are concerned, 'Caius' could as well be Caius, and 'poor Tom' could as well be poor Tom. It does not really matter that Kent and Edgar are unaware of each other's identity or that Lear is unaware of the identities of both. To say so much is not to question the essential power of these climactic scenes, but only to acknowledge that this power owes little to the elaborate machinery of discrepant awarenesses that has been built up. Kent's and Edgar's secrets are major ones in this machinery; their disguises create gaps as broad as those created by the disguises of Rosalind and Viola in the comedies, as broad as those created by Hamlet's 'antic disposition' and Iago's 'honesty' in the tragedies. But the difference is that they do not *matter* to the story of Lear as the others do to their respective central plots, and therefore they do not yield comparable effects.

But of course in these scenes and elsewhere the identity of 'poor Tom' matters to Gloucester, and the initial misunderstanding between them, wrought by Edmund's practice, matters to both; therefore these aspects of their unawareness matter also to us, and accordingly the dramatist exploits them repeatedly. Gloucester first encounters 'poor Tom' in III. iv, outside the hovel, when, approaching with a torch, he is identified by Edgar as 'the foul fiend Flibbertigibbet'. During periods when his father is present the disguised Edgar appears to put on his lunatic act with special fervour, and in his rags and filth, acting out the farthest extreme of lunacy, he maintains a teetering balance between the grotesquely comic and the painfully pathetic. Were he truly the lunatic that he pretends, and behaved as outlandishly, he would doubtless puzzle and embarrass our emotions: we would be at once compelled to laugh and ashamed to do so. Knowing that we do not truly have a wretched madman before us, we feel freer to laugh, yet not entirely free, for we know Edgar's own troubles to be acute and his father's grief to be even more severe. All elements considered, perhaps the final effect of 'poor Tom's' antics is to puzzle our emotions all the more because we know that they are only an act; or perhaps 'puzzle' is an inadequate word. In the complex and anguished situations in which they are exhibited most extravagantly, they do more than confuse the emotions: they stun them.

It is the incidents of direct, pointed exploitation that precipitate the greatest crises in our emotions: when Edgar identifies his father, approaching with a torch, as the foul fiend; when Gloucester, appalled and stirred to compassion by the words, actions, and foul appearance of 'poor Tom', queries Lear, 'What, hath your Grace no better company?' And when Gloucester, with 'Caius' and 'poor Tom' standing before him, grieves for the conditions of Kent and Edgar:

> Ah, that good Kent!
> He said it would be thus, poor banish'd man!

Thou say'st the King grows mad; I'll tell thee, friend,
I am almost mad myself. I had a son,
Now outlaw'd from my blood; he sought my life,
But lately, very late. I lov'd him, friend . . .
(III. iv. 168–73)

Until this point, near the end of a scene committed mainly to exhibiting the deterioration of Lear's mind, the dramatist had avoided direct exploitation of Gloucester's unawareness of the presence of Kent and Edgar, and here the pointedly exploitative remarks suggest Shakespeare's reluctance to let a situation so richly serviceable pass by without being made to yield at least a token payment for his trouble in creating it—but only a token, for extended exploitation of the extraneous Gloucester–Kent–Edgar relationship would too violently wrench our attention from its proper object, Lear's madness. Even so, brief and incidental as it is, perhaps Gloucester's bald speech about Kent and Edgar while 'Caius' and 'poor Tom' look on actually occasions too much of a diversion not only from the Lear story but from the essential nature of the dramatic experience that the king's plight, especially during these searing moments, is designed to thrust upon us; for Gloucester's error, being so palpable, borders on the comic. So, in the next lines, does his embarrassed hesitance to invite such an unsavoury wretch as 'poor Tom' to leave the hovel and come along with the rest to more comfortable quarters.

In general, the awareness gap between Gloucester and Edgar is most effectively exploited in scenes that do not involve Lear, for then exploitation does not affect our concentration upon the mad king's travail. After the terrible scene of his blinding, Gloucester assumes a dramatic importance that rivals Lear's—an importance much exceeding that of earlier scenes when he appears mainly as a functionary who moves in and out, busy with errands. Accordingly, when the disguised Edgar encounters him in IV. i, the awareness gap between them has taken on a larger importance also, with correspondingly increased potential for exploitation. Here, too, the character of the gap has changed drastically, for Gloucester now knows that both he and Edgar were abused by Edmund, but of course remains ignorant that 'poor Tom' is Edgar; Edgar, for his part, remains ignorant that both were abused by Edmund and is ignorant also that, were he to make himself known, his father would now embrace him. Here the irony is that Edgar's reason for his disguise has ceased to exist—but Edgar is, and long remains, unaware of the fact. Twice in the scene the changed situation is exploited through Gloucester's remarks that are unheard by Edgar:

> O dear son Edgar!
> The food of thy abused father's wrath!
> Might I but live to see thee in my touch
> I'd say I had eyes again!
>
> (IV. i. 23–6)

And again,

> I' th' last night's storm I such a fellow saw,
> Which made me think a man a worm. My son
> Came then into my mind, and yet my mind
> Was then scarce friends with him. I have heard more since.
>
> (Ibid. 34–7)

Twice, too, speaking aside, Edgar is made to justify the necessity—for so he sees it—of continuing the masquerade: 'I cannot daub it further. . . . And yet I must.'

Whether directly exploited or not, the awareness gap between Edgar and Gloucester is consistently productive of dramatic effect, hence is far more useful to the dramatist than that between Kent and Lear; indeed, the gap itself is the basic dramatic stuff of their story, whereas the Kent–Lear gap is a mere speck on the periphery of Lear's story. Even so, exploitable as is the 'poor Tom'–Gloucester relationship, the dramatist could not simply fill up whole scenes with Edgar commenting 'I cannot daub it further' and Gloucester endlessly repeating 'O dear son Edgar / . . . Might I but live to see thee in my touch.' If the situation is to avoid such repetition, it must receive occasional infusions of drastically different elements, and in IV. vi, where we next meet the pair, just such elements are introduced. Here Edgar grafts a startling new practice upon the old, first making his father believe that he has ascended to the top of Dover Cliff and now stands at its very edge, then—launching yet another practice that entails his adopting a new identity—convincing him that he has indeed tumbled from the 'dread summit of this chalky bourn' and yet survived. Of all practices in the plays of Shakespeare, this is surely the most bizzarre. 'Why I do trifle thus with his despair', Edgar insists just before Gloucester's fall from the non-existent cliff, 'Is done to cure it.' And, at least for the moment, his drastic practice works; says Gloucester, 'Henceforth I'll bear / Affliction till it do cry out itself "Enough, enough," and die.' Later he wishes that 'the ever-gentle gods' may 'take my breath from me', and later still he invites Oswald to kill him: 'Now let thy friendly hand / Put strength enough to't.' Yet the wish for death is not identical to the active effort to kill himself. 'Let not my worser spirit tempt me again / To die before you please,' he tells his gods, who, he thinks, have miraculously preserved him because they will not suffer their will to be challenged by a mere mortal. He has been humbled by his supposed experience and will not again pridefully take the issue of life or death into his own hands. Actually, of course, since he fell from no cliff, nothing of his gods' will has been demonstrated; he has only been gulled by Edgar's drastic but benign practice and by his bold declaration: '. . . therefore, thou happy father, / Think that the clearest gods, who make them honours / Of men's impossibilities, have preserv'd thee'. Though Edgar's desperate remedy has truly cured his father's despair, it was in the deepest sense a cruel hoax. Before the incident of the cliff, Gloucester voiced a bitter sentiment on the gods' relation to men: 'As flies to wanton boys, are we to th' gods, / They kill us for their sport.' He has now reversed his belief, not because his gods have indeed intervened to save him, but because of Edgar's trick.

Blind, deluded Gloucester and mad Lear are briefly brought together just after the incident of the cliff, and during their conversation Edgar stands aside, remarking on the pity of the spectacle but appropriately drawing no attention to himself as a practiser; but when the king has run off, pursued by his attendants, Edgar resumes his role, not as 'poor Tom' but as 'A most poor man, made tame to fortune's blows'—the same identity, presumably, as that he assumed on first addressing Gloucester after his supposed fall. Then, with no advance notice to us, he adopts still another identity at the approach of Oswald, addressing him in a Southern dialect: ''Chill not let go, zir, without vurther 'casion.' Yet Oswald addresses Edgar as 'bold peasant' even before the latter speaks at all; hence the assumption of the dialect would seem wholly superfluous as an added safeguard adopted to hide Edgar's true identity. 'Go your gait,' says Edgar, 'and let poor volk pass.' But Oswald is not put off, and Edgar continues to use his impromptu accent up to the moment of striking the intruder dead. That Gloucester evinces no surprise at Edgar's sudden shift of accent is itself hardly remarkable, for by now Gloucester is surely accustomed to his companion's habit of changing identities. Edgar's suddenly assumed dialect, in any event, appears gratuitous very much as does Feste's costuming himself as 'Sir Topas' in *Twelfth Night* in order to deceive Malvolio—who, locked in darkness, is quite unable to see his tormentor, costumed or not. But, indeed, Edgar's act seems not so much needed to deceive Oswald, who is already deceived, as designed to exhibit his versatility with an unexpected new addition to his extensive repertoire. Though Edgar's initial practice was forced on him, his successes appear to have caused him to relish masquerading; his final role, the incitement to which is provided by discovery of Goneril's self-incriminating letter on Oswald's body, we shall next examine.

Edgar approaches Albany in v. i, gives him Goneril's letter (or perhaps an explanatory one of his own devising), and departs on a darkly mysterious note: 'Wretched though I seem, / I can produce a champion that will prove / What is avouched there.' Albany urges him to stay until he has read the letter, but Edgar declines with an attitude of even darker mystery—'I was forbid it'—and vanishes. He is now ready to stage his final practice, which, though brief in the acting, proves as spectacular as his lunatic role and as strange as his trickery at Dover cliff. It is also his first practice undertaken in full awareness of his own situation, for not only has he learnt of Edmund's complicity in the wicked sisters' plots, but (as we ourselves are not to learn until later) he has divulged his identity to his father, received a blessing, and presumably been informed by the dying old man of Edmund's initial practices that set father and son at odds. Thus, when the mystery-shrouded 'challenger' replies to the herald and claims the right to do battle with Edmund, it is, at long last, Edgar who is the knowing practiser and the villain Edmund, prime instigator of it all, who is the victim. When Edmund falls wounded, it is Goneril who sharply points the truth: 'This is mere practice, Gloucester.' When the stricken villain demands to know who has triumphed over him, Edgar, dramatically synchronizing word and gesture, raises his helm: 'My name is Edgar, and thy father's son.' Thus the many and varied practices of Edgar, among the most elaborate, durable, and spectacular in all Shakespeare, abruptly end.

Still, shrewdly timed and executed for effect as it is, Edgar's revelation of his identity, together with his triumph over Edmund and his emotion-charged account of the off-stage scene when Gloucester died and the banished Kent suddenly arrived, lacks the moving force of the greatest dénouements. It seems excessive, swelling out of proportion the actual importance of the entire Edgar–Edmund–Gloucester affair just when the affair of Lear is also coming to an end. It is not that, just here, Edgar's busy action lacks relation to the conclusion of Lear's story; on the contrary, his killing of Edmund brings a sudden end to the military triumph of Lear's enemies and actually wipes out all the forces of evil that have tormented Lear—for if Edgar cannot quite claim credit for the death of Regan, already poisoned by Goneril, yet his wounding and exposure of Edmund precipitates Goneril's own immediate suicide and so snuffs out villainy's last flicker. But we remain painfully aware, during all the while that Edgar's affairs are being thus happily concluded, that Lear and Cordelia have been marked for death by Edmund and that even now their executioner should be about his task; with this awareness burdening us, we may have difficulty in witnessing Edgar's final histrionics without a nagging sense of annoyance. And perhaps there is one further bar: our sense that though indeed Edgar, by killing Edmund, destroys Lear's enemies, he does so only incidentally, his own sole interest being to settle his and his father's score with Edmund. Either Albany or Kent might more appropriately than he have destroyed the villains of the Lear plot. In short, the final, spectacular intrusion of Edgar, intent on settling family affairs, into the last moments of the Lear plot gives the scene a discomfiting resemblance to the final massacre in *Titus Andronicus*, where various late survivors are slain by the wrong avengers and for the wrong reasons.

We must now return, but briefly, to bring into perspective the career of the play's most professional practiser, Edmund. Initially, his practices bear no relation to the Lear plot, but concern only himself, his father, and his brother. 'Well, then,' he soliloquizes at the start, 'Legitimate Edgar, I must have your land.' By means of a forged letter to which he draws Gloucester's attention by seeming to hide it, he easily turns Gloucester against Edgar; and with a second practice, as easily accomplished, he persuades Edgar to flee for his life. In the next stage of the double practice he uses Edgar himself as an unwitting accomplice, inducing him to draw his sword while Edmund shouts, 'Yield! Come before my father'—who, of course, is conveniently within earshot. Thereafter he needs few words to supplement the evidence of a self-inflicted minor wound in order to convince his father that Edgar is an incipient parricide; says Gloucester, 'Let him fly far. / Not in this land shall he remain uncaught; /

And found,—dispatch.' With this pronouncement, the purpose of Edmund's initial practices has been achieved.

But the arrival of Cornwall next enables the villain to elevate his sights. Impressed, as we have earlier noted, by Edmund's 'virtue and obedience', Cornwall embraces him as his own: 'You we first seize on.' From this moment Edmund is no longer concerned with the division that he has been at pains to effect between Gloucester and Edgar. Learning, in III. iii, of the French invasion designed to avenge the wrongs done to Lear, he promptly delivers to Cornwall a letter that incriminates Gloucester: 'How malicious is my fortune, that I must repent to be just!' This is the letter which he spoke of, which approves him an intelligent party to the advantages of France. 'O heavens! that this treason were not, or not I the detector!' Though this second major practice betrays Gloucester, it is not in fact related to Edmund's initial practices that set father against son; that step having been completed, his aim now is to gain not merely 'legitimate Edgar's' share of his father's lands, his first goal, but all that his father possesses: 'This seems a fair deserving, and must draw me / That which my father loses.' But, even more importantly, by this manœuvre he further ingratiates himself with Cornwall.

The family resemblance of Edmund to Iago is more than casual. As practisers they share the same sense of centrality of self, the same pleasure in the abuse of others, particularly those who are credulous and noble. Both raise their sights as opportunity arises, or is created by their machinations, and both capitalize on each stray chance. But, even more importantly, they function as the prime movers in their respective dramas. Even as Iago's practices impel the entire action of *Othello*, so Edmund's initial practices set Edgar fleeing for his life and occasion all his histrionics down to his final heroics as 'challenger'. Further, Edmund's more monstrous practice in betraying his father to Cornwall costs Gloucester his eyes and sends him, too, fleeing, as homeless as Edgar. Edmund's practices, thus, though originally unrelated, are alike in their effects, for each sends its victim into the open country where their paths meet and where both cross the path of Lear. These practices are thus key elements, as Edmund himself is, in the fusion of the Lear and Gloucester plots; but not less important is the sheer abundance of dramatic business that they contribute to the total play, for they occasion all of Edgar's subsequent practices—and, indeed, virtually all of the exploitable

awareness–unawareness gaps in the tragedy with the exception of
the gap created by Kent's masquerade.

Unlike Iago, however, Edmund himself, having set a vast
machinery in motion, turns his back on it. Between the time of his
betrayal of Gloucester to Cornwall and his death at Edgar's hands in
the final scene, we are directly reminded only twice, neither time by
Edmund himself, of his responsibility for the woes of Edgar and
Gloucester. The first occasion is the eye-gouging scene:

> *Glou.* All dark and comfortless. Where's my son Edmund?
> Edmund, enkindle all the sparks of nature,
> To quit this horrid act.
> *Reg.*   Out, treacherous villain!
> Thou call'st on him that hates thee. It was he
> That made the overture of thy treasons to us,
> Who is too good to pity thee.
>
> (III. vii. 85–90)

The second is in Lear's maddest speech, when, bedecked with
flowers, he encounters Gloucester and Edgar just after the incident
of Dover cliff:

> Let copulation thrive; for Gloucester's bastard son
> Was kinder to his father than my daughters
> Got 'tween the lawful sheets.
>
> (IV. vi. 116–18)

Lear, of course, is mistaken: Edmund has been no kinder to his
father than Goneril and Regan have been to theirs; here Lear's error
would be recognized by Gloucester, who learned the truth in the
eye-gouging scene, but not by Edgar, who at this point still believes
that Edmund's timely warning saved his life.

After betraying his family, Edmund travels a course that does not
relate directly to either Lear's story or Gloucester's. After the
opening of the eye-gouging scene, when he is dispatched to escort
Goneril home, he does not reappear until IV. ii, when, upon taking
leave at Albany's gate, he is kissed by Regan: 'This kiss, if it durst
speak, / Would stretch thy spirits up into the air. / Conceive, and
fare thee well.' We next hear of but do not see him in IV. v, when
Regan waylays Oswald, tries to get Goneril's letter to Edmund from
him, alludes to the 'strange œillades and most speaking looks' that
Goneril gave Edmund at Gloucester's castle, and finally advises
Oswald to desire his mistress 'call her wisdom to her': Edmund, she
insists, is 'more convenient' for her own hand than for Goneril's.

We next hear of Edmund in IV. vi, when Edgar takes from Oswald's body the letter that incriminates both Goneril and Edmund: '. . . if he return the conqueror; then am I the prisoner, and his bed my gaol; from the loathed warmth whereof deliver me, and supply the place for your labour'.

Edmund, thus, unlike Iago, who as chief practiser is incessantly with us, pursues a shadowy course through a kind of limbo during all the middle part of the play. We hear of him at intervals, seemingly more pursued by the daughters than pursuing them, but we do not hear him speak significantly between III. v, when he betrays his father, and v. i, when Goneril and Regan quarrel over him, each well aware that the other lusts for him, and cast vicious darts between them. Left alone, Edmund then speaks a soliloquy that gives us our first look into his mind and estate since III. v:

> To both these sisters have I sworn my love;
> Each jealous of the other as the stung
> Are of the adder. Which of them shall I take?
> Both? one? or neither? Neither can be enjoy'd,
> If both remain alive. To take the widow
> Exasperates, makes mad her sister Goneril;
> And hardly shall I carry out my side,
> Her husband being alive. Now then we'll use
> His countenance for the battle; which being done,
> Let her who would be rid of him devise
> His speedy taking off. As for the mercy
> Which he intends to Lear and to Cordelia,
> The battle done, and they within our power,
> Shall never see his pardon; for my state
> Stands on me to defend, not to debate.
>
> (v. i. 55–69)

Beside his earliest speeches, most notably the rousing 'Stand up for bastards' soliloquy of I. ii, this speech seems the merest dramatic patchwork, both in its uninspired phrasing and in the awkwardness of its introduction of new and unexpected turns, at this precise moment, just as the decisive battle for Britain begins. As exposition it is of course sorely needed, for it reaffirms Edmund's presence as an active force in events when we have heard nothing directly from him during many scenes; it partly clarifies the ambiguity—by mere assertion—of his relations with the sisters; and, finally, it forewarns us of the deadly practice to be directed at Lear and Cordelia—though indeed all the lines that follow 'As for the mercy' have the sound of afterthought, tacked on by Edmund on the spur of the moment; they give us our first notice of the lofty plane to which his ambition has soared without our having ever been informed of the fact.

Ending as it does with the threat of death to Lear and Cordelia, the soliloquy makes an obvious but patently unsatisfactory effort to connect the activities of Edmund, throughout the play, with the central story of Lear. As we have noted, Edmund's initial practices are not related to Lear's story, and, again as we have noted, Edmund soon turned his back even upon those practices. Between Acts III and V we have heard of him only as a handsome trophy coveted by both the sisters. Their private tug-of-war—which has involved letters, the goings and comings of Oswald as messenger, scarcely veiled insults and rival claims bandied back and forth, and, on the rarest and briefest occasions, actual appearances of one or the other sister with Edmund himself at meeting or parting—has contributed many fragments to the business of scenes, but, like the antic capers of Edgar, it has always seemed rather a side-show than an integral part of Lear's tragedy. Goneril herself, just before the battle, seems almost to make the dramatist's apologies when she reminds that 'these domestic and particular broils / Are not the question here'. At least until the long soliloquy of v. i, Edmund's shadowy doings with the sisters, added to the Lear plot and the Gloucester plot, tend to make *King Lear* unique indeed among the tragedies, a tragedy with not a double but a triple plot.

So, through the patchwork soliloquy, Shakespeare appears to improvise a desperate remedy by relating Edmund's triangular involvement with the sisters to the main Lear plot: Edmund will take one sister or the other, eliminate Albany, Lear, Cordelia, and the remaining sister, and rule Britain. Further, by bringing in Edgar as the mysterious challenger who puts a sudden end to Edmund's fantastic, seemingly spur-of-the-moment aspiration—though indeed Edgar's purpose in killing Edmund has nothing to do with this—the dramatist also, after a fashion, connects the good brother and all his masquerades to Lear's story.

But surely these efforts come too late and too unexpectedly to be wholly satisfying. If in the end the appearance is given that Edmund's shadowy course has led directly to his final practice—the secret order for the deaths of Lear and Cordelia—yet it is true that during the occasional scenes that touch on his relations with the sisters we have had no notice that any such purpose was in his mind, or in the sisters' minds, or, for that matter, in the dramatist's mind. Edmund's dealings with the sisters have appeared as extraneous elements occupying the farther outskirts of the drama while Lear's ordeal moved towards its end. So too with Edgar's affairs: though Edgar stops Edmund's career just short of its goal, yet he does so from private motives and is unaware, when he challenges and wounds Edmund, of his practice on the king's life—which, from our own point of view, is the great reason why Edmund should be slain. On the surface the heavily loaded final scene would seem to offer more than merely enough to involve us deeply: the ceremonious trial by combat, with the trumpet call, the appearance of a mysterious champion, whom we alone know to be the much-wronged Edgar, confronting and overthrowing the cynical villain; the highly wrought tale of Gloucester's pitiful end; the arrival of the noble Kent; the reports of Goneril and Regan's violent deaths and the bringing of their bodies to the stage. The practice-ridden careers of Edmund, Edgar, Kent, and the wicked sisters converge in these tremendous moments before Lear enters with Cordelia's body in his arms; but though they converge their convergence seems rather to scatter than to concentrate the force of the last moments.

In *Hamlet* and *Othello*, to take the nearest cases, the actions of all persons in the closing scenes bear directly upon the centre. It is not so in the final scene of *King Lear*. Though we may respond with morbid pleasure to the gruesome ends of Goneril and Regan, we are yet obliged to recognize that they die not for what they did to their father but because they were rivals for Edmund. Edmund's practices on his father, his brother, and the sisters have been those of a private entrepreneur who expressed no interest in Lear prior to his patch-work soliloquy of v. i. The antics of Edgar as 'poor Tom', his services to Gloucester in successive disguises, and his final appearance as mysterious champion—of precisely what cause?—have from the outset lacked more than accidental relation to Lear. The brothers' courses have loosely paralleled the king's course without affecting it; Lear was never an element in their conflict. As for Kent, though his masquerade is undertaken in direct connection with Lear's story, it begins to lose point after Lear's first rebuff by Goneril and becomes merely superfluous after Regan's rebuff sends the king into the storm. Hence, though the paths of all these are made to converge at last, the convergence lacks relation to Lear's end, and Albany's blunt verdict on Edmund's death, 'That's but a trifle here', seems apt for all.

Possibly, too, our interest in the three principal practisers is compromised not only because their practices are diversionary rather than integral but because we really never come to know them in their own persons as intimately as we know Shakespeare's greatest practisers—the heroines of the romantic comedies, whose special feminine qualities always shine through their masquerades; Hamlet; Iago; and the rest. Edmund begins splendidly, but we see the best of him in his first two soliloquies, after which he is a shadowy figure whom we glimpse only occasionally and of whom we hear from time to time as being concerned with neither the mischief that he initiated nor with Lear; and by the time his final purpose is unexpectedly announced at the tag-end of his soliloquy in v. i, we have to greet him almost as stranger and newcomer. Kent, too, starts with strong promise, first braving Lear at the risk of his head and then getting himself set in the stocks for the king's sake; but whatever the Earl of Kent promised soon vanishes into the changeable, ambiguous 'Caius' on whose nebulous being our affections can take no real hold; and perhaps, even, we resent the fact that, at the critical moment when Lear is seized, imprisoned, and Cordelia hanged, neither the Earl of Kent nor 'Caius' is in the vicinity. Finally, we can hardly applaud this uncertain, unknown Kent when he reappears, too late, and boorishly insists upon inflicting the fact of his true identity upon dying Lear. And as for Edgar, whose role rivals Lear's as the most conspicuous in the play, we never do come to know him personally so that we can care about him as himself. In his first two appearances, when the stage is dominated by the then-glamorous Edmund, Edgar makes no impression at all on the mind. He next enters, inopportunely, when Kent falls asleep in the stocks and we momentarily expect the arrival of Lear, to announce that he will vanish into the person of 'poor Tom': 'Edgar I nothing am,' he asserts with more truth than he knows. Thereafter, lines like 'I cannot daub it further' remind us that the rags of 'poor Tom'—whom we have come to know intimately—presumably hide the real Edgar: but what is the real Edgar? The mind of Edgar, not that of 'poor Tom', invents the drastic device of Dover cliff; but does the device tell us anything of the inventor? After the supposed fall, Edgar assumes the identity of a peasant who witnessed the 'fall' from below, and, later, the peasant abruptly takes on the accent of a visitor from the South for his encounter with Oswald. And, finally, Edgar answers the trumpet call as Albany's visored 'champion'. But who and what is the Edgar behind the visor? When the visor is raised and Edgar's voice emerges—'My name is Edgar, and thy father's son'—we should, perhaps, respond with such a thrill as we experience when Hamlet leaps into Ophelia's grave with Laertes, crying 'This is I, Hamlet the Dane.' But we cannot do so, for while we know Hamlet and care about him, we know Edgar not at all; Edgar is a non-entity, whose original dim image has been obliterated by his successive roles.

Were the practices of Edmund, Edgar, and Kent strictly related to the core of Lear's tragedy, they might involve us deeply even though we perceived the real beings behind the masqueraders only indistinctly; or if the practisers were fully realized persons in their own right, then, even though their devices were only tangentially related to Lear's tragedy, they might still involve us. Being neither integral nor real, they divert rather than concentrate, dissipate rather than focus our final view. Edmund's passing is truly 'but a trifle here'. Albany's impromptu offer of the kingdom to Kent and Edgar falls notably flat—as though Britain were being offered to strangers. Noble Kent rejects the offer on the grounds that he will shortly die. Edgar makes no reply at all, and his silence seems not to matter; not knowing him—though we may indeed have become enamoured of 'poor Tom'—we cannot much care what he might have to say, one way or the other. What tragic force emerges through the last moments of *King Lear*—and of course it is remarkable—does so not because of the three principal practisers, but in spite of them.

CHAPTER VII

The Dramatist as Practiser: *Macbeth*

'I'll charm the air to give a sound . . .'

IN *Romeo and Juliet*, because the Chorus addresses us privately, we gain a secret that is denied to all participants in all scenes, namely, that the lovers are doomed; with the lovers, the Friar, and the Nurse we share a second great secret, that of the marriage; and we share a final secret with only Juliet and the Friar, that of the sleeping potion. Like boxes set one within the other, the three secrets provide the exploitable gaps from which arise the main effects of the tragedy. In *Hamlet* from I. v until v. ii we again hold the great secrets from which the main effects arise: with Claudius, Hamlet, and later Horatio we share the secret of the King's guilt; and with Hamlet and Horatio we share the secret that Hamlet knows the King's guilt. Besides these we share in several subordinate secrets that create effects within the 'endless jar' that is occasioned by the two main ones. In *Othello* we share with Iago alone the one enveloping secret of the play: that he is not, indeed, 'honest Iago'.

In their relation to the three tragedies, these secrets have certain common characteristics: they are deep, central, and enduring. Exploitation of them not merely enhances the dramatic force of each play but virtually *is* the play. They underlie and direct the main and secondary actions and are the very magnets of our attention. In *King Lear*, on the other hand, the initial secret of Lear's error in judging his daughters is central but neither deep nor lasting—for besides ourselves it is known instantly to everyone but Lear, who himself glimpses it before the end of Act I and soon thereafter sees it clearly. Three other major secrets—of Edmund's treachery and Kent's and Edgar's disguised identities—are durable enough but not central; they relate only tropically, we may say, to Lear's story, and they scatter rather than concentrate dramatic interest.

Macbeth begins by offering us two secrets: that Macbeth is first an incipient and then an actual murderer, and that he himself is doomed. The first is central but not at all enduring, the second deep, central, and enduring. But though they provide substance for exploitation during several scenes, neither exercises the kind of unrelenting pull on our minds and emotions that is exercised by the great secrets of *Hamlet* and *Othello*. In this respect *Macbeth* stands somewhere between these two tragedies, on the one hand, and *King Lear*, on the other.

Concerning the first secret, Macbeth's guilt, we may usefully note that in Holinshed no secrecy attend sDuncan's murder; according to the *Chronicles*, the murder is accomplished not covertly but overtly:

> At length therefore, communicating his purposed intent with his trustie friends, amongst whome Banquho was the chiefest, vpon confidence of their promised aid, he slue the king at Enuerns, or (as some say) at Betgesuane, in the sixt yeare of his reigne. Then hauing a companie about him of such as he had made priuie to his enterprise, he caused himselfe to be proclaimed king, and foorthwith went vnto Scone, where (by common consent) he receiued the inuesture of the kingdome according to the accustomed maner.

Rejecting this outright style of murder, Shakespeare modelled his killing on Holinshed's account of the murder of King Duff by Donwald. According to this account, Donwald and his wife first drown King Duff's two chamberlains in drink while the unguarded king lies asleep:

> Then Donwald, though he abhorred the act greatlie in heart, yet through instigation of his wife hee called foure of his seruants vnto him (whome he had made priuie to his wicked intent before, and framed to his purpose with large gifts) and now declaring vnto them, after what sort they should worke with feat, they gladlie obeied his instructions, & speedilie going about the murther, they enter the chamber (in which the king laie) a little before cocks crow, where they secretlie cut his throte. . . .

After the murder the servants dispose of the body in hugger-mugger style, carrying it through a postern gate into the fields, while Donwald himself hastily 'got him amongst them that kept the watch, and so continued in companie with them all the residue of the night'; and in the morning, with the discovery of the blood-spattered chamber, Donwald 'with the watch ran thither, as though he had knowne nothing of the matter, and ... foorthwith slue the chamberleins'.

Shakespeare's substitution of the circumstances surrounding King Duff's murder for those of Duncan's accords perfectly with his long-standing predilection for exploitable situations that result when a gap separates what the audience knows from what certain participants know. As told by Holinshed, Duncan's murder offered no way to develop his favourite dramatic situation; Duff's murder, however, offered an obvious base for scenes to his liking before, during, and after the murder. It is with these scenes that we shall be first concerned.

Our initial advantage is a double one, half ambiguous, half clear. In the opening scene we hear Macbeth's name spoken by the Witches and linked with an ominous paradox that is twice voiced: 'When the battle's lost and won'; 'Fair is foul, and foul is fair'. If the context is not sufficient to make the meaning certain, yet our senses are alerted to the idea that Macbeth moves under a cloud of un- known but dark significance. What the initial Prologue statement does directly for *Romeo and Juliet* the Witches' brief scene does obliquely for *Macbeth*: the first gives us an explicit advantage over the lovers, the second an inexplicit but foreboding one over Macbeth. But the second half of the advantage that we gain over Macbeth before we meet him in person is quite clear. In I. ii Duncan—before echoing the Witches' ominous paradox with his 'What he hath lost, noble Macbeth hath won'—orders Ross to hail the warrior with the traitor Cawdor's title. Macbeth remains ignorant of his new title until the Second Witch hails him by it, and then remains doubtful of it until Ross reports Duncan's command: 'He bade me, from him, call thee thane of Cawdor.' At this point Banquo's astonished 'What, can the devil speak true?' exploits the yet-insecure advantage that we hold over Macbeth, one that still amounts to little more than an uneasy sense that the Witches are engaged in some dark plot involving both Macbeth and Duncan.

Whatever ambitious thoughts he may have had in the past, the idea of murder that is to culminate in actual commission of the deed first seizes on Macbeth's mind in his first meeting with the Witches and is directly prompted by their prophecies. Perhaps it first touches his imagination as he hears these, for during his noticeable silence directly afterwards Banquo remarks that Macbeth 'seems rapt withal'. Between the Third Witch's 'All hail, Macbeth, that shalt be King hereafter!' and the next point at which the idea may have struck him, Macbeth learns from Ross that he is indeed thane of Cawdor. The truth of two prophecies has now been confirmed, and perhaps we should understand that the first thought of murder strikes him when he perceives this fact: 'Glamis, and thane of Cawdor! / The greatest is behind.' At the latest, we know that the idea has taken hold when he next speaks aside:

> ... why do I yield to that suggestion
> Whose horrid image doth unfix my hair
> And make my seated heart knock at my ribs,
> Against the use of nature?
>
> (I. iii. 134–37)

In any event, when we hear these lines the dramatist has an exploitable condition in hand that he wastes no time before exploiting; thus Banquo's 'Look how our partner's rapt', spoken even as Macbeth is engaging in self-debate on the question whether the fates expect him to act in order to gain what has been promised or only to wait until it is thrust upon him. Our best advantage here is over Ross, who, not having heard the Witches' prophecies, has no notion why Macbeth is 'rapt'; but Banquo, perhaps even now, entertains suspicions, and his running remarks to Ross suggest that he may be seeking to shield Macbeth from scrutiny. When Macbeth then urges that they find a time to speak 'Our free hearts each to other'—a proposition suggestively phrased—Banquo's quick assent betrays nothing about his own thought; but we can assume both that he guesses what his partner has been brooding on and that Macbeth is aware of the fact.

Shakespeare's heavy-handed exploitation of the gap opened with our awareness that the idea of murder has invaded Macbeth's mind begins in the next scene, with the entrance of Duncan, to whom the idea will matter most. The situation is exactly to Shakespeare's taste, and he would be unlike himself if he failed to lay bold hands upon it. Having heard the report of the traitor Cawdor's execution, Duncan tenders a solemn observation on men and life:

> There's no art
> To find the mind's construction in the face.
> He was a gentleman on whom I built
> An absolute trust.
>
> (I. iv. 11–14)

It is at this precisely calculated instant that the new Cawdor—Macbeth, with murder brewing in his head—enters, and Duncan rushes to embrace him with arms and golden praise: 'O worthiest cousin! / The sin of my ingratitude even now / Was heavy on me. Thou art so far before / That swiftest wing of recompense is slow / To overtake thee.' On such occasions, when he has taken pains to devise a situation the potentiality of which can be made to erupt in flares of irony, Shakespeare never was and never would become a subtle workman. As early as *Henry VI* and as late as *Henry VIII* he seized such chances with both hands. In *Richard III* he could make Clarence plead for his life in the name of the brother who has ordered his death, and in *King Lear* he could make Gloucester threaten Cornwall with the vengeance of the very son who has betrayed him. Just so, in *Macbeth*, he hesitates not at all to have Duncan, with prodigious solemnity, repeat with the new Cawdor the same error that he had made with the old. 'O worthiest cousin!' he cries to Macbeth, whose mind is boiling with 'black and deep desires', and so continues throughout the scene with protestations of trust and affection.

In corresponding situations in the comedies, when bursts of irony illuminate the gap between awarenesses—Orlando making love to 'Ganymede' as though to his 'very Rosalind'; the right Antipholus beating the wrong Dromio; Falstaff brazenly exposing to 'Master Brook' his scheme to make a cuckold of Ford; Malvolio exuberantly displaying his cross-gartered yellow stockings to Olivia while his tormentors watch unseen—the effects are comic and demand laughter from our point of observation. The dramatist's purpose in devising such exploitable moments in the comedies is thus always self-evident. But what is the result when the identical formula is applied in situations that involve life and death? Here the primary purpose cannot be to evoke laughter—though indeed the effect borders on the comic when the unaware speaker's error is as palpable as Duncan's extravagant praise of Macbeth; so does it when Clarence appeals for aid in the name of his brother Richard, and when Gloucester threatens Cornwall with Edmund's displeasure. When a situation is essentially comic, as when 'Ganymede' teases Orlando or Olivia notices Malvolio's dress, the flashes of irony mark the direct hits scored upon dead centre of all that is incongruous in the situation. But when the situation is deadly, as in Duncan's case, the flashes accentuate its deadliness, the boldest of them setting the situation into such sharp relief that it turns grotesque—as when Othello, at the most acute stage of his deception, puts his trust in 'Honest, honest Iago!' So, too, the deadliness of a situation is grotesquely accentuated until it borders on the comic when Duncan, speaking directly after we have heard Macbeth's private and sinister 'Let not light see my black and deep desires', overflows with joyous

affection: 'Let's after him, / Whose care is gone before to bid us
welcome. / It is a peerless kinsman.'

In the next scene, when the Macbeths easily reach an initial
accord on the murder of their guest, the way is prepared for yet
bolder and more sustained exploitation of Duncan's unawareness.
This scene characterizes the interior of Macbeth's castle as reeking
with evil[1], Lady Macbeth's chilling invocation to the spirits to unsex
her and fill her 'from the crown to the toe top-full / Of direst
cruelty' infects the atmosphere with poison. Her husband is ordered
to 'look like the innocent flower, / But be the serpent under't', for
'He that's coming / Must be provided for'. The very atmosphere is
saturated with evil, and it is with these images of horror still assailing
our senses that we next hear Duncan at the castle entry:

> This castle hath a pleasant seat; the air
> Nimbly and sweetly recommends itself
> Unto our gentle senses.
>
> (I. vi. 1–3)

Even Banquo—who surely lacks Duncan's absolute obliviousness of
the danger that lurks beyond the threshold—is made to contribute
his bit to the contradiction that has been established in our minds
between appearance and reality, between the exterior and the
interior of the castle: 'The temple-haunting martlet', he observes
mildly, has taken up residence here, and 'Where they most breed
and haunt, I have observ'd / The air is delicate.' Yet gaudier are the
flashes that erupt from Duncan's conversation with his hostess. The
one joyfully ignorant, the other hyper-conscious, as are we, of the
reality, they vie in expressing extravagant compliments: 'Your
servants ever / Have theirs, themselves, and what is theirs, in
compt,' says Lady Macbeth at the end of their elaborate exchange,
'To make their audit at your highness' pleasure, / Still to return
your own.' And with these gracious words the hostess guides her
guest into the castle and his death.

During these scenes, from Macbeth's first meeting with Duncan
after the idea of murder has infected him until the King's entrance
into the castle, exploitation of the gap between our awareness and
Duncan's has been unceasing, shooting sparks of irony from the
speeches of both victim and assassin. The power of many scenes in
Macbeth to exercise a spellbinding effect is famous, and if we find
ourselves hypnotized by these particular scenes perhaps we do so
because of the very vividness of the flashes. Perhaps, too, the sheer

intensity of the exploitation owes something to the dramatist's realization that this useful gap will not last long; to make it yield its utmost before the actual murder closes it, he loads each line spoken between victim and assassin with special meaning for us.

The gap, in fact, is as good as closed with Duncan's entrance into the castle, for we never see the innocent victim again. We do, however, hear of him once more before the murder, when Macbeth, going about his sinister business in the dead of night, encounters Banquo walking near Duncan's chamber door. Here the situation is virtually self-exploiting because we have just heard Macbeth conclude his debate with his wife by pronouncing an unequivocal decision: 'I am settled, and bend up / Each corporal agent to this terrible feat.' We know not only that he is resolved to kill the King, but that he is even now on the way to do so. Meeting just at this point, neither Macbeth nor Banquo can speak a line without producing a shower of irony—irony, however, the quality of which contrasts markedly with that which accompanied the conversations of Macbeth and Lady Macbeth with Duncan shortly before. In the previous scenes, Duncan was oblivious of the truth that we shared with the Macbeths, whereas in the present scene Banquo—whatever may be the exact degree of his awareness of Macbeth's immediate intent—is certainly not oblivious. Though exhausted, he has resisted sleep: 'Merciful powers, / Restrain in me the thoughts that nature / Gives way to in repose.' What are these thoughts? Of private ambition, inspired by the Witches' promise that he would get kings but not be one? Or of Macbeth, and Duncan's immediate danger from that quarter? Pointedly, he mentions the 'great largess' just bestowed by Duncan upon Macbeth's servants; and, more pointedly still, he displays a diamond that is Duncan's gift to Lady Macbeth. The oddness of the timing makes the incident suspect: should we not have expected Duncan himself, during the evening or the next morning, to have presented the diamond to his hostess? Perhaps Shakespeare has stretched probability to emphasize a point: Banquo has been given the diamond (not so much by Duncan as by the dramatist) in order to display it at this unlikely hour, after midnight, in the corridor outside Duncan's chamber, just when Macbeth, his mind made up, is on his way to kill the King. It is as though Banquo, with a kind of bribe, would stay the assassin's hand at the last instant. We may imagine that, after this exchange, he goes directly to bed and falls sound asleep, confident that Macbeth, shamed by this added evidence of the King's generosity, cannot possibly bring himself to proceed with murder.

In any event, we must certainly assume that Banquo suspects something, and therefore his conversation with Macbeth evinces a different quality from that noted in the earlier conversations of the Macbeths with Duncan. Here, the effect is that of a fencing match in which the adversaries, knowing each other's habits, anticipate each stroke and parry. 'I dreamt last night of the three weird sisters,' Banquo tentatively opens; and Macbeth parries, 'I think not of them.' But then he thrusts by inviting Banquo—'when we can entreat an hour to serve'—to discuss the matter with him; Banquo assents blandly, and Macbeth thrusts again: 'If you would cleave to my consent . . . / It shall make honour for you.' Banquo's carefully ambiguous reply at once breaks off the match and leaves the way open for later resumption: 'So I lose none / In seeking to augment it, but still keep / My bosom franchis'd and allegiance clear, / I shall be counsell'd.' But that the positions of the opponents are reversed, the closing remarks might have been made by Hamlet and Claudius just before Hamlet is dispatched to England. 'Good,' says Hamlet; and Claudius replies, 'So is it, if thou knew'st our purposes.' Hamlet answers, 'I see a cherub that sees them.' Thus Macbeth: 'Good repose the while'; and Banquo: 'Thanks, sir; the like to you.' Though here the intent of each speaker is privately ironic, the depth of his irony is not only hidden from the other but also not fully appreciated by himself. The future will afford no rest for either.

From the meeting with the Witches until his dying moment, the degree of Banquo's awareness of Macbeth's intent remains ambiguous. Shakespeare had at least three choices: to follow Holinshed in making Banquo privy to the murder; to make him oblivious of Macbeth's purposes beforehand and guilt afterwards; or to hedge, portraying him with deliberate ambiguity both before and after the murder. Holinshed's way would have been dramatically inferior because it would have compromised Macbeth's centrality as tragic hero; dramatic focus and intensity would have been sacrificed had Macbeth and Banquo shared the guilt. The second choice would have strained our credulity overmuch: Banquo heard the Witches tempt Macbeth and saw his partner standing 'rapt'; the dramatist would have had to make him blind and stupid in order to convince us that he could remain oblivious thereafter. Conceivably politics entered the dramatist's calculations also: with the tragedy ultimately designed as a compliment to King James, Shakespeare would not wish to portray his sovereign's legendary ancestor as either a blockhead or a murderer. But there were also compelling dramatic reasons for choosing the way of ambiguity, and these will emerge as we examine the aftermath of the murder.

With the commission of the actual murder, the gap between awarenesses shifts: where formerly it stood between Duncan, unaware that he was to be murdered, and the Macbeths, it stands next between those unaware that a murder has been committed, and the Macbeths. The scene of the murder itself offers no occasion for exploitation of this second gap, since during this time we see and hear only the guilty couple. But exploitation resumes soon after, with the arrival of someone who does not know: the knocking at the gate signals the approach of 'the unknowing world'. The Porter who answers knows neither who is without nor what has happened within, and the breadth of the gap between his awareness and ours is exploited with no great subtlety by his taking it into his badly hungover head to fancy himself porter of hell-gate—as, in the special sense known to us, he is. But the boldest exploitation, matching that which just preceded the murder, begins with the entrance of Macduff and Lennox from without and the return of Macbeth from his own chambers within. 'Is the King stirring, worthy thane?' inquires Macduff. 'Not yet,' replies Macbeth. While Macduff has gone to wake the King, Lennox casually asks, 'Goes the King hence today?' 'He does,' answers Macbeth—and adds after a pause the significance of which is of course lost on Lennox, 'he did appoint so.' When Lennox has described some phenomena of the night—screams of death, fearful prophecies, the owl's clamourings, the shaking of the earth—Macbeth blandly concurs: ''Twas a rough night.'

With Macduff's return from the murder chamber crying 'Horror, horror!' the nature of the gap shifts again: where formerly it was between the Macbeths and those ignorant that a murder has been committed, it now stands between the Macbeths and those ignorant of the murderer's identity. It seems reasonable, at this point, to expect this to become at once and to continue to be the single most important, most exploitable and most exploited, gap in the tragedy. But, as we shall see, of this crucial discrepancy Shakespeare makes only slight and incidental use. *Macbeth* is not at all a 'mystery' play, a play of suspense, for the question of who killed King Duncan hardly even figures as an issue. In fact, such exploitation of the discrepancy between our knowledge, shared with the Macbeths, and the world's ignorance of the murderer's identity as is dramatically significant is confined to the remainder of the very scene in which the fact of the murder is discovered. Before the murder, Lady Macbeth had assured her husband that none could suspect them, for the blame would obviously fall upon the King's chamberlains: 'Who dares receive it other, / As we shall make our griefs and clamour roar / Upon his death?' The 'griefs and clamour' they now mix as naturally as they can into the general uproar that follows discovery of Duncan's body: 'What is't you say? The life?' cries Macbeth; and Lady Macbeth, entering, contrives just that appearance of concern that should be correct for a considerate hostess whose household guests have been disturbed by some hideous noise below stairs: 'What's the business, / That such a hideous trumpet calls to parley / The sleepers of the house? Speak, speak!' She is by far the abler dissembler; not only are her first words shrewdly calculated to suit the occasion, but her second outcry, after Macduff's 'Our royal master's murder'd!' cues her, is a socio-psychological masterstroke: 'Woe, alas! / What, in our house?' It is a sly imitation of what the proper hostess would be expected to say when a guest has been murdered under her roof, calculated to expose a mind as innocent of guilt as Gertrude's in *Hamlet* with her gem of purely disinterested dramatic criticism, 'The lady protests too much, methinks.'

Even so, the climactic act of her performance is her fainting spell. Of course the spell is not, as some have insisted, a real one, induced either by horror at the deed, accentuated by Macbeth's overly vivid description of the gory bedchamber, or by late reaction, just now striking home after her brave struggle to be strong for her husband's sake. It is partly, but only partly, the timing of the fainting fit that disproves its genuineness. Carried away by his fevered imagination, Macbeth has just wallowed in a descriptive orgy; ever since returning from the murder scene, he has mouthed extravagances like an inspired drunk. That he is not in command of his faculties must be all too apparent to his wife, and one plausible explanation of her feigned swooning is her sense of the need to take the spotlight away from him before he betrays them both, as one actor rushes to the aid of another in desperate trouble with his lines. But a second explanation is at least as plausible. Twice already Lady Macbeth has spoken the proper, the 'expected' sentiment—first demanding the cause of the uproar that has disturbed her guests, then evincing the first, appropriately social, concern of the good hostess for what others may think of her breach of good manners in allowing a guest to be murdered: 'What, in our house?' Now, for her finale, she takes her cue from Macduff's masculine concern for her feminine sensibilities: 'O gentle lady, / 'Tis not for you to hear what I can speak; / The repetition in a woman's ear / Would murder as it fell.' Following Macduff's comment, she next hears that Duncan has in fact been murdered and then hears her husband's overwrought description of the bloody scene. A 'gentle lady' would be expected to respond in only one manner to these horrors. And so she faints.

But her controlled performance is not matched by that of her partner in crime: whatever may be Macbeth's capabilities, a talent for dissembling is not among them. His physical reaction to the Witches' prophecies was so pronounced that Banquo marked it: 'Why do you start and seem to fear / Things that do sound so fair?' Twice, once to the Witches and again to Ross, Banquo remarked that Macbeth was 'rapt'. When he first confronted his wife after hearing the prophecies, his face betrayed him: 'Your face, my thane, is as a book where men / May read strange matters.' Sternly, she ordered him to 'look up clear', warning that 'To alter favour ever is to fear'. Thereafter we have had no further view of Macbeth in company until just before the murder when he meets Banquo, for it is Lady Macbeth, not he, who greets Duncan at the door; and during the evening's entertainment of the King we are not shown how well or ill he manages to 'look like the innocent flower / But be the serpent under't'. On the way to kill Duncan, he appears at ease, though an alert Banquo must note a contradiction in his remarks: 'I think not of them', he replies to Banquo's mention of the Witches, but adds in the next breath that '. . . when we can entreat an hour to serve, / We would spend it in some words upon that business'. But it is in the moments just after Duncan's body is discovered that he most exhibits his ineptness in the art of seeming. Returning from the murder chamber with Lennox and Ross, he launches first one and then another flight of nervous poetry; it is at the end of the second that Lady Macbeth finds it expedient to faint—and perhaps it is her action that saves the day, for Macbeth, growing ever wilder, might soon have so incriminated himself that no wit of his wife's could rescue him.

What he does speak is not directly incriminating, but its manner is such as to arouse rather than allay suspicion. His reply to Macduff's peremptory 'Wherefore did you so?' is floridly excessive, the speech of one who has much to hide. Hearing it, Malcolm and Donalbain quickly exchange words and looks of fear and suspicion, each remarking that, though they 'most may claim this argument for ours', yet neither has given way to an emotional outburst. 'Our tears are not yet brew'd,' says Donalbain, and Malcolm pointedly adds that 'To show an unfelt sorrow is an office / Which the false man does easy.' To assert that at the end of this, the very scene in which the murder is discovered, the Macbeths' guilt is known throughout the castle is to assert too much; but to say that at this point Shakespeare considered their secret to be no longer useful for exploitation is only to report what the text verifies. If Acts I and II have prepared us to expect the remainder of the play to represent Scotland's search for unknown murderers, they have misled us, for after Act II the dramatist turns quite away from this issue; indeed, except for passing allusions, Duncan's murder itself appears to have been forgotten by everyone but Lady Macbeth, who broods on it, reviews it in her sleep, goes mad, and finally dies because of it. But the Macbeths' guilty secret is not pursued by others: when Malcolm and Macduff raise an army to destroy Macbeth, it is not because he killed Duncan but because he has become a bloody tyrant. The Macbeths' guilt is never debated, proved, or published; it appears, rather, to be taken for granted.

Banquo and Macduff, who seem from the first to guess the truth, do not pursue their suspicions. Replying to Ross's 'Is't known who did this more than bloody deed?' Macduff baldly states, 'Those that Macbeth hath slain.' Perhaps this is an assertion of Macduff's true belief; more likely it is irony. In any event, the conversation shows clearly that, already, Macduff's primary concern is not with the question of who killed Duncan but with Scotland's future under Macbeth's rule. Banquo opens Act III by remarking 'Thou hast it now: King, Cawdor, Glamis, all, / As the weird women promis'd, and, I fear, / Thou play'dst most foully for't'—and momentarily his emphasis on 'fear' and 'most foully' may suggest to us that he means to pursue the question. But in fact Banquo, like Macduff, exhibits a curious indifference to the murder itself; neither here nor elsewhere does he express interest in calling the murderer of the King to justice. In this scene (III. i) the best that can be said of him is that he means to do nothing, one way or the other, but will merely stand and wait to see whether the fates will or will not make him founder of a royal line; the worst that can be said is that he means to blackmail Macbeth: 'Let your Highness / Command upon me,' he tells his former partner, 'to the which my duties / Are with a most indissoluble tie / For ever knit.' It is a blunt reminder that they share a dark secret, and it suggests to Macbeth a more direct threat than perhaps Banquo quite intends, for it prompts a swift succession of ominous questions: 'Ride you this afternoon?' . . . 'Is't far you ride?' . . . 'Goes Fleance with you?' Banquo appears not to catch Macbeth's drift, and here the dramatist has quickly created and as quickly exploited a deadly discrepancy between awarenesses, ours and Macbeth's on the knowing side, Banquo's on the other.

The incident of Banquo's murder gives us also another passing advantage that we share only with Macbeth. Lady Macbeth, for the first time, is not privy to her husband's plans, and throughout III. ii her ignorance of what we have witnessed in the preceding scene, in which Macbeth instructs the Murderers, provides an exploitable gap of richer potential than that which yielded only a few sudden sparks in Macbeth's interrogation of Banquo. In the earlier scene we were not ourselves advised in advance, so that our awareness of Macbeth's true intent remained uncertain until the final lines. In the scene with Lady Macbeth, however, Shakespeare's more typical method is restored, for III. i, Macbeth's scene with the Murderers, has given us all the light we need, and the last words of the scene focus the light precisely: 'Banquo, thy soul's flight, / If it find heaven, must find it out to-night.' Quite unaware of her husband's arrangement, Lady Macbeth opens the scene with what may or may not be an innocent query: 'Is Banquo gone from court?' But when Macbeth enters it is he, not she, who first mentions Banquo—for though he has just come from ordering his death and has thus deceived his wife, it is clear that his basic insecurity requires her approbation. To his tentative 'Thou know'st that Banquo and his Fleance lives', she replies with a pointed 'But in them nature's copy's not eterne'—a remark that conveys her attitude unmistakably even to Macbeth's essentially slow understanding. Her oblique assent not merely reassures but exhilarates him and occasions an exuberant flight of imagery:

> Be innocent of the knowledge, dearest chuck,
> Till you applaud the deed. Come, seeling night,

Scarf up the tender eye of pitiful day,
And with thy bloody and invisible hand
Cancel and tear to pieces that great bond
Which keeps me pale!

(III. ii. 45–50)

Thus he continues for as many more ecstatic lines uttered as though a naughty boy who has conspired to commit a wicked deed should find, to his great relief, that his mother thoroughly approves; at the end of the scene, surely the actor of Macbeth should exit gleefully, carrying his 'chuck' aloft, and kicking up his heels.

Aside from the fact that his method gives us a brief advantage over Lady Macbeth, it is uncertain why Shakespeare chose to have Macbeth keep his wife ignorant of his scheme for eliminating Banquo. As with Duncan's murder, he might have presented it as a joint enterprise, with Lady Macbeth present to help instruct the Murderers. The dramatist's purpose can hardly have been to demonstrate Macbeth's bad judgement in acting independently, since it is quickly shown that Lady Macbeth also wishes Banquo dead and would surely have approved the scheme. Presumably her abrupt 'Is Banquo gone from court?' which opens the scene implies that she, too, had been thinking of his assassination. It has sometimes been argued that Macbeth denies his wife knowledge of the intended murder because he has already recognized signs of her future breakdown and wishes to spare her the ravages of additional guilt; but nothing that we have yet seen in Macbeth—and nothing that we will see afterwards—suggests that he possesses the least sensitivity to his wife's troubles, or to those of any other but himself. One fact, at least, is clear: the scene marks a turning-point in the partners' relationship. Hitherto it was Lady Macbeth who seized and held the initiative: she reversed her husband's mind after he had decided not to kill Duncan; she outlined the precise means; she returned the bloody daggers to the chamber when the distraught Macbeth refused to go again; she sought to calm and reassure him; she even rescued him by seeming to swoon when all eyes were fixed on him. Until now her will, mind, and hand have managed every-thing; but now, with Banquo's murder, Macbeth has taken the reins. One immediate consequence of his doing so seems, for the moment, fortunate for them both: not knowing of the murder, she does not, like her husband, see the blood-boltered ghost which shortly appalls Macbeth at the banquet table, hence is able, after a fashion, to cover for him and to empty the room of the astonished guests. But of course Macbeth could not have foreseen this need of her when he ordered the murder and left his wife innocent of the plan.

But, now, of what actual importance is the banquet scene itself as a factor in the exposure of the Macbeths' guilty secret? It has been usual to assign it much importance—even to assert that here one of the tragedy's more conspicuous ironies operates: Macbeth has Banquo murdered in order to gain security, but his violent outburst at sight of Banquo's ghost arouses universal suspicion and loses him whatever security he formerly had. On first consideration this view seems corroborated by a later scene in which Lennox's conversation with 'another Lord' serves to epitomize the world's opinion; says Lennox,

> The gracious Duncan
> Was pitied of Macbeth; marry, he was dead.
> And the right valiant Banquo walk'd too late;
> Whom, you may say, if't please you, Fleance kill'd,
> For Fleance fled; men must not walk too late.
> (III. vi. 3–7)

Continuing with the same sarcasm, Lennox summarizes what we are presumably to take as the prevailing opinion of the nobility: that Macbeth killed Duncan, killed Banquo, and would kill Malcolm, Donalbain, and Fleance if he had them 'under his key'. But this opinion is not one that emerged suddenly after Macbeth's wild conduct at the banquet. As we noted earlier, strong suspicions were voiced by Malcolm, Donalbain, and others immediately after Duncan's body was discovered, and these were then repeated in bolder terms by Macduff and Ross in the next brief scene. In its main function, which is expository, the Lennox–Lord scene merely parallels the much earlier Ross–Macduff scene and adds little to it: suspicion was as strong immediately after Duncan's murder as after the banquet. Further, it is noteworthy that Macduff, who denied his presence at the banquet, had earlier denied it also at Macbeth's coronation. Lennox and the Lord do not make clear whether Macduff had already left Scotland for England before the banquet, but in any event his departure cannot have been precipitated by Macbeth's strange behaviour there, since he was not present to witness that outburst. Also noteworthy is the fact that Macbeth and his wife, once the banquet guests have been dismissed, show no alarm lest Macbeth's behaviour should have betrayed their guilt;

neither even mentions the possibility. Macbeth immediately expresses anger that Macduff was not present and resolves to deal with him at once; and Lady Macbeth, having consumed her waning energies in covering for her husband, seems privately dispirited and is mainly silent. And, finally, it is noteworthy that the scene in England with Malcolm and Macduff, shortly to follow, includes no mention of the disruption caused by Macbeth's conduct at the banquet; Ross, who was present on that occasion, describes in vivid detail Scotland's condition under the tyrant but says nothing of the banquet. In all, it must appear that Macbeth's guilt in Duncan's murder was so generally accepted long before the banquet that his wild manner then revealed nothing new to the observers.

But before turning away from the hardly existent, hence dram   unserviceable gap between Scotland's and our own awaren   he Macbeths' initial guilt in order to consider the effects of a more important discrepancy, we must examine one further scene, Lady Macbeth's sleepwalking (v. i) as it relates to the first gap. With most of what we have noted in tracing the history of the Macbeths' 'secret' this scene seems in curious conflict. Much of its potent force derives, obviously, from its vivid delineation of Lady Macbeth's guilt-ravaged conscience; but possibly even more of its force arises from our involuntary fears that, having no control of her tongue, she will divulge all and thus absolutely betray her husband and herself before witnesses. As in the banquet scene, where Macbeth teetered on the brink, seemingly in danger of exposing their secret, when in fact there was none left to expose, so here Shakespeare very shrewdly capitalizes on what, though non-existent, remains somehow exploitable for spectacular effect. To this end he sets the stage with a pair of eavesdroppers, the Doctor and the Gentlewoman, quickly creates the illusion, by means of their hushed and nervous conversation, that there truly are unheard-of secrets in imminent danger of being divulged, then brings in Lady Macbeth to walk before them, uttering incoherent scraps and pieces that, to our informed minds, obviously pertain to the murders of Duncan, Banquo, and Macduff's family. Spellbound by her words and actions and prodded by the eavesdroppers' expressions of shock and amazement, we cannot easily avoid supposing that her admissions really are dangerous to herself and her husband. To the Doctor's hushed 'what, at any time, have you heard her say?' the Gentlewoman replies, 'That, sir, which I will not report after her.' Says the

Doctor, midway through the scene, 'Go to, go to; you have known what you should not,' and the Gentlewoman replies, 'She has spoke what she should not, I am sure of that.' When the sleepwalker has returned to her bed, the Doctor confides that 'Foul whisp'rings are abroad', and closes the scene with 'I think, but dare not speak.' Perhaps the eavesdroppers have indeed, hitherto, heard no more than rumours, and perhaps Lady Macbeth's involuntary confessions are truly news to them; but at this late point it is difficult to believe that they would be news of any significance to the Scottish nobility. For the counter-action is already well under way, without need of further testimony: Malcolm and Macduff are even now invading with an army from England. Were the Doctor to take heart, flee the castle, seek them out, and breathlessly disclose his suspicions that the Macbeths are murderers, he would astonish no hearers. But it is not to be denied that Shakespeare's device has served his turn: by exploiting a non-existent gap as though it were real, he has enormously increased the spellbinding potential of the sleepwalking scene; and if the scene uses dramatic chicanery, it is no less remarkable for that.

To this point we have been concerned primarily with Shakespeare's exploitation of the awarenesses that are shared between the Macbeths and ourselves. In the scenes that precede Duncan's murder we have noted that this exploitation is typically direct and obvious, producing effects of irony so glaring that they border on the comic. After the murder, with our advantage fast vanishing because the Macbeths' secret is no real one at all, we have marked only brief moments of similarly broad exploitation, as in Macbeth's pointed 'Ride you this afternoon?' interrogation of Banquo. The fact is that except for the scenes that just precede or follow the initial murder, the Macbeths hold no such advantage over other participants as that held, for example, by Hamlet or Iago. Dramatic interest centres only briefly on their secret, even as, in *King Lear*, it centres only briefly on the wicked daughters' secret as deceivers of their father. Unlike the hero Hamlet or the villain Iago, Macbeth, who is at once hero and villain, is never the proprietor of a great central secret. It is, thus, not the advantage that we share briefly with the Macbeths that matters to the drama as a whole, but the advantage that we hold over *them*, and especially over Macbeth. Just what this advantage is, when and how we acquire it, and what the dramatic consequences are of our holding it are the questions to which we now turn.

We come very early to hold a tremendous advantage over Macbeth. Do we do so because of some quality in his nature that impairs his judgement, as Brutus's addiction to honour impairs his? Or do we gain advantage as we do over Othello, because we are privy to the plots of a cunning practiser? Our first advantage is insignificant and brief and has already been noted: we learn before Macbeth that Cawdor has been condemned and that his title is now Macbeth's. Macbeth hears himself hailed as thane of Cawdor by the Witches but does not immediately believe them; hence our advantage lasts until Ross hails him similarly. Though our advantage here is unexploited, it does carry a special importance: we learn, before Macbeth, that the Witches have spoken the truth and are accordingly conditioned to receive their next pronouncements as truths also. Macbeth will be King, and Banquo will get kings. It is worthy of note in passing that our confidence in the Witches' prophecies proves to be justified, for in fact the sisters never break faith with us: Macbeth *is* thane of Glamis and Cawdor and does become King; Banquo's children will be kings; Macbeth is safe until Birnam wood comes to Dunsinane; he should indeed beware Macduff; and no man 'born' of woman harms him.

If it is always truth that the Witches tell us, so is it truth that they tell Macbeth, for they tell us only what they tell him. We do not gain advantage over him because we are privy to their counsel as we are privy to Iago's; they do not tell us privately facts that contradict what they tell Macbeth. He hears the prophecies that we hear and also hears Banquo's warning as we do:

> ... oftentimes, to win us to our harm,
> The instruments of darkness tell us truths,
> Win us with honest trifles, to betray's
> In deepest consequence.
>
> (I. iii. 123–6)

In heeding this warning we gain advantage over Macbeth, though it proves at last to be not quite the same advantage we think. Banquo is not strictly accurate in his prediction, for the Witches do not, in fact, first tell Macbeth truths in order to gull him with later falsehoods; they continue to tell him nothing but truths right up to the end. It may be useful here to compare Macbeth's situation to Lear's at a similar moment. Goneril and Regan lie to Lear about loving him, and, with the aid of Cordelia and Kent, we recognize that they do so, while Lear, though he, too, hears Kent, continues to take lies for truths and refuses to alter his decision. Thus we gain a vital advantage over him, and later events prove our first judgement to have been accurate, for the wicked daughters do turn on him, Cordelia stays loyal, and Kent's warning proves true. But the Witches, unlike Lear's daughters, foretell what is to come, and what they foretell does in fact come to pass. Because, on two occasions, their vindictive character is exposed to us privately (their threat of vengeance, for example, for the sailor's wife's refusal to give the First Witch chestnuts), we incline to distrust them and would doubtless advise Macbeth, if we could and as Banquo does, to believe none of their prophecies even though they have already spoken truth in hailing him as thane of Cawdor. Here, then, curiously, we hold an advantage, which the dramatist exploits, that proves in the end to have been false: we distrust the Witches and would have Macbeth do so, yet in the end we must acknowledge that they never speak an untruth to Macbeth, whose naïve acceptance of their prophecies proves righter than our own more sophisticated suspicion. Curiously, too, in the end Macbeth's and our views of the Witches become reversed. In his last hours he calls them 'juggling fiends', equivocators 'that palter with us in a double sense', fiends 'that lie like truth'; but we, having seen the truth of all their utterances borne out, recognize that Macbeth's bitter complaints are that same 'admirable evasion of whoremaster man' noted by the bastard Edmund— whoremaster man who lays the fault for his personal disasters on the sun, the moon, the stars, on anything and everything but himself. Like the doomed hero of a morality play, Macbeth warns us never to believe in such powers of darkness, wishes that the curse of Cassandra may fall upon them, and neither recognizes nor admits that any fault in his own character has brought about his wretched end.

If, then, our initial advantage over Macbeth does not, after all, derive from our truer assessment of the Witches, how do we gain it? For assuredly, by the end of the scene in which he first meets the Witches, we are conscious of a very considerable advantage, and essentially this same advantage continues to function as a gnawing certainty in our consciousness throughout the tragedy. Possibly the start of our advantage occurs with Banquo's question that directly follows the Witches' first round of prophecies: 'Good sir, why do you start, and seem to fear / Things that do sound so fair?'—a startling echo of their choral utterance in the eleventh line of the play, 'Fair is foul, and foul is fair', which Macbeth's own first words spoken in our hearing have already echoed only a few lines before Banquo's: 'So foul and fair a day I have not seen.' But it is not only this obvious linkage of Macbeth with the Witches that promotes our advantage; what is more significant is that Macbeth has reacted physically to the prophecies, and so visibly as to attract Banquo's attention. It is our very first sign of a slumbering imperfection in his character that will soon show itself fully. Our sense of this imperfection grows with his lines spoken aside just after Angus has confirmed the Witches' word that he is thane of Cawdor. 'Glamis and thane of Cawdor! / The greatest is behind,' he muses, and then continues:

> ... why do I yield to that suggestion
> Whose horrid image doth unfix my hair
> And make my seated heart knock at my ribs,
> Against the use of nature?
>
> (I. iii. 134–7)

'That suggestion', we learn in his next lines, is murder. We have noted earlier that with his use of this word we gained instant advantage over Duncan, the unsuspecting victim; but, though Shakespeare exploits that advantage heavily, as we have seen, our great advantage gained here is over Macbeth himself.

To suppose that it consists only in the fact that we, with Banquo, distrust the Witches' prophecies whereas Macbeth does not is to miss the real point; what sets him most significantly below our level of vision is the fact that the idea of murder has invaded his mind. It is noteworthy that the Witches have not mentioned murder or suggested any wrongful act. Neither do they ever suggest, even by the obscurest hint, that Macbeth commit a wrong in order to gain the throne: the idea of murder is exclusively his own, and the fact that his mind conceives and embraces it is the source of our greatest and most enduring advantage. For with this first thought of murder he drops to a plane of moral awareness below our own.

His plane is in fact not one of moral awareness, but of moral ignorance; and from this plane he never rises. From beginning to end of the action he remains oblivious of murder as a moral fault. The wrongness of murder is a subject that he neither debates in soliloquy nor discusses with his wife, for he is oblivious of the issue itself. Subsequent experience, as we shall see, teaches him much: it teaches him that escaping punishment for murder is harder than his wife had made him think; it teaches him that one murderous act is but a step that must be followed by others; it teaches him also, as he laments in a moment of whining self-pity, that a murderer cannot look to have 'that which should accompany old age, / As honour, love, obedience, troops of friends', any more than he can expect to find respite from the fear of being caught; it teaches him, in short, that murder brings all sorts of inconveniences upon the murderer. But of 'right' and 'wrong' it teaches him nothing at all. If, in the end, he were given a second chance to decide whether to kill Duncan, he would decide not to do so; but his decision would again be based, as before, not on moral grounds but on consideration of consequences to himself.

What he gains from his experience is only explicit confirmation of the very fears that he expressed in the soliloquy of I. vii, 'If it were done when 'tis done', where Shakespeare anatomizes the mental cast of a criminal mind as he does nowhere else in the canon. Here Macbeth raises the only arguments against murder that his imagination can conceive—and when he does so our advantage over him becomes complete, for we then gain clear sight of his fatal limitation: his lack of moral sense. Not once in this appalling soliloquy does he raise a moral objection to murder, argue it down, and finally reject it; he simply does not raise it at all, for he is unacquainted with any such idea. He is aware of a risk to his immortal soul if he should kill Duncan—but fear of punishment after death is no equivalent to moral scruple; had he been dissuaded from murder by his awareness of this risk, instead of shrugging it off with a casual 'We'd jump the life to come,' it would yet have been fear, not moral sense, that deterred him. He is keenly aware of the practical danger that he faces 'upon this bank and shoal of time', and it is finally his fear of earthly consequences only that restrains him. 'But in these cases', he says,

> We still have judgement here, that we but teach
> Bloody instructions, which, being taught, return
> To plague th' inventor.
>
> (I. vii. 8–10)

Fear of heaven's punishment he brushes off; it is the fear of man's justice that stays him. He reasons, further, that in this instance man's justice will be particularly dangerous because of the special circumstances of the murder: he is Duncan's kinsman, his subject, and his host; he is conscious that men will expect him, as host, to shut the door against an intruder, not bear the knife himself. What is more, he reasons, Duncan has been a *good* king, 'meek', and 'clear in his great office'. Here, for a fleeting instant, we may hope that, after all, at least a semi-moral thought has struck him, for he seems about to argue that it is *especially* wrong to murder a *good* man, a *good* king. But then our expectation is abruptly quenched by his statement of *why* it is worse to kill a good king than a bad one:

> . . . his virtues
> Will plead like angels, trumpet-tongu'd, against
> The deep damnation of his taking-off;
> And pity . . . .
> Shall blow the horrid deed in every eye . . .
>
> (Ibid. 18–24)

In short, it is not particularly *wrong* to kill a good king; it is only particularly risky, for men will be the more enraged and the likelihood of apprehension and punishment all the greater.

If we distill Macbeth's reasoning in the soliloquy, we get the following:

> If I could kill this man and avoid earthly justice, I would risk whatever the hereafter may bring. But there is danger of earthly punishment for murder, and in this case it is increased because Duncan is my king, kinsman, and guest, and is, besides, a *good* king, whose murder will offend people extremely and increase their zeal to catch and punish me.

With this reasoning he talks himself out of doing the murder. 'We will proceed no further in this business,' he flatly informs his wife when she comes to him; and then he adds what she instantly recognizes as mere sham:

> He hath honour'd me of late; and I have bought
> Golden opinions from all sorts of people,
> Which would be worn now in their newest gloss,
> Not cast aside so soon.
>
> (Ibid. 32–5)

A superlative wife who reads her husband like a primer, she wastes no word in countering his lame and whining rationalization, but goes straight to the real matter, charging him with letting 'I dare not' wait upon 'I would'. His next pretext—'I dare do all that may become a man; / Who dares do more is none'—she brushes aside with biting contempt. Having now exhausted his meagre store of excuses, he blurts out the plain truth: 'If we should fail?' On hearing this admission of what really causes him to balk, she darts into the breach, swiftly and explicitly detailing how they can do the deed and *not* fail. When she has finished, Macbeth responds with no further lame objections or hesitant assent, but with a burst of exuberance:

> Bring forth men-children only,
> For thy undaunted mettle should compose
> Nothing but males. Will it not be receiv'd,
> When we have mark'd with blood those sleepy two
> Of his own chamber and us'd their very daggers,
> That they have done't?

(Ibid. 72–7)

By showing him how to do it and not be caught, she has demolished his sole restraint.

Perhaps we should recapitulate at this crucial point by marking three distinct, though obviously interrelated, advantages that we hold over Macbeth. First, though Lady Macbeth has easily overcome his fears and convinced him that he can murder Duncan and escape punishment, she has certainly not convinced *us*. Though Shakespeare has not directly supplied us with evidence that detection and punishment will follow—unless Banquo's strong suspicions may be counted as such evidence—yet our dramatic sophistication and our traditional habits of thought about crime and punishment coincide and prevail with us against Lady Macbeth's too-easy assurances. So deeply is the idea that punishment will follow crime embedded in our consciousness that we would lay odds of a thousand to one that if Macbeth murders Duncan he will be caught and punished. Second, for similar reasons, at this point we strongly distrust the Witches, as Macbeth evidently does not. In part our doubts have been planted dramatically, for we have noted the ominous linking of these powers of evil with Macbeth from the first scene on through the device of repetitive phrasing; at the end of Act I we might have counted some twenty-eight verbal plays on the theme first struck by the Witches' 'Fair is foul, and foul is fair'. Conspicuous among these powerful conditioners of our awareness is Macbeth's own first line, 'So foul and fair a day I have not seen,' but in fact the seven scenes of Act I are laced with variations—some obvious, like Banquo's '. . . why do you start, and seem to fear / Things that do sound so fair?' and some less obvious, like Lady Macbeth's 'look like the innocent flower, / But be the serpent under't' and 'So from that spring when comfort seem'd to come, / Discomfort swells'. We can hardly have avoided being affected, consciously or unconsciously, by the heavy repetition, in various wordings, of this basic paradox. It is a dramatically shrewd device, calculated to infect us with a mood of unrest, of deep uneasiness. Perhaps even without this elaborate pre-conditioning, and without Banquo's pointed warning to beware, we might yet have distrusted the Witches; still, the insistent hammering on the single theme clearly deepens our sense of misgiving.

But neither our certainty that Macbeth will not escape justice nor our distrust of the Witches is the great advantage that we hold over him. If these were our main advantages, *Macbeth* might stand as a suspenseful melodrama, but never as the profound tragedy that it is. Holding only these advantages, we might warn Macbeth thus: 'Do not kill the King, Macbeth, for if you do you will be caught and punished,' and, again, 'Do not trust the Witches, Macbeth, for witches are never to be trusted.' But these are shallow injunctions and do not touch the deeper issue, which is a moral one. The crucial gap between Macbeth's awareness and ours is simply that we have a moral sense and he has none. If we could warn him before the murder of Duncan, 'Do not do it, Macbeth, for murder is morally wrong,' he would not heed our warning because he would not understand our meaning. He does not know what 'wrong' means before the murder, and he never comes to know it afterwards. That murderers may be caught and punished, he understands; that men disapprove of killing a kinsman, a sovereign, a guest, he understands. But to understand only so much and no more is to be deeply flawed. Macbeth's flaw is not a defective moral mechanism that gives in to his overweening ambition; it is rather his total lack of a moral mechanism.

Prudence, if nothing else, requires acknowledgement here that few students of the play have characterized Macbeth in these terms. Most have found him initially admirable—indeed, even moral—and only corrupted by forces that penetrate and corrode his innate goodness. Two broad categories encompass the major traditional and current views: first, a majority of critics have found Macbeth initially an 'essentially good' man who succumbs to strong temptation despite a deep moral repugnance; second, a smaller number of critics acknowledge that he reasons consciously on non-moral grounds, but at the same time insist that beneath this outward show lies a moral consciousness that the verbal arguments seek to hide or control—that Macbeth, in the key soliloquy, *tells* himself that only practical consequences such as detection and punishment need to be considered, while his true moral sense that lies within writhes in torment. What is common to these two views is the idea that Macbeth is a moral man. According to the first, he is a moral man who *does* argue on moral grounds against murder; according to the second he is a moral man who subdues his moral revulsion by pretending that other considerations are all that matter. The second view appears to have more to recommend it because it offers a subtle way around the hard evidence supplied by the key soliloquy, whereas the first view offers no way around at all except blindness to the words that Shakespeare wrote. After the murder has been done, the two views become a single one: each sees the tragedy as a study of moral decline, the fall of a man who was essentially good into the abyss of evil; further, each finds Macbeth, once the fall has begun, agonized by his conscience, aware of and appalled by the evil that he has done, and plunging into new acts of wickedness in a vain effort to smother the cries of his tormented conscience.

Again, lest the foregoing digest be perceived as a distortion of critical views, prudence suggests the need here of a random but liberal sampling of the precise words by which Macbeth and his tragedy have been characterized in standard criticism of the past two centuries; all the following quoted words and phrases are taken directly from this criticism. Before the much-noted 'moral decline' sets in (just after Duncan's murder), Macbeth is said to possess 'initial nobility', 'initial moral stature', 'every potentiality for goodness', 'native love of goodness', 'great goodness', 'essential goodness', 'potentiality for greatness'; he is an 'admirable man'. Before his decision to murder Duncan he is said to debate (i.e., in the key soliloquy) 'the moral issues involved' and to 'sacrifice his moral beliefs', finding the deed 'abhorrent to his better feelings'. What temporarily dissuades him is his realization of 'the hideous vileness of the deed'. Once the murder has been done, reference is made to his 'prior goodness', his 'prior virtue'; but now his 'inner being is convulsed by conscience', he has an agonizing 'consciousness of guilt', he is in 'the throes of remorse', his 'conscience is terrifyingly alive', he utters a 'conscience-stricken lament'. If it is somewhat lamely acknowledged that he never does 'advance from remorse to repentance', yet the idea most steadily insisted upon is that from the very first he 'was aware of the wickedness of his act'.

But surely it would be impossible to find, in the entire vocabulary of English, words less suited to the character of Macbeth than these;

if confronted with them, Macbeth himself would be incapable of understanding any of them. That he is a valiant warrior the text makes clear, as it does also that he recognizes the enormity of the risk involved in murder. But it is one thing to recognize how monstrous the act of murdering a king must appear to the state, and quite another thing, of a quite different order, to abhor murder as a moral wrong, regardless of whether it occasions any unpleasantness at all for the murderer. Macbeth is never, either at the outset, in the middle, or at the end of his career, 'aware of the wickedness of his act' *per se*. He is aware only that murdering a good king will stir up a tempest that will threaten his own security. It is the distinction between these understandings, his and ours, that makes the true tragic gulf between Macbeth's awareness and ours.

This gulf continues throughout the action and is unbridged at the end. But to say so much is not to say that Macbeth learns nothing from his experience. He eventually learns enough to overcome the two advantages, other than the moral one, that we gained over him during the early scenes—the advantage derived from our conditioned or inherited sense that, despite Lady Macbeth's easy assurances, he will not avoid justice, and the advantage derived from our expectation that the Witches, though they tell only truth, are not his friends. We begin to lose the first of these during the very scene of the murder, for Macbeth, as we learn when he emerges from the death chamber, there experienced the first disagreeable consequences of committing murder in the very act of committing it. These unpleasant effects came as a surprise to him, for they were of a kind that he had not foreseen in the soliloquy, when his imagination was busily conjuring up fearful ideas of bloody instructions returning to plague the inventor and poisonous ingredients being put to the lips of the poisoner. Lady Macbeth had then assured him that he could escape detection, and that if he could but escape that, he would escape all such undesirable effects of murder as his limited awareness could fathom. But in the very act of murder he heard voices saying that he would sleep no more, found that 'amen' stuck in his throat, and underwent such a mental and physical convulsion that he rushed from the chamber with the incriminating daggers in his hands. He is next agitated by the sound of knocking at the gate, shaken by the sight of blood on his hands, and ends the scene by crying, 'Wake Duncan with thy knocking! I would thou couldst!'

Thus his experience of the initial consequences has made him a wiser man already than he was in the soliloquy: he has learned that there are unpleasantnesses attendant upon murder besides those he had earlier feared. But we are not to confuse this new understanding with any such thing as a sense of moral wrong. His shouted wish that Duncan might be waked is no doubt sincere; but it signifies only his desire to halt the disagreeable sensations by which he is being visited at the moment. He is no more a moral man than before the soliloquy; he is only a man who has learned that murder produces frightening symptoms in the murderer.

During the action that follows, what he experiences is a painful extension of these same manifestations. He is unable to sleep—or, when he sleeps, is shaken by nightmares. He is driven by his terrors upon a career of violent acts, of which the murders of Banquo and Macduff's family are only the major examples; Scotland becomes a land 'Where sighs and groans and shrieks that rend the air / Are made, not mark'd'. Soon he finds himself envying the dead Duncan, who 'sleeps well'. His mind is 'full of scorpions'. Dread of the here-after, which once he was willing to 'jump', gnaws at him after all, for he believes that his 'immortal jewel' has been surrendered to hell's eternal punishment. In his final days he complains of his isolation, his loss of all the goods that should attend a man in his declining years: '. . . that which should accompany old age, / As honour, love, obedience, troops of friends, / I must not look to have'. He loses his wife to violent, untimely death—a casualty, ultimately, of that initial fault, Duncan's murder; and now it seems to him that life itself is but a mockery, '. . . a tale / Told by an idiot, full of sound and fury, / Signifying nothing'. And finally he loses his own life; his 'cursed head' is cut off, and there is none to speak one good word of him: he is only a 'dead butcher'. The seed of Banquo will flourish in Scotland after all, while his own soul lies forever in the toils of 'the common enemy of man'.

But though, through direct experience of them, Macbeth comes to awareness of these grim consequences of his act, he remains morally oblivious, as before; as a moral being he is neither better nor worse than when he debated the advisability of killing Duncan. Though he orders the deaths of Banquo and Macduff's family more easily than he decided to kill Duncan, it is not because he has suffered a moral decline that he can do so, for they were not moral considerations that formerly gave him pause. He becomes an increasingly worse menace to the state not because he has declined morally but because he wishes to hold on to what he gained by the initial murder. *Macbeth* is not the tragedy of a good man's moral deterioration, but of a man who lives and dies without knowing what moral sense means.

To this conclusion all the evidence of the play steadily points. To learn, as at last Macbeth does, that crime does not pay, and to learn only that, is not to gain a moral sense. To regret—because it did not bring him what his ambition coveted but instead brought only terror, loss, and finally death—that he committed the first murder is not to experience remorse, or anything like remorse. When Macbeth first voices the wish that Duncan were yet alive, he does so because he is experiencing sensations that frighten him: every noise appals him; he sees blood on his hands that he fears will stick forever. His agonized outbursts that rise to a crescendo thereafter, as action continues, are not the cries of conscience, but only responses to present and future dangers. Thus, for example, when he contemplates the murder of Banquo, he does so for two explicit reasons: 'to be safely thus', and to prevent Banquo's heirs from inheriting the crown. Bitterly he laments that

> For Banquo's issue have I fil'd my mind;
> For them the gracious Duncan have I murder'd;
> Put rancours in the vessel of my peace
> Only for them . . .
>
> (III. i. 65–8)

He has not gained all that he sought: Duncan's killing has brought him only discomforts that are intolerable unless Destiny can be altered and the crown secured to his own heirs.

The fact is that it is never once for the wrong of his past action, in itself, that Macbeth agonizes; it is always for the present and the future. Lady Macbeth, possessed of the capability of moral awareness, erroneously assumes that her husband feels what she herself feels and seeks vainly to console him: 'Things without all remedy / Should be without regard; what's done is done.' But his thoughts are not at all on 'what's done'. How grossly she has mistaken the cause of his torment is marked by his reply:

> We have scotch'd the snake, not kill'd it;
> She'll close and be herself, whilst our poor malice
> Remains in danger of her former tooth.
>
> (III. ii. 13–15)

At this point he has already ordered Banquo's death; but because she is ignorant of the fact, Lady Macbeth continues to suppose that his anguish is for Duncan's murder: 'You must leave this', she urges him. Again Macbeth bursts out in pain:

> O, full of scorpions is my mind, dear wife!
> Thou know'st that Banquo and his Fleance lives.
>
> (Ibid. 36–7)

Commentators who would be strictly honest should never separate this pair of lines, as they sometimes do in arguing the reality of Macbeth's 'conscience'. The phrase 'full of scorpions' by itself suggests the pricks of conscience: but these are not the scorpions of conscience. The pain they inflict is not for the dead Duncan at all, but for the living Banquo, and, worse, the living heir, Fleance. In the next instant, as we have already noted, having satisfied himself that his wife is with him in the decision to murder both, he is quite free of scorpions, playfully calls her his 'dearest chuck', and disgorges a burst of poetic images that are nothing if not euphoric. 'Thou marvell'st at my words,' he cries, when the flight of images has passed, 'but hold thee still; / Things bad begun make strong themselves by ill.' It is just here that the actor of Macbeth should exit dancing, kicking his heels together in the air.

Macbeth's euphoria continues into his next scene, when, greeting the banquet guests, he is as jovial as Capulet at the ball. Turning aside to speak with the Murderers, he is elated even more at their first words of success: Banquo has been taken off. But then, as suddenly again, he is stricken by news that Fleance has escaped: 'Then comes my fit again,' he cries,

> I had else been perfect,
> Whole as the marble, founded as the rock,
> As broad and general as the casing air;
> But now I am cabin'd, cribb'd, confin'd, bound in
> To saucy doubts and fears.
>
> (III. iv. 21–5)

And 'saucy doubts and fears' aptly describes the range of his torment, hitherto, now, and hereafter. His agitation proceeds from no such phenomenon as conscience; there are no pangs for either dead Duncan or dead Banquo: his 'fit' is only for the escape of Fleance. When he has convinced himself that there is no immediate danger from 'the worm that's fled', he is again jovial, and in his high good humour spices his chatter with a bold dash of irony: 'Here had we now our country's honour roof'd, / Were the grac'd person of our Banquo present.'

His high spirits are abruptly dampened by the appearance of the Ghost of Banquo, and as the scene goes on his agitation on being twice confronted by it in quick succession is the most violent since that he experienced during and just after Duncan's murder. That his agony here is real, intense, and deeply felt there is no question; but to confuse his fearsome paroxysms, here or elsewhere, with 'agonies of conscience' is to abuse the term. Whether the Ghost is real or imagined, a point often debated, is irrelevant: what matters is that its appearance must be laid to Macbeth's fear for himself in the present and the future, and to no such thing as remorse for past deeds. His first words to the Ghost, at its return, make plain what unnerves him: 'Thou canst not say I did it; never shake / Thy gory locks at me.' Not conscience but raw, superstitious terror shakes him. When Ghost and guests have gone, his fit continues, his incoherent chatter betraying not conscience but only terror for himself. But that the Ghost returned to affright him, he would no more have thought again of Banquo than of Duncan; he frets not for what has been, but only for what is and will be. His very next words to his wife, when they are alone, concern the man who, now that Banquo is dead, stands first in line as a threat to him: 'How say'st thou that Macduff denies his person / At our great bidding?' And he continues, 'For mine own good / All causes shall give way.' He feels no prick of conscience for those already slain; it is only to satisfy himself, for better or worse, about the future that he will immediately seek out the Witches.

As for the Witches, they know well enough what torments him. 'He knows thy thought,' the First Witch warns him: 'Hear his speech, but say thou nought.' When the First Apparition cautions him to beware Macduff, he replies, 'Thou hast harp'd my fear aright.' When the Second Apparition assures him that 'none of woman born' shall harm him, he first decides that Macduff shall live, then changes his mind: 'I'll make assurance double sure / And take a bond of fate'—for by doing so, he believes, he can 'tell pale-hearted fear it lies, / And sleep in spite of thunder'. The thunder that has disturbed his sleep is not, and has never been, for those already slain, but for those who threaten his future; with the Witches' final assurance that he will not be vanquished until Birnam wood comes to Dunsinane, his spirits once more soar high. But then he voices the thought that most torments him:

> ... my heart
> Throbs to know one thing: tell me, if your art
> Can tell so much, shall Banquo's issue ever
> Reign in this kingdom?
>
> (IV. i. 100–3)

On being shown that Banquo will indeed found a long succession of kings, he responds with ungoverned fury, berating the informants themselves—who are, after all, only revealers, not causers of events— as 'Filthy hags!'

Thus in his second interview with the Witches his thought and manner run true to form: he is concerned solely with his own prosperity, present and future. Neither before Duncan's murder, nor after it, nor before Banquo's murder, nor after it, nor in this final interview has a syllable that suggests 'essential goodness', or moral sense, or conscience escaped his lips. Nor do the remaining scenes of the tragedy show him in any better light.

After the interview, he resolves on the immediate slaughter of Macduff's family. Then we see him no more until V. iii, the first of the frantic final scenes, when, besieged, he is appalled by the advance of enemy forces, enraged by the revolt of his own subjects, distressed that his wife—just at this time—should be perilously stricken. He strives to cover his terror by raging at those about him and by reminding himself of the Witches' promise: 'Till Birnam wood remove to Dunsinane, / I cannot taint with fear.' It is here, too, that he laments, with a fresh surge of imagery, his lack of 'that which should accompany old age, / As honour, love, obedience, troops of friends'. But this outburst of self-pity, saturated with malice towards all those who deny him these comforts, is not to be confused with remorse or anything akin to remorse; at best he regrets what he has done because it has brought him to this lonely stand, and he blames others rather than himself even for that.

Macbeth's much-quoted expressions of concern, here and shortly after, for his wife's health seem, because of their eloquence, their golden sounds, to do him more credit as a humane being than do any other passages in the play; and, indeed, perhaps they do. But even here three considerations render his humanity suspect. First, his expressions of concern for his wife by no means dominate the scene, but take a minor place beside those that concern his own comfort and safety. Though the Doctor is present at the start of the scene, it is only after thirty-seven lines that Macbeth inquires 'How does your patient, doctor?' Then follows an intense speech in which he entreats the Doctor to cure her by any means—but in the next lines Lady Macbeth loses place to the thanes 'who fly from me' and to the advancing enemy, which, with a gross image, he wishes might be scoured from his ailing kingdom even as his wife's body might be purged. At the end of the scene his thought is exclusively for himself: 'I will not be afraid of death and bane, / Till Birnam forest come to Dunsinane.' Second, even his few lines that directly concern his wife's illness—in all, a mere thirteen in a scene of sixty-two—lack feeling for her primarily: what dominates is the idea that she has failed him just when his thanes fly from him, when he lacks 'troops of thanes'. Third, and most damaging to the idea that he is capable of moral sense, is the fact that none of his words hints that, for all his teeming imagination that can suggest every other sort of fancy to him, he conceives any connection at all between Duncan's murder and his wife's affliction. The Doctor outlines the general character of Lady Macbeth's illness, and for us, who know that conscience is killing her, the clues are blatant; but Macbeth cannot guess the truth because he is unacquainted with the idea of conscience. So he speaks only in general terms of 'a mind diseas'd', a 'rooted sorrow', and 'that perilous stuff / Which weighs upon the heart'. Like the merchant of Venice with his melancholy, he cannot conceive how his wife caught, found, or came by her affliction. Lamely, wide of the mark, he demands that the Doctor 'cast / The water of my land' in order to find the cause and effect her cure.

If this climactic scene offers no reason to contend that, even in his concern for his wife, Macbeth exhibits an 'essential goodness', neither does the scene which brings him word of her death. At the sudden cry of women, he asks what the noise is, then comments that night-shrieks no longer startle him—though we might have expected this particular cry, coming from the chamber where his wife lies stricken, to convey an instant significance and to elicit a special response, however much inured he has become to horrors. But he lumps it with any and all other cries: 'Direness', he asserts, 'cannot once start me.' When told that his wife is dead, he replies with the famously ambiguous 'She should have died hereafter'. Though one of the possible readings ('She would have had to die some time') appears to do him less discredit than another ('This is an

inconvenient time for that'), neither shows him capable of compassion for any other than himself. He feels his wife's death as an
added vexation placed unfairly on him at a bad moment, and cites it
as proof that all of life is an empty cheat. He has, in truth, no word at
all for his dead wife, even as he had no word for the dead Duncan or
the dead Banquo. He has no comprehension of what killed her; the
thought that the murder of Duncan might bear a causal relation to
her death is beyond him. His splendidly imaginative disquisition on
the utter meaningless of life, which follows, is no requiem for her;
for all its impressively moral sound, its sense is hollow and selfish.
It is only one more item to document his history of whining self-pity,
like his earlier lament that in the 'yellow leaf' of age he is denied
honour, love, obedience, and troops of friends, and, indeed, like all
the other utterances he has made when events annoyed him; it was
even so when he found that he could not say 'amen' at a moment
when a blessing would have been welcome.

Hemmed about, his ill-gotten kingdom falling about his ears, he
loses most of the smooth, golden poetic sound that formerly obscured
the brutish thought within; but harsh, vivid images fly from his
mouth in abundance: 'Upon the next tree shalt thou hang alive, /
Till famine cling thee'; 'They have tied me to a stake; I cannot fly, /
But bear-like I must fight the course'; 'Why should I play the
Roman fool, and die / On mine own sword? Whiles I see lives, the
gashes / Do better upon them.' Unlike other tragic heroes, he does
not seem to 'grow' in his final hour, not even in brutishness; but the
mask of glittering poetry has slipped away, disclosing more nakedly
the beast that has always lurked behind it. Neither, on the other
hand, is there any sudden genesis of a moral attitude that might, even
at the last moment, cause us to reassess his moral potential; during
these last scenes his thoughts are as they have always been—wholly
for himself: 'I gin to be aweary of the sun' is of a kind with the
earlier complaint that he lacked honour and friends, of a kind with
his declaration that life is a tale told by an idiot. His thought is for
himself, not for Duncan, or Banquo, or Macduff's family, or his
own wife. 'But get thee back,' he tells Macduff; 'my soul is too much
charg'd / With blood of thine already.' If this warning sounds com-
passionate, even moral, yet it has nothing of compassion or morality
in it: it merely expresses his fear that the eternal punishment he was
once willing to 'jump' may be all the more terrible if Macduff's
blood is added to his account. Last-moment fear of punishment

evinces no more moral awareness than any other fear evinces; the thought that, because it is too heavily weighted with blood, his soul may never make its flight to heaven is not a moral thought, but, like all his thoughts, a thought for himself.

Nor, finally, is there anything moral in the show of valour that he makes in his last moment. A demonstration of mere physical courage—if it were courage used to defend some noble person or principle, or even to defend one's own life against a wrongful blow—might argue, if ever so slightly, on the side of moral capability. But Macbeth fights only to save his wicked self from a well-deserved fate. What is more, the quality even of his physical courage is compromised. Before Birnam wood came to Dunsinane, he cheered himself with the Witches' assurance that he need fear nothing; after it came, he cheered himself with their assurance that none born of woman should harm him. Yet even while he sought to bolster himself with these assurances, he remained frantic with fear. When Macduff then strips away the last false shield, he makes his stand not, indeed, from courage, but from hatred and from fear of a fate worse than death: 'I will not yield, / To kiss the ground before young Malcolm's feet / And to be baited with the rabble's curse.' He flings himself at Macduff in hate, rage, and fear. If to do so is somehow more admirable than to flee, yet there is nothing moral in the act.

It is otherwise with Lady Macbeth, over whom we hold no such advantage as that we hold over her husband. It is remembrance of Duncan's murder that torments, sickens, and finally kills her. From the first, whereas Macbeth feared Duncan's murder only as an act that might bring repercussions dangerous to himself, Lady Macbeth recognized the moral wrong. Though she assured her husband that 'A little water clears us of this deed'—and may even, for the moment, have convinced herself that she spoke the truth—yet she also pleaded with the spirits to 'Stop up th' access and passage to remorse' in herself. Soon enough she finds that her entreaty was futile, for 'compunctious visitings of nature' prevent her from killing Duncan herself; and thereafter, though indeed a little water washes off the literal blood, the blood of guilt penetrates brain and heart and kills her. But the blood that Macbeth feared would incarnadine the seas was, for him, only literal blood, which he washes off with ease, then turns his attention to present fears. After the murder neither of them can sleep, but from different causes: she because of her sense of guilt, he because of fear for himself and his throne. If, with respect to sense of moral wrong, we hold any advantage over Lady Macbeth, it consists in our recognizing what she never comes to see: that her husband has none. Her sleepwalking scene evinces what she has suffered from during the days and nights that follow the initial murder: she has suffered from guilt, and would suffer neither more nor less if there were no Banquo or Macduff or any other threat to her royal security; and, finally, she would have died of it just the same had the thanes not revolted or the English forces besieged the castle. But her husband, if both Banquo and Fleance had died and the forces of justice had not hounded him, would have lived happily ever after, sleeping not 'in spite of thunder', but quite without any thunder. For the only thunder that he has heard since Duncan's murder has been that of his fears for himself and his throne.

It is, then, the gap between our moral awareness and Macbeth's moral obliviousness that, after initial, lesser gaps have closed, remains open and exploitable throughout the tragedy. But exploitation of this particular gap is no easy matter, for, having neither moral sense nor awareness of its existence, Macbeth cannot logically be made to speak about it even in scorn. In his obliviousness he exceeds even Antonio, the remorseless villain of *The Tempest*:

> *Seb.* But, for your conscience?
> *Ant.* Ay, sir, where lies that? If 'twere a kibe,
> 'Twould put me to my slipper; but I feel not
> This deity in my bosom. Twenty consciences,
> That stand 'twixt me and Milan, candied be they
> And melt ere they molest.
> (*Temp.* II. i. 275–80)

Antonio at least knows enough to scoff at conscience; he has heard of this phenomenon and its alleged force even though he has had no direct experience of it himself. But Shakespeare cannot make Macbeth even allude to conscience for the purpose of scoffing at it— and indeed no line ever suggests that this villain-hero is acquainted with the idea of conscience even as a hypothetical commodity. Had he heard of it, but regarded it as mythical, we might have expected him to raise it as a possible argument against killing Duncan, and then to reject it summarily, just as he does the argument of eternal punishment: 'We'd jump the life to come'. But he does not raise the issue at all.

Shakespeare easily exploits the gap between Othello's ignorance and our knowledge of Iago's dishonesty, needing only to make Othello interject, at a moment of crisis, 'Honest Iago!'—and the gap is exploited. But since in Macbeth he means to represent a mind that is too unacquainted with conscience to make any reference to it, he must exploit the gap indirectly. He does so not by having Macbeth flout conscience but by having him itemize the various inferior substitutes, conspicuously omitting everything with a moral connotation. Thus, before the murder, he debates the advisability of murder but raises only the expedient objections to it; then, after the murder, he begins to suffer, and suffers with rising intensity thereafter until his death—but for all the wrong reasons: he fears detection and punishment, fears for his 'immortal jewel', fears for his throne, fears Banquo and Fleance, fears Banquo's Ghost and Fleance's flight, fears Birnam wood, fears Macduff. Each time he is made to express the wrong reasons for not committing murder, each time he is shown in a paroxysm of fear and frustration rather than—like Lady Macbeth—in an agony of conscience, the gap between his awareness and ours of moral right and wrong is exploited obliquely.

As the end nears, Shakespeare uses Macbeth increasingly in the capacity of the moralist, whose example serves as a warning: 'Look at what has happened to me because of my wicked acts,' he seems to say, 'and judge whether you would choose to do as I have done.' Other persons, too, are used to point up his warning function: 'Now does he feel / His secret murders sticking on his hand,' remarks Lennox, and continues,

> Those he commands move only in command,
> Nothing in love. Now does he feel his title
> Hang loose about him, like a giant's robe
> Upon a dwarfish thief.
>
> (v. ii. 19–22)

Pointedly, Macbeth advises that one like him must not look to have honour, love, obedience, troops of friends, but must expect to end his days joylessly. He laments his loss of normal feelings; because he has 'supp'd full with horrors', his senses are dulled. 'Look at me,' he seems to say again: 'to such a one as I have become, life is only a tale told by an idiot, signifying nothing.' And, finally, learning that Macduff was from his mother's womb 'Untimely ripp'd', he figuratively shakes a finger at us in direct warning:

And be these juggling fiends no more believ'd
That palter with us in a double sense,
That keep the word of promise to our ear,
And break it to our hope.

(v. viii. 19–22)

These and other palpably didactic speeches express the sum of what he has learned from his experience. Were he to speak the Epilogue, he would tell us not to kill a good man, especially one who is our king, kinsman, and guest, because if we do so we shall certainly be caught and punished, first on earth and afterwards eternally. 'Crime', he would admonish us, 'does not pay.'

Macbeth is the tragedy of one whose experience taught him nothing better to teach us than this shallow admonition. But it is only his admonition, not Shakespeare's; to confuse Macbeth's 'lesson' with Shakespeare's would be to cheapen both the tragedy and ourselves, for it would mean setting our own moral level, and Shakespeare's, at Macbeth's level. To imagine that Shakespeare, like Macbeth, made no distinction between moral sense and the mere expedient arguments that Macbeth's limited capability was able to bring against murder is of course absurd, for consciousness of conscience is everywhere in the plays, from first to last. The consciences of Henry IV, Henry V, and Henry VI are real enough, as are those even of villains like Claudius, Edmund, and Richard III. In *Measure for Measure*, a year or two before *Macbeth*, the disguised Duke anatomizes moral sense in contradistinction to any inferior substitute as he interrogates pregnant Juliet:

Jul. I do confess it, and repent it, father.
Duke. 'Tis meet so, daughter, but lest you do repent
As that the sin hath brought you to this shame,—
Which sorrow is always towards ourselves, not heaven,
Showing we would not spare heaven as we love it,
But as we stand in fear,—
Jul. I do repent me, as it is an evil,
And take the shame with joy.
Duke. There rest.

(*Meas.* II. iii. 29–36)

Macbeth's sorrow is indeed always towards himself, not heaven, and indeed he would not spare heaven as he loves it, but as he stands in fear; and fear for oneself is not conscience.

Macbeth's tragedy is that of one who, because his imagination can

supply only shoddy arguments against wrong, succumbs to temptation and never comes to recognize the true error of his ways. After
killing Duncan he is denied even such leavening anguish as that
which Claudius, for example, suffers; instead, he endures only the
meaner torment of one who has stolen what he sought and fears
losing it. Our hardest question, at last, is how it can be that *Macbeth*
has so regularly been characterized as the tragedy of an 'essentially
good man' whose principles give way to overmastering ambition
and who thereafter undergoes moral deterioration, experiencing all
the while those agonies of conscience that, indeed, only an essentially
good man can experience. This pervasive view, which acknowledges
no gap between Macbeth's moral awareness and our own, and which
does not distinguish between the prick of conscience and mere fear of
losing what murder has won, is contradicted, as the foregoing pages
have shown, by all the evidence of Shakespeare's text.

Two explanations suggest themselves. The first has to do with the
quality of the poetry that Shakespeare gives Macbeth to speak; the
second concerns our possibly unconscious zeal to prove Macbeth's
right to the high title of 'tragic hero'.

It is needless to cite here a wide representation of passages that
evince Macbeth's extraordinary poetic gift, which has long been
acknowledged and much admired. Macbeth's language is everywhere richly imaginative, vivid in the extreme. It teems with figures
of every sort and with ideas that erupt, stage pyrotechnical displays,
then yield to new eruptions—all with such rapidity that the effects
dazzle eye, ear, and mind alike. Though indeed Macbeth grows in
poetic power as his experience of life continues, yet he is prodigiously
endowed at the very start. Nor is his habit of poeticizing everything
confined to major, extended passages; it is indulged as startlingly in
quick, passing phrases and half-line darts as well. He sees, feels,
thinks, and speaks habitually as a poet—not, indeed, like one who
self-consciously coins clever phrases and devises extravagant
conceits, or like one who, as Richard II does, narcissistically
exploits his precious lyric voice for the beauty of its sound, but like
one who is literally possessed by a poetic demon, who could never
speak dully if he tried. His is relentlessly, from start to finish, the
imagination of that very *sauvage ivre* that Voltaire improperly
charged Shakespeare with being; but of course this drunken savage
is not Shakespeare himself, but his creation, whose perfervid
imagination is integral to the creation. Only once does Macbeth

deliberately call on his imagination to produce what he thinks he needs at the moment—in the flashy, glittery, false-sounding 'Here lay Duncan' speech; elsewhere his utterances wildly spill from his intoxicated brain because it is filled to overflowing. His imagination is not under his control; he is its creature. It drives him not only to pour out striking images and splendid sounds in profusion but to see visions, hear voices, conceive bizarre and inexplicable ideas: 'Had I three ears I'd hear thee.' Had not Banquo also seen the Witches, we might be tempted to assume that they and their prophecies were merely bodied forth by Macbeth's fantastically active imagination.

But for all its charm, Macbeth's poetic power does not fool Macduff, who is not even 'in' on his secret, as we are; and perhaps it should never have deceived students of the play. Yet there can be no question but that Macbeth's spellbinding power of language has contributed much to the deception of generations of critics who have found him better than he is—to be, indeed, possessed of moral sense and inner goodness that in plain fact he lacks. Totally self-centred, judging all that lies in past, present, and future in terms of consequence to himself only, he has thrust his sensations upon us in such dazzling images that we have been hoodwinked; so finely are some of his most reprehensible utterances dressed that they have passed for moral sentiments that proclaim the speaker to be nothing if not 'essentially good'. For many generations young students were required to memorize such passages as 'If it were done when 'tis done', 'Is this a dagger', and 'To-morrow, and to-morrow, and to-morrow' and to intone them solemnly as though they were moral profundities. College students, asked to paraphrase the first of these passages in plain prose, often get all the words right, but then, being asked to state on what grounds Macbeth argues himself out of killing Duncan, reply in chorus: 'On moral grounds; he cannot bring himself to kill a good old man who is his king, guest, and kinsman.' So awesome is the tone of the soliloquy that it has bewitched the ablest readers: nothing that sounds so nobly reasoned can be anything but noble. But of course what the dramatist made his villain-hero actually say here is neither noble nor moral: 'I dare not kill this good old king, for if I do I shall have to pay for it.'

And so it is throughout the tragedy. When Macbeth delivers his brief apostrophe to sleep 'that knits up the ravell'd sleeve of care', his images are so golden that he seems a good man merely by uttering them. Much later, complaining of being denied the usual joys and comforts of age, he whines so becomingly that his sentiments seem nothing if not moral. Yet these ingratiating words do not say 'Do not kill, for it is wrong to do so.' They say only, 'Do not kill, for you will be unable to sleep and will be deprived of benefits that should be yours.' When he sums up his experience of life in the magnificent 'To-morrow, and to-morrow, and to-morrow', he does so with such grace that the passage is regularly cited among Shakespeare's noblest sentiments; and yet in fact Macbeth here ignobly blames life itself as a cheat when it is only he who has cheated. Finally, when he confronts Macduff and refuses to fight because 'My soul is too much charg'd / With blood of thine already', his wording makes the thought seem no less than magnanimous; but in fact the thought is not at all noble: 'It will be all the rougher with me in the hereafter if I kill you in addition to your wife and children.'

The poetry that Macbeth speaks, thus, is everywhere deceptive, its very sound making the speaker seem better than he is; poetry is his mask, like Iago's mask of honesty. But Macbeth wears his mask with a difference, and it is the only such mask in Shakespeare. Claudius and Iago, to take the immediate contrasts, use their masks to deceive other participants: the 'smiling, damned villain' deceives all Denmark, and 'honest Iago' deceives all Venice and Cyprus. But their masks do not hide Claudius and Iago from *us*; while other participants see only the masks, Shakespeare unfailingly exhibits their naked countenances to us. But Macbeth constantly holds his glittering mask of poetry between his true character and us—and the history of *Macbeth* criticism leaves no doubt that he has done a better job of deceiving the most discriminating among us than of deceiving his peers in the tragedy.

Macbeth himself, of course, can have no possible interest in deceiving us. Hence all the deception must be a practice devised by the dramatist; and if we now ask why Shakespeare should have wished to deceive us by equipping his protagonist with such a blinding gift of words that we should mistake his character utterly, we come to the edge of our second reason why critics, in spite of the text's unequivocal evidence, have, almost universally, pronounced Macbeth an 'essentially good' man whose morality suffers corrosion. By means of the witchcraft of poetry, Shakespeare has managed to foist upon us a protagonist who, as 'tragic hero', is a cheat and a fraud; and, conscious of the dignity that attends the high title of 'tragic hero', critics have been more than willing, actually eager, to be seduced. Whatever, in recent years, we have found to balk at in Aristotle's definition of the tragic hero, we have not balked at one self-evident point: if his action and his disaster are to move us with the effects of true tragedy, the tragic hero must not be beyond the pale of sympathy—so bad that our sympathy is denied him utterly. And precisely there is the rub: how can we sympathize with one who is incapable of moral reasoning, immune to moral feeling, unacquainted with conscience, remorse, repentance, even as empty words?

Yet to deny *Macbeth* the name of tragedy and Macbeth the title of tragic hero is obviously unthinkable: the work has to stand as an absolute masterpiece even in Shakespearian competition. Confronted with a critical dilemma, we have eagerly taken the way that the dramatist, perpetrating for once his own monumental practice upon us, has seductively invited; we have harkened to the sounds of spellbinding poetry and have bestowed on Macbeth the requisite moral sense and agony of conscience that any proper tragic hero must have. Though Shakespeare assiduously denied his villain-hero a single moral impulse, a single thread of moral restraint, a single pang of conscience, a single sigh of repentance, we have done our best to upgrade him with substitutes, citing the Sergeant's stirring account of his bravery in battle, noting that his hair rose at the first thought of murder, repeating Duncan's glowing (and quite mistaken) early praise, citing his 'desolation' at his wife's death; we have even slandered Lady Macbeth, making *her* the one who knows no moral restraint.

With a single stroke like Claudius's 'How smart a lash that speech doth give my conscience!' Shakespeare could have opened the way to a view of Macbeth as capable of moral feeling. But he made no such stroke, and this hero never deviates into a moral thought. Since the dramatist evidently proceeded deliberately to prevent even a momentary lapse, we do violence to the play by asserting that its tragic hero is moral and therefore worthy of the title. The main problem here is with sympathy as the due of a tragic hero. If we do manage to sympathize with Macbeth, it may be either because we have been misled about his true character or because we recognize him for what he is, a moral cripple, and pity him for his lack of the one attribute that could save him.

Perhaps Shakespeare wagered with himself that he could create a brute with no moral sense and no virtue but the physical, exhibit him in a succession of murders, and yet bless him with such a gift of poetry that we would be moved to sympathy for 'a great soul in torment'. The alternative is to suppose that, though perpetrating a practice on us, Shakespeare meant us to see through it, see his villain-hero for what he is behind his poetic mask—*and, even so, extend some measure of sympathy*. Perhaps what we should hear, above all, in Macbeth's poetry, is the reminder of his tragic lack, for this poetry is the dramatist's primary means of exploiting the gap between our secure moral sense and Macbeth's moral obliviousness. In this way, then, the poetry does indeed serve to move our sympathy, not by masking the moral void, but by magnifying it. Once we have become aware of Macbeth's fatal deficiency, the very brilliance of his language continues to illuminate it, for his imagination supplies his tongue with everything to say except the one right thing, which he dies without ever learning.

In the end he has learned no more than that the only consequences he knew to fear at the beginning are indeed to be feared and that evil spirits are bad counsellors because in following them one comes to disaster. To extend him, on these terms, a measure of the pity due a tragic hero is doubtless harder than to extend it on the usual terms; hence Macbeth is likely to gain far less of our sympathy if we take him for what he is, a moral idiot, than if we take him for what his poetry has so long deceived us into thinking him, a moral man gone wrong. Sympathy for the latter is easy because it is given to one more or less like ourselves, who might also, under certain conditions, succumb, act wrongfully, and suffer the special torments that afflict moral beings who have done wrong. But to sympathize with Macbeth as a moral idiot is to sympathize with one unlike ourselves, powerless to become like us, doomed by lacking what we take for granted. Surely it entails no dimunition of the tragedy if we see Macbeth so. Shakespeare created no other tragic hero who so much tests our capacity for compassion; he does not even, at the last gasp, make it easier for us by expending a kind word or two on his villain-hero's behalf. The deaths of other tragic heroes are accompanied, or immediately followed, by showers of gracious praise—'Good-night, sweet prince, / And flights of angels sing thee to thy rest!'—calculated to stimulate a fresh, final surge of emotion in our breasts; Macbeth's death, quite deservedly, is celebrated only with Macduff's jubilant shout: 'Behold where stands / Th' usurper's cursed head.' To see him only so, and yet to extend a measure of compassion: this is the true burden the tragedy imposes.

CHAPTER VIII

The High Professionals: *Antony and Cleopatra*

'... his quails ever
Beat mine, inhoop'd, at odds.'

CONSIDERED from any of many possible points of view, *Antony and Cleopatra* represents a drastic departure from the dramatic methods of the preceding tragedies. Prior to this change, the most severe shift in the line occurred between *Romeo and Juliet* and *Julius Caesar*, when external fate or circumstance yielded to human character as the principal agent of catastrophe. *Antony and Cleopatra* brings nothing new in this respect; like its predecessors since *Romeo and Juliet* and like *Coriolanus* afterwards, it is nothing if not a tragedy of character. Yet the change in method is conspicuous, and so pervasive as to rival the earlier shift.

In part—though this difference is only incidental to our special concern in these pages—what we notice in *Antony and Cleopatra* is an essentially static situation rather than the usual action. In all the earlier tragedies, action 'goes somewhere'; the heroes undertake some challenge which, for better or worse, they eventually meet head on, and die. Brutus, moved by causes both private and public, assassinates Caesar; his act tumbles him into a flood of subsequent events, which flood sweeps him to defeat and death. Hamlet accepts the duty of removing his usurper uncle in order to restore health to the state; the greater part of the tragedy exhibits his tortuous progress towards achievement of this task, which achievement occasions also his own death. Othello, more acted upon than acting, is manœuvred along an unhappy course from joyous bridegroom to wife-killer. Lear absurdly gives away his kingdom to the wrong beneficiaries, endures a long, dark night of the soul, and dies a purged and better man. Macbeth kills his king, strives to avoid the unpleasant consequences by additional murders, and finally dies the death of a mad dog. Though the routes of these heroes are dissimilar, they have in common a definite forward movement; situations arise, are met, yield to new ones, and so on until the final one, which is always far removed from the first. Much *happens* between the time Hamlet seals an oath to avenge his father's death and the moment of his triumph; between the time Lear casts off his kingdom and the time he holds dead Cordelia in his arms; between the time Othello eloquently describes his wooing of Desdemona and the time he kills her; between the time Macbeth chooses to kill Duncan and the time Macduff kills him.

In contrast, *Antony and Cleopatra* appears to end almost where it began. Forty scenes stand between I. i and v. ii, yet do not essentially alter the initial situation: there appears no compelling reason why Octavius should not have invaded Egypt immediately after I. i and achieved the same débâcle that he accomplishes at the end of v. ii. In the opening scene he recognizes Antony's dotage as complete, and the time for destroying him is no riper five acts later. Action during the intervening scenes has not advanced, but, 'Like to a vagabond flag upon the stream', has only gone 'to and fro, lackeying the varying tide'. Instead of the usual action, these scenes are devoted to what might be called 'illustration' of character and situation—a dramatic phenomenon that sharply distinguishes the play from its predecessors and relates it more closely to its successors; *Timon*, in particular, is given almost wholly to illustration in place of action, and, though obviously action occurs in *Coriolanus*, yet many of its scenes, too, are largely illustrative.

During Act I, while Antony remains in Egypt, successive scenes illustrate his Alexandrian life, with its restless tug-of-war between pleasure and duty. Throughout Act II and the first part of Act III, when he is in Rome and Athens, some scenes illustrate his relations with Caesar and the affairs of Empire, while others represent Cleopatra marking time during the great void of his absence. Before the middle of Act III, Antony is again in Egypt, and the remaining scenes illustrate his trials of love and war, with successive choppy incidents signifying the manœuvrings of armies and fleets, the fluctuations in his wasted fortunes as he deals now with Octavius, now with Cleopatra, and, as we shall note, always with himself. But for a slow drift towards disaster, the basic situation remains unchanged: instead of action and plot development, we witness repeated illustration of the initial situation—a fallen, enslaved

Antony, always tormented by the image of what he was yet powerless not to be what he has become. So the situation stands until death ends it.

To say so much of the tragedy is by no means to disparage it, but only to mark a basic way in which it differs from its predecessors.

But the play differs also in another way that is our primary interest here. *Antony and Cleopatra* is the only tragedy in which Shakespeare seems actually to prefer inexplicitness to explicitness in his management of awarenesses. Where elsewhere he takes pains to advise us beforehand about important actions so as to exploit them for such effects as they can be made to yield, here he shows no clear inclination to do so; instead of regularly enabling us to hold advantage over participants, here he often allows them to hold advantage over us. As much as any other, it is this drastic change in method that differentiates the tragedy from its predecessors, and because, in this and other ways, the play confronts us with peculiarly difficult problems, it seems wise to approach it somewhat differently. We shall first briefly survey the instances of discrepant awarenesses, then review our first view.

1. *First View*

Only eleven of the forty-two scenes set our awareness above that of participants—a low proportion that contrasts sharply with that in *Romeo and Juliet*, where all participants stand below our vantage point in all scenes, and that in *Othello*, where we hold advantage in all but the opening scene. Even so, it is not the low proportion alone that marks the change of method, but the fact that in none of these eleven scenes is our advantage of much importance; in most it exists only incidentally and briefly, often during only part of the scene. In no scene does Shakespeare seem to have taken pains to prepare a discrepancy with the primary purpose of creating significant effect through its exploitation.

The first gap between awarenesses occurs in I. iii after we have learned two momentous facts in I. ii: that Fulvia is dead and that Antony will return to Rome. Knowing neither fact, Cleopatra comes seeking Antony, quickly divines that he means to leave her, and promptly begins using the feminine wiles about which she had just been arguing with Charmian. Plying them in earnest, to hold her man, she will not let Antony speak, and thus a usable gap exists between her awareness and ours for about forty lines until he manages to tell her of Fulvia's death. But Shakespeare makes no use of the gap as a principal source of effect; rather, that source is the very exhibition of Cleopatra's wiles. Her unawareness only gives occasion for this exhibition, for she would otherwise have had no need to sting Antony with such barbs:

> What says the married woman? You may go.
> Would she had never given you leave to come!
> Let her not say 'tis I that keep you here;
> I have no power upon you; hers you are.
>
> (I. iii. 20–3)

So spectacular is her display of real or simulated fury that it obliterates our sense of the main fact of which she is unaware; and then, when she has learned of Fulvia's death, her continuing performance is again the source of dramatic effect:

> O most false love!
> Where be the sacred vials thou shouldst fill
> With sorrowful water? Now I see, I see,
> In Fulvia's death, how mine receiv'd shall be.
>
> (Ibid. 62–5)

Given such a performer as Cleopatra, perhaps Shakespeare recognized that he no longer needed an exploitable gap in order to create effects.

The second instance of a minor gap occurs when, in I. iv, directly after we have learned that Antony will return to Rome, Octavius and Lepidus meet to assess the immediate dangers to the Empire: Pompey's popularity is rising; he is 'strong at sea'; the 'famous pirates', Menas and Menecrates, 'Makes the sea serve them, which they ear and wound / With keels of every kind'. The need for Antony is urgent, and the effective emphasis of the scene is rather upon the urgency than upon Caesar's unawareness that he is even now homeward bound. Caesar plainly relishes the Empire's perilous situation as an occasion to assail the errant Antony for neglect of duty; he begins the conversation with a caustic account of Antony's revelry in Alexandria, while Lepidus feebly seeks to mitigate Antony's fault and is chided like a schoolboy: 'You are too indulgent.' Three speeches by Caesar simultaneously excoriate Antony and plead for his sudden return: 'Let his shames quickly / Drive him to Rome.' Though throughout the scene the two remain ignorant of what we

know from the preceding scene, that Antony will be on hand to bear
his part against Pompey, Shakespeare, concentrating this time on
exhibiting the overwhelming presence of Caesar, lets the exploitable
gap lie unused.

Next, in II. i, it is briefly the turn of Pompey, with Menecrates and
Menas, to stand below our vantage point: 'Mark Antony / In Egypt
sits at dinner,' he knowingly assures his subordinates, 'and will
make / No wars without-doors.' Ignorant that Antony has already
embarked for Rome, he offers what is perhaps the play's finest
summary of the forces that, supposedly, will detain him forever
in Egypt:

> But all the charms of love,
> Salt Cleopatra, soften thy (wan'd) lip!
> Let witchcraft join with beauty, lust with both!
> Tie up the libertine in a field of feasts,
> Keep his brain fuming; Epicurean cooks
> Sharpen with cloyless sauce his appetite,
> That sleep and feeding may prorogue his honour
> Even till a Lethe'd dulness!

(II. i. 20–7)

Pompey's brief period of only indirectly exploited unawareness ends
with Varrius's announcement: 'Mark Antony is every hour in
Rome / Expected.'

Next, in II. v, just after we have learned of Antony's agreement to
marry Octavia, Cleopatra is shown in a state of unawareness during
a space of about forty lines. Restless, she toys with the notion of
engaging first in one and then another entertainment—music,
billiards, fishing—while her longing for Antony makes each unsatis-
fying. At line 23 a Messenger enters with the news known to us, but
it is not until line 60 that he manages to deliver it to Cleopatra:
'Madam, he's married to Octavia.' The intervening lines represent
perhaps the most concentrated brief exploitation in the play, for
here, almost uniquely in this play, Shakespeare makes the gap
between our awareness and Cleopatra's the central dramatic fact.
Apprehending at a glance that the Messenger bears ill news, she
frantically postpones its hearing even while she alternately berates
the Messenger—'I have a mind to strike thee ere thou speak'st'—
and rewards each tiny morsel of good news that she manages to
extort from him: 'Make thee a fortune from me.' Even so, the duration of this exploitation is so brief that in the total scene it takes
second place to the effects created once again by the display of

Cleopatra's emotional reactions on learning, at last, of the marriage.

Next, in II. vii, Shakespeare pointedly, and properly, omits an opportunity for exploitation on a tremendous scale. This is the magnificent drinking scene with Caesar, Antony, and Lepidus aboard Pompey's ship. Midway through the scene, Menas draws his master aside and urges him to '. . . let me cut the cable; / And, when we are put off, fall to their throats. / All there is thine.' Had he wished, the dramatist might easily have introduced the scene with these lines even before the world's triumvirs came aboard; but in that event the glorious scene, acted under its black cloud, would have been totally changed. The effect would have been potent in the extreme, but quite irrelevant to the drama as a whole. Earlier in his career—say at any point up to *Macbeth*—Shakespeare might have been unable to resist the temptation; but now he confines the incident to a space of a few lines and abruptly ends it with Pompey's harsh order to Menas, 'Desist, and drink.' Perhaps no single incident better exemplifies the shift that *Antony and Cleopatra* represents in the succession of tragedies.

In II. iii we hear a flat pronouncement made by Antony: 'I will to Egypt; / And though I make this marriage for my peace, / I' th' East my pleasure lies.' Act III includes three scenes over which our private knowledge of this statement casts a shadow. The most telling exploitation occurs in III. ii, as Antony and Octavia prepare to leave Rome for Athens; says Caesar,

> Most noble Antony,
> Let not the piece of virtue which is set
> Betwixt us as the cement of our love,
> To keep it builded, be the ram to batter
> The fortress of it; for better might we
> Have lov'd without this mean, if on both parts
> This be not cherish'd.
>
> (III. ii. 27–33)

'You shall not find, / Though you be therein curious,' blandly states Antony, 'the least cause / For what you seem to fear.' Perhaps no line ever does him less credit than this, when we know that his intentions directly contradict his protestation. He flatly lies, and in the whole of the parting scene betrays no sign of his true intent, but seems, instead, honestly moved as he watches the parting of brother and sister—appears, even, to feel an instant of genuine affection for his new bride: 'The April's in her eyes; it is love's spring, / And these the showers to bring it on.' He embraces Caesar at parting, and but for the heavy emphasis of his earlier statement we could not guess that he means to betray them both.

Again in III. iv Shakespeare sets aside his usual way and lets us learn too late the facts that would have influenced our view of the scene. Here Antony castigates Caesar for wronging him by waging wars on Pompey and speaking slightingly of Antony himself. Torn between husband and brother, Octavia insists on returning to Rome in an effort at reconciliation. Here, as in the parting scene at Rome, no hint is given us of what Antony's mind truly holds; he represents himself as blameless for the rupture that has occurred, calls his wife 'Gentle Octavia', appears honestly grateful for her attempt at reconciliation, and is most generous with offers for the journey to Rome: 'Choose your own company, and command what cost / Your heart has mind to.' And during her absence, he tells her, he will raise 'the preparation of a war / Shall stain your brother' in the event that reconciliation fails. What Shakespeare holds from us is that Octavia's departure for Rome is Antony's cue for immediate departure to Egypt, his irrevocable break with Caesar and Empire. Had he followed his earlier way, he might have prefaced the parting of Antony and Octavia with a soliloquy in which Antony advised us that he would leave for Egypt as soon as his wife was aboard ship. In the very next scene, III. v, Enobarbus states that 'Our great navy's rigg'd', and Eros pointedly adds, 'For Italy and Caesar.' Antony is said to be walking in the garden, furious at the news that Caesar has made new wars on Pompey, then imprisoned Lepidus for treason. Ambiguous as it is, the brief scene still leaves us to suppose that Antony will do as he had promised his wife—wait in Athens until word comes of her success.

But then, in III. vi, we are subjected to such a shock as the dramatist's more typical method of dealing openly with us would have forbidden. Caesar opens the scene, speaking to his lieutenants:

> Contemning Rome, he has done all this and more
> In Alexandria. Here's the manner of 't:
> I' th' market place, on a tribunal silver'd,
> Cleopatra and himself in chairs of gold
> Were publicly enthron'd. At the feet sat
> Caesarion, whom they call my father's son,
> And all the unlawful issue that their lust
> Since then hath made between them. Unto her

> He gave the stablishment of Egypt; made her
> Of lower Syria, Cyprus, Lybia,
> Absolute queen.
>
> (III. vi. 1–11)

At no point in any other tragedy of Shakespeare's is our awareness so jolted, for the two immediately preceding scenes had made us suppose that Antony was still in Athens awaiting word from his wife. Shocking as it is to us, Caesar's speech gives us instant advantage over Octavia, arriving a moment later, unaware that her husband has reverted to Cleopatra. 'Where is he now?' asks Caesar; and his sister replies, 'My lord, in Athens.' In blunt words Caesar then tells her the truth; we have held advantage over her for a space of only twenty-five lines. Doubtless our image of Antony suffers severely during this period, both because of the harsh, unexpected account that Caesar has given of him and because of the pain that is brutally inflicted upon Octavia. Without our knowing it at the time, Antony had coldly lied to her at parting; learning only thereafter that he lied, we can hardly be expected to think as well of him as while we watched the tender parting scene under the impression that he was dealing honestly with his wife. Yet presumably Shakespeare counted on our thinking less harshly of him than had we known, at the time the parting was in progress, that he was lying. Though we then witnessed his perfidy, we did not know what we were witnessing.

Such finely exploitable situations as that of Octavia's arrival in Rome unaware that Antony has deserted her are not only infrequent in the play, but momentary and incidental as well. The fact of her unawareness creates a considerable passing effect but lacks significance for the course of the tragedy—though of course the desertion itself bears terrible significance because it offends Caesar; it was not idly that Shakespeare earlier emphasized Caesar's closeness to his sister and expressed repeatedly the ominous warning that the marriage might eventually bring about the reverse of the purpose for which it was intended. In contrast, Antony's unawareness in III. vii, the next case in point, though it creates little passing effect, is of first importance to the course of the tragedy. Here Enobarbus quarrels with Cleopatra over her presence at the scene of the approaching wars. Caesar, with incredible speed, has swept both land and sea to engulf Antony's powers at Actium. We ourselves, at the moment, lack means to know whether it is better for Antony to fight by land or by sea; hence we hold no advantage when he first announces his decision: 'Canidius, we / Will fight with him by sea.' Cleopatra, who has not been silenced by Enobarbus's blunt objection to her presence, agrees at once; but in quick succession Canidius, Enobarbus, and a common soldier tell him outright that he is wrong. But three times, in as quick succession, Antony rejects their advice: 'I'll fight by sea.' To Enobarbus's persuasive analysis Antony offers no rebuttal at all, but clings to his decision merely because '. . . he dares us to't'. Before the scene is half finished, we know beyond doubt that Antony's decision is an error.

But so ungeneral-like does Shakespeare make Antony appear here that question is inevitable: does the dramatist mean to imply that Antony, after all, is aware that the decision to fight at sea is wrong? Are we really expected to find it credible that Canidius, Enobarbus, and the soldier can readily perceive the truth whereas the great campaigner is blind to it? Are we to understand that Antony, though knowing his folly, nevertheless insists upon it because Cleopatra will have it so? Or is there yet some other and even more compelling reason for his determination to fight Caesar on Caesar's own terms despite the odds? The scene itself offers no clues to resolve these questions, which for the present we must leave suspended.

If the state of Antony's awareness is ambiguous here, it is more so when we next encounter a possible discrepancy. In III. xiii Antony becomes enraged when Thyreus takes liberties with Cleopatra's hand, has him whipped, and returns him to Caesar with insulting words. But, indeed, ambiguity persists throughout the scene. Cleopatra receives Thyreus, who says that Caesar knows 'that you embrace not Antony / As you did love, but as you fear'd him', and therefore pities the scars upon her honour 'as constrained blemishes, / Not as deserv'd'. To this she replies,

> He is a god and knows
> What is most right. Mine honour was not yielded,
> But conquer'd merely.
>
> (III. xiii. 60–2)

Knowing Cleopatra, we cannot doubt that her sentence is loaded with irony; and yet, inexplicably, Enobarbus—who should know her better than we—quickly speaks aside:

> To be sure of that,
> I will ask Antony. Sir, sir, thou art so leaky

> That we must leave thee to thy sinking, for
> Thy dearest quit thee.
>
> (Ibid. 62–5)

The first sentence appears like typical Enobarbian irony; but the second evinces that Enobarbus believes Cleopatra to have left Antony like the rat that leaves the sinking ship. Enobarbus goes out, and shortly returns with Antony—to whom he has evidently communicated his impression as fact; for when Antony surprises Thyreus kissing Cleopatra's hand he first orders him whipped, then turns his fury on Cleopatra, calling her viler names than any used against her in Rome, charging her brutally with the basest of instincts:

> I found you as a morsel cold upon
> Dead Caesar's trencher; nay, you were a fragment
> Of Cneius Pompey's; besides what hotter hours,
> Unregister'd in vulgar fame, you have
> Luxuriously pick'd out; for, I am sure,
> Though you can guess what temperance should be,
> You know not what it is.
>
> (Ibid. 116–22)

Four speeches, each of some length, make up his tirade, the savagery of which suggests that it marks an eruption of feelings lying just below the surface of his daily manner; the tongue-lashing that he administers, compact of vile terms, gross images, and venomous emotion, coming suddenly as it does, can hardly be understood but as just such an eruption of a long-festering sore. When Antony pauses at last, after nearly a hundred lines of name-calling, Cleopatra responds with a reprimand so mild that it should freeze him in his tracks: 'Not know me yet?' After this she speaks a few soothing words, and Antony says 'I am satisfied.' Therewith he abruptly changes the subject, and the scene resumes as though the incident of Thyreus had never been.

But it is just here that Enobarbus says, 'I will seek / Some means to leave him', and, as a whole, the scene appears to mark the point at which Antony has slipped the whole way down—a judgement confirmed in the next brief scene, when Caesar, reporting the treatment that was accorded Thyreus, derides Antony's chiding 'as he had power / To beat me out of Egypt', off-handedly refers to 'the old ruffian', and ends with a word of contempt thinly coated with pity: 'Poor Antony.'

The dozen scenes that stand between IV. i and IV. xiv include none

that are relevant here. Though Scene vii exhibits Antony, with
Scarus, in a brief victory, and though Scene viii shows the old
campaigner so jubilant that he gives Scarus Cleopatra's hand to kiss,
yet we are not to suppose that we hold advantage over him in our
awareness of what the ultimate outcome must be. If he displays the
same swaggering manner earlier remarked by Enobarbus—'Now
he'll outstare the lightning'—yet neither he nor Cleopatra nor any
other who appears in these scenes is truly deceived. Our own
expectations, of course, have long since become certainties: if we
did not know from history how the combat of Antony with Caesar
is to end, we would know it from the uncommonly full and varied
prophetic signs given us by the dramatist. These indications and
manifestations are expressed not only through deserters like
Enobarbus and Canidius, but by Antony and Cleopatra also, as in
IV. ii, when, shamelessly, Antony so manipulates the emotions of his
servants as to make them weep. They are expressed also through the
most obvious of dramatic devices, as in IV. iii, when the soldiers on
guard hear supernatural noises that they identify as the sounds of
Hercules abandoning Antony to his fate. But perhaps most of all our
sense of the inevitable is deepened by a steadily increasing emphasis
on the idea of Caesar's invincibility.

To imagine, then, that Antony is deceived by his brief victory is
to suppose him as naïve as Brutus, as blind to his fate as is Othello to
Iago. But Antony, whatever his faults, shares nothing of the blindness of these protagonists; like Hamlet, he is a thoroughly aware hero,
and we can never claim to know more about his fate than he himself
knows. With the first battle he has gained a momentary reprieve,
earned yet 'one other gaudy night', but he is not deceived. Returning
to the palace from the field, he swells with seeming confidence:

> Had our great palace the capacity
> To camp this host, we all would sup together
> And drink carouses to the next day's fate,
> Which promises royal peril. Trumpeters,
> With brazen din blast you the city's ear,
> Make mingle with our rattling tabourines,
> That heaven and earth may strike their sounds together,
> Applauding our approach.

(IV. viii. 32–9)

But by its very excess, his show of confidence betrays him; to
borrow Romeo's words, it is only 'the lightening before death',
which Antony recognizes as clearly as we.

We hold, in fact, no exploitable advantage over him until, in IV. xiv, Mardian brings the false news that Cleopatra has killed herself. Though this word precipitates his death, it rather hastens than actually causes it, as the false word of Juliet's death causes Romeo's. In IV. xii, having witnessed the defection of his fleet, he told us that he must die: 'O sun, thy uprise shall I see no more: / Fortune and Antony part here; even here / Do we shake hands.' Further, at the beginning of the present scene, prior to Mardian's arrival with the false news, he had sought to console the weeping Eros: 'Nay, weep not, gentle Eros; there is left us / Ourselves to end ourselves.' He would die in any event, by his own or Eros's hand—but not before he had slain Cleopatra. In IV. xii, after the defection of the fleet, he asserted that he would be 'reveng'd upon my charm'; then, when Cleopatra entered, he drove her off with a furious account of the fate that he wished for her at Caesar's hands. But after she had run away in terror, he then returned to his former vow: 'The witch shall die. . . . She dies for't. . . . She hath betray'd me and shall die the death.'

On learning of Cleopatra's death, Antony appears to forget his rage suddenly and totally; and then, on learning—having just given himself a mortal wound—that he has been hoaxed, he appears to bear no grudge, says no unkind word, but quietly demands to be taken to her. Twice earlier Shakespeare had exhibited a similar phenomenon: first, after Cleopatra's flight at Actium and Antony's ignominious pursuit, he had been angry enough to kill her—then had abruptly relented at her mere plea for pardon; next, in III. xiii, after the hand-kissing incident with Thyreus, when he had reviled her with phrases unfit for the basest strumpet, Cleopatra cooled his wrath with a single speech: 'I am satisfied,' he then said, and turned to other matters. Thus merely sending word that she has slain herself proves as effective in winning Antony back as actually killing herself; he appears to accept the token as proof once more of her fidelity. But we shall soon need to look again at these curious occasions of Antony's towering rage followed by abrupt and seemingly complete forgiveness.

Very briefly and unimportantly in V. i Caesar is unaware of Antony's death, Dercetas breaking the news to him after a few lines at start of the scene. Except for the Guardsman who unwittingly allows the basket of asps to be carried in to Cleopatra, no person stands in a condition of exploitable unawareness during Act V.

Nevertheless, all the closing scenes continue to exploit what might be called the conflict of 'rival awarenesses' that is the characteristic mode of *Antony and Cleopatra* throughout. In representing this special kind of conflict, Shakespeare quite rightly chose to be more inexplicit than explicit. In the preceding pages we have been concerned with the relatively explicit; we now re-survey the play, this time with an eye to the inexplicit and the game of rival awarenesses.

2. *Second View*

Shakespeare did not, of course, suddenly invent this game as a dramatic replacement, in *Antony and Cleopatra*, for the familiar structure of unequal awarenesses. The game of wits between equals or near equals, which is closely akin to this conflict, is a feature of all the comedies and, at one point or another, of the earlier tragedies—indeed, the histories, too. Often the game is purely verbal, and is played between clown and clown, clown and master, clown and heroine; between Beatrice and Benedick, Hal and Falstaff, Romeo and Mercutio, Claudius's pair of spies and Hamlet. The matches are so prevalent and so prominent as to need no review here. The object of the game is usually the same—to put the opponent down and to make great fun in the doing. More often than not the game of rival wits is played as an interlude, sometimes standing apart from the main issues of the plot, which we hold in mind while contestants engage in their sparring match. Of course the game is not always played for fun, nor is it always wholly verbal; that which runs throughout *Hamlet* between the 'mighty opposites' is a deadly version, and one that comes nearer to what we find in *Antony and Cleopatra*.

Yet it is not quite the same. If we grant that Hamlet and Claudius are fairly well matched intellectually, yet their actual awarenesses are not at all matched, for the Ghost's disclosure gives Hamlet advantage. What is more, we fully share Hamlet's advantage, having been present both when the Ghost reveals Claudius's guilt and when Hamlet states his intent to play the lunatic. Thus the game that runs throughout *Hamlet* actually takes place within the usual frame of unequal awarenesses that structures all the earlier tragedies. But that which runs throughout *Antony and Cleopatra* is set in no such frame. Neither Antony nor Caesar holds such advantage over the other as Hamlet over Claudius or Iago over Othello; neither, certainly, do we hold any such advantage over either as we hold over Othello; rather, we occupy the same level, for the most part, as the three persons whose involvement in the conflict comprises the heart of the drama.

What we have previously sketched—the dozen or so instances of exploitable gaps between the participants' and our awareness—reveals nothing that is either central or productive of great effect; if the dramatic interest of the play depended upon such effects, *Antony and Cleopatra* would be a dull chronicle indeed. What we shall now examine in detail is the conflict of awarenesses among the three great persons. These are highly sophisticated awarenesses, fiercely competitive, jealous of advantage. Further, the relative states of the awarenesses are never announced to us openly, as they regularly are in earlier plays, but are hidden behind the physical action, between the very lines that the characters speak, so that it is always difficult and often impossible to fix the exact status of one awareness in relation to another with which it competes. Where Shakespeare's usually abundant and explicit expository devices are applied at all in this play, they are used in connection with matters of minor importance and deliberately omitted in situations of great importance; indeed, in some crucial situations not only are true guides lacking, but deceptive ones are provided, often with heavy emphasis. In earlier plays, if Shakespeare may be said to have erred expositorily, it was on the side of too much explicitness; never did he dress a heroine in male disguise without letting us know beyond doubt; the fact of Iago's duplicity is so heavily emphasized that no dull spectator, though he alternately slept and waked, could possibly mistake it. We were made to know the exact state of each participant's awareness at any given moment, so that, with only rare exceptions, in the actions of plays that precede *Antony and Cleopatra*, had we chosen to do so, we might have tabulated all of the aspects of a situation as they were known to some and unknown to other participants.

In earlier plays we cannot mistake who is practiser, what his practice is, who his victims are; but in *Antony and Cleopatra* practices are invented and set in motion without announcement, proceed without reminders such as soliloquies and asides, and finally terminated without direct notice being given that they occurred. Though in this play the dramatist still employs his usual complement of practices, practisers, and practisees, he has so far deviated from the usual expository means of keeping us advised of these that we can easily survey the play without becoming aware of their presence. What is more, it is in the areas that lie about the core of the tragedy that Shakespeare has been most inexplicit.

We are thus confronted in the opening scene of the play by a problem of interpretation which, when it is resolved in one way, yields a drastically different view of the tragedy from that sketched in our preliminary survey. But neither in the theatre nor in the study do we immediately recognize even the existence of this problem, for the dramatist does not cue us either before the action begins or while it is under way; hence, for a considerable period, we must watch and hear without occasion to question whether what we see and hear is in fact what it seems. The opening scene uses a technique so familiar from the first plays onward that we are unlikely to suspect anything unusual. The first speaker, and our first informant, Philo, speaks thirteen lines of which the thesis is 'this dotage of our general's'. The erstwhile formidable captain has become a slave of passion. At the end of his speech Philo pointedly directs Demetrius— and us, of course—to mark the proof of what he speaks: 'Take but good note, and you shall see in him / The triple pillar of the world transform'd / Into a strumpet's fool. Behold and see.' The 'proof' is then set before us: the pillar and the strumpet enter, perform exactly as Philo foretold, and depart. When this initial illustration of Antony's 'dotage' has been completed, Demetrius restates its message as pointedly as Philo first stated it: 'I am full sorry / That he approves the common liar, who / Thus speaks of him at Rome.'

From first to last, the scene that we have witnessed, standing at what is consistently for Shakespeare the most vital expository post in a play, its beginning, appears to expose only unqualified truth; no word or action hints that it is other than a straightforward introduction to the actual situation and to subsequent action. It appears to serve, that is to say, exactly the function of initial passages that in earlier plays cast their rays faithfully ahead as our guide to what succeeds. What is more, it must be acknowledged here that nothing in the rest of the play ever overtly contradicts the impression given by this first exposition; on the contrary, most that is overt afterwards seems to confirm the rightness of the initial impression. Thus from Antony himself in the very next scene we hear, 'These strong Egyptian fetters I must break, / Or lose myself in dotage'; and, again, 'I must from this enchanting queen break off', for 'She is cunning past man's thought', and 'Would I had never seen her'.

Later, in Rome, on the eve of his marriage to Octavia, he says in soliloquy, '. . . though I make this marriage for my peace, / I' th' East my pleasure lies'. Soon thereafter he does in fact return to Cleopatra while the unsuspecting Octavia travels to Rome. Then, contemning all advice but Cleopatra's, he fights Caesar at sea, and, when Cleopatra 'Hoists sails and flies', he executes the most ignominious act of his life: 'Claps on his sea wing, and like a doting mallard, / Leaving the fight in height, flies after her'. And, finally, brought false news of Cleopatra's suicide, he mortally wounds himself, and then, finding that the news was false, orders himself carried to Cleopatra to die with his head in her lap.

The accuracy of Philo's opening lines is seemingly verified also, and repeatedly, by the remarks of other observers at later moments. These subsequent comments, indeed, appear to be strategically distributed throughout the play as if we must not be left for long without a fresh reminder of Antony's enslavement. Thus the first words that we hear Caesar speak concern his addiction to the delights of the East: 'He fishes, drinks, and wastes / The lamps of night in revel.' He apostrophizes the absent triple pillar with 'Leave thy lascivious wassails', but entertains no real expectation that Antony can or will do so. In all, the impression we gain from Caesar confirms that we had from Philo; and but a scene intervenes before we hear the same again, this time from Pompey, in a long passage earlier quoted: 'Let witchcraft join with beauty, lust with both! / Tie up the libertine in a field of feasts, / Keep his brain fuming.' Indeed, it may be Pompey's brilliantly succinct summary that best sustains the image first presented us by Philo. In any event, Pompey's version of Antony's condition matches Caesar's, even as Caesar's matches Philo's, and all seem perfectly corroborated by Antony's own complaints of the 'strong Egyptian fetters' and the 'enchanting queen' that hold him in Egypt. It would therefore seem all but superfluous that to the mass of such early evidence is added the testimonial of the play's professional informant, Enobarbus, who speaks unequivocally to Maecenas when, on the eve of Antony's marriage to Octavia, Maecenas has opined that now Antony must leave Cleopatra utterly: 'Never. He will not'; and, finally, the same spokesman, directly after the marriage, informs us flatly that Antony 'will to his Egyptian dish again'.

All such evidence appears to support the assessment of Antony's condition that is made by Philo and thus to fix the nature of the tragedy as that of a man enslaved and ultimately destroyed by passion: enslaved by Cleopatra, Antony neglects his imperial duty; momentarily he breaks his chains, returns to duty, binds himself to Caesar by marrying his sister; but then he succumbs to temptation, abandons Octavia, returns to Cleopatra, and brings upon himself the final wrath of Caesar; yielding to Cleopatra, he abandons sane military judgement and confronts Caesar on his own terms, then loses the crucial battle by turning tail to follow his fleeing mistress; and, finally, he kills himself for Cleopatra and dies with his head in her lap. Such is the outline of the tragedy, and such, according to the overt evidence, is its cause; though critics have expressed varying opinions on particular aspects of the play, including sharply divergent views of Cleopatra, the history of criticism shows no dissent from the main outline.

But, as we have stated, *Antony and Cleopatra*, all but devoid of the usual expository devices, is a play in which we must suspect that more is hidden than shown. And if ever Shakespeare had proper occasion to reverse his characteristic dramatic habit of laying motives bare to the audience and at the same time hiding them from participants, it is here, in a play that is of politics all compact: Antony, Caesar, and Cleopatra all are high professionals engaged in policy of a rare order, and the dramatist's method is shrewdly altered to fit their characters and the magnitude of their game.

In seeking what lies beneath the surface, we must begin with Antony—but with a line not from his play but from Macbeth's; says Macbeth of Banquo,

> ... under him,
> My Genius is rebuk'd, as, it is said,
> Mark Antony's was by Caesar.
> (*Macbeth* III. i. 55–7)

Shakespeare's habit of directly or obliquely referring, during the writing of one play, to characters or incidents that he would use in the next, or had just used in that immediately completed, is too well known to require elaboration. He may, of course, have known Plutarch's life of Antony long before he came to *Macbeth*—as early as *Julius Caesar*, or earlier; but at the time of writing *Macbeth* it is clear that the subject of his next tragedy was being turned over in his mind—and, indeed, not merely the general subject of the tragedy but the very key to its treatment. For it is Macbeth's statement that, at the climactic moment, the very hinge of the tragedy, he elaborates in striking detail. Here Antony has just agreed, seemingly in good faith, to marry Octavia, giving tender assurances to her that, regardless of any past blemishes his character may have acquired in the world's view, all shall hereafter 'be done by the rule'. We are given no cause to doubt his honesty at the moment he makes these assurances. But, then, immediately afterwards, being left alone with the Soothsayer, he asks a monumental question: whether his own or Caesar's fortunes shall rise higher. 'Caesar's,' replies the Soothsayer, and, quite without invitation, continues:

> Therefore, O Antony, stay not by his side.
> Thy demon, that thy spirit which keeps thee, is
> Noble, courageous, high, unmatchable,
> Where Caesar's is not; but, near him, thy angel
> Becomes a fear, as being o'erpower'd: therefore
> Make space enough between you.
>
> (II. iii. 18–23)

'Make space enough between you': from what instantly follows, it is plain that the Soothsayer has struck a nerve at the core of Antony's being. 'Speak this no more', he commands, with an urgency like that with which Hamlet swears to silence his friends who have seen the Ghost. But the Soothsayer, having touched the sore spot, mercilessly turns the knife:

> To none but thee; no more, but when to thee.
> If thou dost play with him at any game,
> Thou art sure to lose; and, of that natural luck,
> He beats thee 'gainst the odds. Thy lustre thickens
> When he shines by. I say again, thy spirit
> Is all afraid to govern thee near him;
> But he away, 'tis noble.
>
> (Ibid. 24–30)

'Get thee gone', Antony orders—then brushes the subject aside as if it were trivial: 'Say to Ventidius I would speak with him.'

Left alone, however, he speaks the one true soliloquy that Shakespeare allows him in the entire play; here for once, without policy or self-deception, he bares his soul to both himself and us:

> Be it art or hap,
> He hath spoken true. The very dice obey him,
> And in our sports my better cunning faints
> Under his chance. If we draw lots, he speeds;
> His cocks do win the battle still of mine,
> When it is all to nought; and his quails ever

Beat mine, inhoop'd, at odds. I will to Egypt;
And though I make this marriage for my peace,
I' th' East my pleasure lies.

(Ibid. 32–40)

'Make space enough between you': it is to the Soothsayer's admonition that his closing lines respond: 'I will to Egypt'.

If the Soothsayer's words shock Antony, they do so not by disclosing what was previously unknown to himself, but by revealing that another, whether by 'art or hap', has pierced the wall and recognized his deepest secret: *what had kept him long in Egypt and will now drive him there again is not his body's lust for Cleopatra but his spirit's terror of Caesar.* Cleopatra is but his excuse, the convenient lie behind which to hide the truth from the world.

Just how completely he has deceived the world we have already noted: Philo and Demetrius first express the world's false view in the opening scene; Caesar marks its acceptance at the highest level; Pompey subscribes to it utterly. Enobarbus, closest of all men to Antony, nevertheless has no notion of the truth, and has been unknowingly instrumental in establishing the legend: it is he who, with golden words, describes for Agrippa and other gaping Romans Antony's first meeting with Cleopatra 'upon the river of Cydnus'; it is he who counters Maecenas's argument that Antony's marriage will cause him to leave Cleopatra with a memorable summary of her charms: 'Age cannot wither her, nor custom stale / Her infinite variety'; and it is he who flatly asserts, when the marriage to Octavia has been solemnly sealed, that Antony 'will to his Egyptian dish again'. And so, indeed, Antony will—but for no such reason as Enobarbus thinks. He will return to Egypt, but not for love.

Cleopatra is Antony's screen, blocking from the world's view the truth that his own heart recognizes as shameful and that he seeks ever to hide from himself as well as from the world. 'I must from this enchanting queen break off', he tells Enobarbus, and 'These strong Egyptian fetters I must break / Or lose myself in dotage.' But these are loaded words borrowed from the world's conception of his condition, for the tragedy of Antony is not that of a man enslaved by passion but of a man desperately fleeing, lying to himself and the world. On command, prompted by Cleopatra, and with the world as audience, he grandly proclaims that Rome can melt into the Tiber and the whole wide-ranging Empire fall: but these and all the others are *public* boasts, ostentations designed for the world's eyes and ears;

and in reporting them as truth, Philo and Demetrius are deceived, and would unwittingly deceive us as well. Such desperate hope as Antony has hangs upon his success in having the legend taken for truth, that passion makes him keep space between himself and Caesar: to be known as the slave of passion is a lesser blemish on his honour than to be known as one whose soul shrinks before Caesar.

When Antony leaves Cleopatra for Rome, his argument is that duty calls and can no longer be denied, for the Empire is endangered and Caesar needs him; and indeed it is precisely because Caesar needs him that he *can* return, for while Caesar urgently needs him against Pompey and the pirates his Genius need not cower; it is then safe enough. It is noteworthy that all signs of his well advertised passion for Cleopatra vanish as soon as he conceives himself safe at Rome, momentarily secure with Caesar. Even so, the old, unequal relationship persists, and Shakespeare illustrates it with awesome economy:

Caes.	Welcome to Rome.
Ant.	Thank you.
Caes.	Sit.
Ant.	Sit, sir.
Caes.	Nay, then.

(II. ii. 28)

And Caesar sits; in the single pentameter line the whole tale is told: Caesar sits first because he *can* sit first at Antony's bidding, not Antony first at his. Caesar's quails have beaten his again.

As the conversation continues that begins thus inauspiciously, it is noteworthy that Antony makes no disclaimer of reports about his life in Egypt; on the contrary, he boasts of excesses about which, in the presence of Caesar, we might have expected him at least to be silent: 'Three kings I had newly feasted, and did want / Of what I was i' the morning.' And, again, 'Neglected rather, / And then when poisoned hours had bound me up / From mine own knowledge.' Here, and later in conversation with Pompey, his way is to exaggerate, not diminish, the reports of his riotous life with Cleopatra. And it is significant that Shakespeare does not once show him undergoing the slightest degree of conflict, let alone soul-struggle, over the rival claims of duty and passion. His solitary references to the 'enchanting queen' are expressed rather as matters of fact than as the effects of inner struggle. His final interview with Cleopatra before leaving for Rome is studded with professions of love and assurances of fidelity and quick return, but it contains no hint that the parting is difficult for him; his ringing protestations arise here, as elsewhere, as responses to her prompting, not his passion. It is her insistence that produces his final declaration: 'Our separation so abides and flies / That thou residing here go'st yet with me, / And I hence fleeting here remain with thee.' But this is not, for Shakespeare, and never was at any point in his career, the language of passion; it is merest make-do, language of the lips.

That no passion seethes in his parting speech is confirmed by his total silence, during the whole of his time in Rome and up to his desertion of Octavia in Athens, on the subject of Cleopatra. His sole allusion to her occurs in II. iii, at the end of his interview with the Soothsayer, earlier quoted: 'I' th' East my pleasure lies.' But this, as we have noted, is no confession of yearning, but rather an involuntary, or conditioned, response to the Soothsayer's warning that he must 'Make space enough' between himself and Caesar. For the rest, the many scenes that show him in Rome and Athens give no sign that Cleopatra ever comes near his thoughts. In his first meeting with Pompey in II. vi, Enobarbus tells Menas that Antony '. . . will to his Egyptian dish again', but Antony himself speaks no word of longing; on the contrary, he boasts to Pompey like a leering boy sharing a salacious joke with another: 'The beds i' th' East are soft.' We may even imagine that he accompanies the remark with a knowing elbow-nudge. Aboard Pompey's galley, he engages loudly in tilting at Lepidus, drinking, and vainly soliciting Caesar to doff the world aside and join the celebrants; failing utterly with Caesar, he joins hands with Pompey and Enobarbus to dance the Egyptian Bacchanals, 'Till that the conquering wine hath steeped our sense / In soft and delicate Lethe'. His riotous wassailing throughout this scene betrays a desperate need to drown something in drink—but it is not his longing for Cleopatra; it is rather the same old gnawing pain: '. . . his quails beat mine, inhoop'd, at odds'. Caesar has easily resisted all his efforts to get him drunk, imbibing only enough to suit the politics of the occasion. And in the end it is Caesar who abruptly breaks off the Bacchanals: 'Our graver business / Frowns at this levity.' Led soddenly from the galley, Antony has lost again.

Like those that precede it, Antony's final scene in Rome, when he and Octavia part from Caesar, is devoid of a suggestion that Cleopatra occupies any space in his mind or heart. Instead, all his thought here seems to be of Octavia. When Caesar warns that he must never let this marriage prove the means of division—meaning, none too subtly, that he must not desert Octavia for Cleopatra— Antony replies, 'Make me not offended / In your distrust' and assures Caesar that he will never find cause 'For what you seem to fear'. Then, while Octavia bids a long farewell to her brother, Antony twice speaks of her with a tenderness that he shows nowhere else in the play—and certainly never for Cleopatra: 'The April's in her eyes. It is love's spring', and, again, of the delicate balance which this parting has wrought in his bride's emotions, 'the swan's-down feather / That stands upon the swell at full of tide / And neither way inclines'. These words are composed of tenderness mixed with awe, and, hearing them, we are unlikely to guess that even now his mind harbours a deceptive purpose, an intent to return to Cleopatra; if it does harbour the thought, then Antony is vile indeed, and not only so, but infinitely clever, guileful, dishonourable, capable of masking the blackest purpose with an appearance of deep sincerity. At no point does Shakespeare make him look more honourable than here, when the sight of 'love's spring' in Octavia's eyes affects his heart; here indeed he appears not only to have forgotten Cleopatra utterly but to have escaped even from his private agony, his consciousness of Caesar's infallible, effortless supremacy. His bearing throughout the scene suggests not only that he would wish never to return to Cleopatra but that, naïvely, he imagines the old and unequal competition with Caesar to have ceased with the marriage. Thus much is implied by his last embrace and final words to Caesar: 'Come, sir; come, / I'll wrestle with you in my strength of love. / Look, here I have you, thus I let you go, / And give you to the gods.' Perhaps at this moment he truly believes that the unequal contest is ended and that he has escaped his fate; but, as we shall note hereafter, even in the instant of embracing his illusion, he has lost the match again.

The parting from Caesar is separated from the parting of Antony and Octavia in Athens by a single scene. As it stands, isolated, with no connecting bridge either before or after it, this second parting scene appears at first to bear no particular significance. Ostensibly it represents only a temporary separation: Octavia will go to Rome in an effort to reconcile brother and husband, and during her absence Antony will prepare for war with Caesar in the event of her failure. Yet in the perspective of the total play, the brief scene proves the very hinge of the tragedy, for it marks the final, irreparable rupture of Antony's relations with Caesar. But Shakespeare, whether deliberately or carelessly, gives us no clue to its significance at the time it is acted. Though, many scenes earlier, we heard Enobarbus state that Antony would leave Octavia for Cleopatra, and heard Antony himself, just after his marriage, say that he would return to Egypt, these assertions are too widely separated from the immediate scene to carry any weight, especially in view of Antony's seeming tenderness and the appearance of a splendid relationship with his wife. Of course there is no question how we are to understand Octavia's mind as the two part: she loves both husband and brother, and there is nothing devious in her plan to reconcile them.

But what is really in Antony's mind during this parting? Does he know that all juggling with Caesar is already at an end? Does he—even as he parts from Octavia with such seeming tenderness—cold-bloodedly deceive her, meaning only to get her off to Rome so that he can flee to Egypt? In any earlier play we would have expected the dramatist to be explicit, especially about so momentous an event, and here his omission of any private prompting of our awareness is one of the extreme instances of the inexplicitness that everywhere characterizes the method of *Antony and Cleopatra*. Nor does he provide us with advice in the next scene, where Eros first tells Enobarbus that Caesar has made wars on Pompey. 'This is old,' replies Enobarbus. To Enobarbus's query concerning the outcome of the wars, Eros replies, speaking of Lepidus, that Caesar, having used him in the wars, 'presently denied him rivality', made accusations against him, and finally imprisoned him 'till death enlarge his confine'. 'Then, world,' says Enobarbus, 'thou hast a pair of chaps, no more, / And throw between them all the food thou hast, / They'll grind the one the other.' The scene ends with flat statements that Antony's navy is rigged 'For Italy and Caesar', and all indications are that Antony, having heard of Lepidus's imprisonment and Pompey's death, will now move directly against his ancient rival.

Therefore we can hardly be less than astonished when we learn at the opening of the next scene that he has neither waited in Athens nor sailed against Caesar, but has fled to Egypt. It is at the opening of this scene that Caesar tells how Antony, 'Contemning Rome', has made a great show in Alexandria, when 'Cleopatra and himself in chairs of gold / Were publicly enthron'd'. 'This in the public eye?' asks Maecenas; and Caesar, 'I' th' common show-place, where they exercise.' It is the boldest advertisement yet made by Antony of the passion for Cleopatra that he would have the world accept as fact.

And indeed it is so accepted. When Octavia arrives at Rome, Caesar inquires, 'Where is he now?' 'My lord, in Athens,' she replies. 'No, my most wronged sister; Cleopatra / Hath nodded him to her. He hath given his empire / Up to a whore.' To her insistence that it was not Antony who sent her to Rome, but herself who 'begg'd / His pardon for return', Caesar replies, 'Which soon he granted, / Being an obstruct 'tween his lust and him.' Maecenas then re-sums the world's view of Antony's conduct and its cause:

> Only th' adulterous Antony, most large
> In his abominations, turns you off,
> And gives his potent regiment to a trull
> That noises it against us.
>
> (III. vi. 93–6)

Antony's lust for Cleopatra and the delights of the East caused him to abandon Octavia and Empire; and this 'truth' Antony himself has blatantly flaunted by his recent ostentation in Alexandria, writing his 'abominations' large in the eye of the world.

But that is Caesar's view, with which our own cannot accord, for neither the scene of Antony's parting from Octavia nor that which immediately follows it supports this understanding. Antony's change of mind, which Shakespeare neither shows nor reports, must have occurred after his wife's departure, when news was brought him that Caesar had eliminated both Pompey and Lepidus, leaving the world with but one 'pair of chaps' that must now 'grind the one the other'. At that news, says Eros, Antony walked in the garden kicking at rushes on the floor and crying 'Fool Lepidus!' Only then, in that offstage scene, could he have made his decision to go to Egypt. If yearning for Cleopatra influenced the decision, the fact is neither expressed nor implied; moreover, he could have had no space for Cleopatra in his thoughts at the moment when he heard the news of Pompey and Lepidus. While Pompey lived, he was a threat to Caesar and a buffer between Caesar and Antony; while Lepidus lived, he served as buffer, mediator, and sop. With them alive, Antony did not stand alone; moving relentlessly toward making himself sole sir of all the world, Caesar first had three rivals; then two; and now but one. And soon there must be none. This fact Antony grasps only too well: intimidated by Caesar even while the others lived, driven even then to make space between himself and his invincible adversary, he now knows not only that the two chaps will indeed grind each other but that Caesar will win, as always:

'His quails beat mine.' All that is left is to flee to Egypt and shout aloud for all the world to hear that passion for Cleopatra drew him there.

Through all that follows, Antony's use of Cleopatra as his screen against the world's eye grows increasingly desperate. In III. vii he insists on fighting Caesar by sea, though the advice of all but Cleopatra is against doing so. 'Canidius,' he says, 'we / Will fight with him by sea', and offers an irrational argument: 'For that he dares us to 't.' It cannot be that he is blind to his disadvantage here, when the evidence is as obvious to him as to Enobarbus and Canidius. Incredibly swift and relentless, Caesar's approach by sea has been ominous, bordering on the supernatural; Antony can have no expectation of stopping it. Why, then, does he insist?

His error, he knows, will be attributed to his enslavement by Cleopatra, and will advertise that enslavement once again throughout the world. Said Enobarbus, at the start of the scene, opposing Cleopatra's presence in the wars,

> Your presence needs must puzzle Antony,
> Take from his heart, take from his brain, from's time,
> What should not then be spar'd. He is already
> Traduc'd for levity; and 'tis said in Rome
> That Photinus an eunuch and your maids
> Manage this war.
>
> (III. vii. 11–16)

Enobarbus cannot guess, of course, that Antony's best hope is just this: to have his foolish error and defeat attributed to his dotage. After the debate, when all arguments have failed to dissuade Antony, Canidius—who knows as little of the truth as Enobarbus—explains to a soldier why the general has clung to a militarily indefensible decision: '. . . his whole action grows / Not in the power on't. So our leader's led, / And we are women's men.' And for Antony's purpose no explanation could be more salutary. His defeat by Caesar, he knows, is inevitable whether he fights by sea or land. To be defeated by land would be to forfeit the comfort of his abiding and precious excuse, his supposed enslavement; to be defeated by sea, when all the world will say—and does say—'our leader's led', will be to enlarge the legend that his honour and his pride require.

What is more, the actual sea battle works out even more usefully, from his private point of view, than he could have anticipated. At the battle's height Cleopatra—

When vantage like a pair of twins appear'd,
Both as the same, or rather ours the elder,
The breese upon her, like a cow in June,
Hoists sails and flies.

(III. x. 12–15)

Plutarch states that Cleopatra insisted on fighting at sea because the sea afforded opportunity for her own escape. But our question is not the wherefore of her flight, but the wherefore of Antony's; reports Scarus,

She once being loof'd,
The noble ruin of her magic, Antony,
Claps on his sea-wing, and, like a doting mallard,
Leaving the fight in height, flies after her.

(Ibid. 18–21)

'Experience, manhood, honour, ne'er before / Did violate so itself,' Scarus adds, understating rather than exaggerating the infamy of Antony's flight. That Antony should twice have turned his back on duty and Empire, that he should have deserted first Fulvia and then Octavia, is one order of shame; but that the glorious captain who had often fronted the enemy under such conditions as to win even the praise of Caesar should desert his forces at the very height of battle is quite another. That such a warrior, in such a circumstance, would actually surrender his judgement and his honour to passion is an inconceivable and wholly unacceptable explanation of his flight.

But Antony himself—as by now we should have come to expect—loses no time in loudly attributing his monstrous act to passion: 'I follow'd that I blush to look upon,' he announces to the Attendants when we first see him after the flight. His brown hairs, he adds, reprove the white ones 'For fear and doting'. And, moments later, with Eros and the Attendants still within earshot, he shouts at his mistress: 'O, whither hast thou led me, Egypt?' and goes on:

Egypt, thou knew'st too well
My heart was to thy rudder tied by the strings,
And thou shouldst tow me after. O'er my spirit
Thy full supremacy thou knew'st, and that
Thy beck might from the bidding of the gods
Command me.

(III. xi. 56–61)

And again,

> You did know
> How much you were my conqueror, and that
> My sword, made weak by my affection, would
> Obey it on all cause.
>
> (Ibid. 65–8)

In these vehement and repeated protestations what sounds loudest is Antony's obsession with making the world believe that his flight did indeed owe all to passion; and, perhaps, if he could shout it long and loud enough, he might even make himself believe. But in his heart he knows well why he fled Caesar; in the entire scene Antony is given but one private speech, and this suggests the truth that the public protestations are meant to hide:

> . . . he at Philippi kept
> His sword e'en like a dancer, while I struck
> The lean and wrinkled Cassius; and 'twas I
> That the mad Brutus ended. He alone
> Dealt on lieutenantry and no practice had
> In the brave squares of war; yet now—No matter.
>
> (Ibid. 35–40)

When Eros directs his attention to Cleopatra, remarking that 'death will seize her' unless Antony comforts her, he directly assails her with 'O, whither hast thou led me, Egypt?' and comforts her not at all. And then, after raging at her through his next three speeches—all asserting and reasserting his boast to the world that his fault was one of passion—at her cry of 'Pardon, pardon!' he abruptly drops all his simulated rage:

> Fall not a tear, I say; one of them rates
> All that is won and lost. Give me a kiss.
> Even this repays me.
>
> (Ibid. 69–71)

Having publicly shifted the blame to her and his passion for her, he has made his point; there is no longer need to rage. The sea-battle, after all, has worked out splendidly; for it brought him a rich windfall of opportunity to show the world a heart so chained to Cleopatra that it could only trail after her even from the height of battle.

The same abruptness that marks his change of mood here also marks that at the end of the highly curious and nearly inexplicable Thyreus scene that next follows. Here, after flying into unprecedented rage at sight of Thyreus kissing Cleopatra's hand, he declares abruptly to her, 'I am satisfied'—and changes the subject as if there had been no such incident. We have been given a guide for our view of this scene in Caesar's order to Thyreus, just before: 'Observe how Antony becomes his flaw', and mainly the scene illustrates just this—the frenzy of a man cornered, with his nemesis swiftly closing in. But so very much more is involved in this focal scene than meets the eye that it is now necessary once again to back up in order to set the incident in perspective.

Caesar's admonition to Thyreus prepares us only for what the scene exhibits on the surface. More sharply than perhaps any other, the Thyreus incident brings into focus an aspect of Antony's unawareness about which we have had hitherto no occasion to say anything, though several earlier scenes have partially exposed it. For all the while that Antony practises upon the world with his pretence that passion for Cleopatra has caused his neglect of Rome, his abandonment of two wives, and his flight from the sea-battle, he is also the victim of his own blindness to Cleopatra's true quality and the depth of her passion for him. Previously we have been concerned only with the falseness of his much-advertised passion for her; but the other side of the matter is equally remarkable. From the outset of the action until the instant of his death, Antony's perception of Cleopatra goes only as deep as the world's; to him as to it, she is the super-subtle enchantress of the East, 'cunning past man's thought', the same voluptuous siren whose charms had captivated Julius Caesar and Cneius Pompey. He holds of her, in short, exactly the image that suits his private purpose: her reputation is his shield against prying eyes, none of which—except the Soothsayer's—must ever divine his deepest secret. But Shakespeare's Cleopatra is deeper than Antony's, and it should be Shakespeare's, not Antony's, that is ours.

That she is wily is undeniable: the play everywhere exhibits her wiles; or, where they are not being exhibited, they are being discussed with something akin to awe. Thus, in the opening scene, when Philo has directed us to note that Antony has been transformed into a strumpet's fool, she is exhibited in performance: 'If it be love indeed,' she prompts Antony, 'tell me how much.' When messengers from Rome are announced, she takes occasion to demand Antony's protestation—for the thousandth time, no doubt—of disdain for Rome, for Caesar, for Fulvia, and of unlimited devotion to herself. When Antony, on cue, has obliged with a ringing declaration—'Let Rome in Tiber melt'—she baits him to even loftier rhetoric, which again he supplies. Her point scored, she grandly exits with Antony in quest of pleasure; but moments later we learn that a 'Roman thought' has struck him and taken him from her side. Then, though she has come to seek him, at his return she instantly turns her back: 'We will not look upon him.' It is Antony who must then seek her, satisfy her demands for the usual signs of doting; Enobarbus here offers us an analysis of her repertory:

> Cleopatra, catching but the least noise of this, dies instantly. I have seen her die twenty times upon far poorer moment . . . her passions are made of nothing but the finest part of pure love. We cannot call her winds and waters sighs and tears; they are greater storms and tempests than almanacs can report. This cannot be cunning in her; if it be, she makes a shower of rain as well as Jove.
>
> (I. ii. 144–57)

And with this assessment Antony, his soul obsessed by the need to have the world believe him sunk in dotage, readily agrees: 'She is cunning past man's thought.'

As the world perceives them, Cleopatra's are shallow wiles that show the shallowness of the woman herself; and, in the world's view, the style of life to which Antony has succumbed is as tawdry as the wiles. To Philo, Antony's function is that of 'cooling a gypsy's lust', of being a 'strumpet's fool'. To Caesar, his life consists of tumbling on the bed of Ptolemy and giving up a kingdom 'for a mirth', a life of 'lightness', 'voluptuousness', for 'Salt Cleopatra' with witchcraft and lust has tied up the 'libertine'. And when Antony has turned his back on Rome for the last time, Caesar bluntly advises Octavia that he has given his empire up 'to a whore'. A whore's wiles: such, in the world's view, is the sum of what Cleopatra has offered and to which Antony has succumbed.

To this estimate of her not even Caesar gives more extravagant testimony than Antony himself, especially when, on three occasions, his fury—or at least what he would have the world believe is fury—rises beyond bounds. The first, as we have noted, directly followed the sea-battle, when Cleopatra fled and Antony ignominiously pursued, then loudly proclaimed that her flight was a deliberate trick, like all her others, to draw him after. Second is his succession of outbursts at sight of Thyreus kissing her hand. 'You were half-blasted ere I knew you', he roars, and 'I found you as a morsel cold upon / Dead Caesar's trencher. Nay, you were a fragment / Of Cneius Pompey's, besides what hotter hours / Unregister'd to vulgar frame, you have / Luxuriously pick'd out.' Third, with which we shall deal later, is his final rage after the surrender of his fleet, when he cries that the 'Triple-turn'd whore' has sold him out to Caesar.

Of course Antony does not always speak so of her; these are occasions of special stress. But his quickness, during each of them, in venting accusation and vilification upon her suggests how near the surface lies his true estimate. He has not, in these moments of stress, reversed his opinion of her; she has always been, and always remains, in his eyes, just what the Roman world calls her, neither better nor worse than a wily temptress, a shallow trickster cunning past man's thought.

It would be preposterous to suggest that she is not in fact a wily temptress; we cannot fail to mark her cunning wits at work from our first view of her until the last, when she outwits Caesar. But it would be as faulty for us to suppose that she is only such and no more. It is one thing for the participants, including Enobarbus and Antony himself, to underestimate her by seeing only her guile; but it is quite another for us, who are privileged to hold a higher vantage point, to mistake her by blindly accepting the Roman image of her. For certainly, though indeed he has not used his typically explicit means of doing so, Shakespeare has enabled us to assess her qualities more accurately than Rome and Antony can. The time may well have been, in her salad days with Pompey and Julius Caesar, when she was no more than the super-subtle seductress that she still seems to the world of *Antony and Cleopatra*; indeed, even in the early period of her relations with Antony, at which the play sometimes takes a backward glance, she may have been only so much. But during the segment of her life that is set before us in this play her wiles serve primarily as Antony's supposed passion serves—*as a cover that hides the truth*. It is only this cover that the Romans and Antony see; but what Shakespeare leaves us occasional openings to glimpse is the depth of the true passion for Antony that lies beneath the cover of her wiles. Where Antony's practice is his continuous pretence of being passion's slave, hers is the unceasing display of artifice by which she masks the genuineness of her passion—and her deep insecurity. For it is not, in fact, she who, as the world believes, has captured Antony and transformed him to a strumpet's fool, but Antony who, desperate to keep his own secret from the world, has captured her and made himself her whole existence. Thus the desperation of each complements that of the other: her compelling need is indeed to enslave Antony, to claim and hold his mind and heart wholly; his is to show the world that she has done so.

In her effort to have Antony hers, Cleopatra's notorious wiles, the very same with which she strives to hold him, are her worst enemies, for Antony's eyes never see through them. Obsessed with his own problem, blinded by it, he remains until he dies, with his head in her lap, oblivious that, like Othello's base Indian, he has cast a pearl away richer than all his tribe—richer in fact than the whole 'wide arch / Of the rang'd empire' that he so loudly boasts he has given up for her. Against his blindness Cleopatra wages a lost battle, knowing all the while that she has neither his mind nor his heart. For all her wiles, she is never able to claim more than his superficial attention, his loud, shallow protestations; he recognizes only that she is cunning past men's thought, and his spirit remains aloof, preoccupied with its own problem. Her best wiles succeed only in getting him to play the role of the doting lover, which, for his own purpose, not hers, he is eager to play; but it is an actor's part, as both know. Such is their game in the first scene, which Philo's world quite mistakes for reality: 'If it be love indeed,' her first words prompt him, 'tell me how much'; and he responds on cue, 'There's beggary in the love that can be reckon'd.' She prompts again, and yet again, and each time he responds with more extravagant protestations; and so the scene goes until they grandly withdraw from the public view. Philo has instructed us, with Demetrius, to mark the 'strumpet's fool' who slavishly responds to an enchantress composed wholly of wiles; but what the scene in fact exhibits is an Antony who responds on cue as an actor does, and a Cleopatra whose wiles superficially obscure an insecure and desperate passion. Moments later, the actor Antony, having abruptly abandoned his public role of enslaved lover, is busily seeking out Caesar's messengers and declaring his intent to return to Rome; but Cleopatra we next see frantically seeking him, and simultaneously plying the wiles that hide her desperation: 'We will not look upon him.'

It is through the conspicuous contrast between the manner of their parting and the manner of each during their separation that Shakespeare most sharply differentiates the feelings of the pair for each other. Antony, as we have noted, undergoes no soul-struggle at parting, evinces no sign of consideration for her pain at his leaving, but expects only to have to weather another shower of her tricks and tears. And indeed she does oppose his going with all her cunning, as we have noted, recalling their happy days when 'Eternity was in our lips and eyes' and finally, perhaps genuinely, seeming to lose her covering mask and to lapse into speechlessness—'Something it is I would,— / O, my oblivion is a very Antony, / And I am all forgotten.' Yet Antony, attuned only to her artifice, fails utterly to understand her final artlessness and charges her harshly with being 'idleness itself'. Again her reply is from the heart:

> 'Tis sweating labour
> To bear such idleness so near the heart
> As Cleopatra this. But, sir, forgive me,
> Since my becomings kill me when they do not
> Eye well to you.
>
> (I. iii. 93–7)

Confessing artlessly that she has used her best arts for him and that they have failed, she might be thirteen-year-old Juliet speaking from her window, unaware of Romeo's presence; but Antony sees no more in her confession than another of her tricks, and his parting words are perfunctory and empty.

More vivid still is Shakespeare's contrast of their feelings during the long separation. As we have noted, neither in Rome nor in Athens does Antony give a sign that he thinks of her at all. It is not so with Cleopatra. 'Give me to drink mandragora,' she tells Charmian after Antony's departure, 'That I might sleep out this great gap of time'. Says Charmian, 'You think of him too much'—and is chided: 'O, 'tis treason.' And thus it is in each scene that shows her in Antony's absence: she thinks, speaks, feels nothing but him. If she attempts light jesting with her eunuch—'I take no pleasure / In aught a eunuch has'—she does so absent-mindedly, then abruptly surrenders again to her constant preoccupation:

> O Charmian,
> Where think'st thou he is now? Stands he, or sits he?
> Or does he walk? Or is he on his horse?
> O happy horse, to bear the weight of Antony!
>
> (I. v. 18–21)

Alternately she practices self-deception—

> He's speaking now,
> Or murmuring, 'Where's my serpent of old Nile?'

—and as quickly undeceives herself:

> Now I feed myself
> With most delicious poison. Think on me,
> That am with Phoebus' amorous pinches black,
> And wrinkled deep in time?
>
> (Ibid. 24-9)

When Alexas arrives with word of Antony, her instant thought is of comparison: 'How much unlike art thou Mark Antony!' Alexas's recital of Antony's mechanical, patently public greetings she seems not to heed, for when the recital is done she begs for a single personal glimpse of the man behind the formal salutation: 'What, was he sad or merry?' Being told that he was neither, but stood between the extremes, she rationalizes his mood so as to suck comfort from it:

> He was not sad, for he would shine on those
> That make their looks by his; he was not merry,
> Which seem'd to tell them his remembrance lay
> In Egypt with his joy; but between both.
> O heavenly mingle!
>
> (Ibid. 55-9)

Had Alexas reported Antony's mood as sad, she would have interpreted his sadness as marking his pain at their separation; had it been reported as merry, she would have interpreted it as affected, put on to reassure her in his absence. Thus deluding herself, she might be Malvolio assuring himself that 'M, O, A, I' signifies none but himself. But of course the analogy is imperfect: Malvolio, being a fool, fools himself absolutely; but Cleopatra, no fool, succeeds not at all.

Helpless to find solace in self-deception, she pours her energies into chatter about Antony, and into a wild excess of correspondence: 'Who's born that day / When I forget to send to Antony', she declares, 'Shall die a beggar'; and, a moment after, 'He shall have every day a several greeting, / Or I'll unpeople Egypt.' Shakespeare avoids mention of Antony's reaction to the flow of letters; but Alexas, who has met twenty messengers on the way from Rome, perhaps reflects his master's attitude: 'Why do you send so thick?' Her immersion in thoughts of Antony continues during the second of her scenes in his absence. 'Give me some music,' she cries—and, before the musicians can begin: 'Let it alone; let's to billiards.' Mardian the eunuch stands ready to play 'As well as I can', but Cleopatra has appetite for neither billiards nor the bawdy talk that 'play' gives rise to: 'I'll none now. Give me mine angle.' But thoughts of fishing give no respite from her constant obsession: 'My bended hook shall pierce / Their slimy jaws, and as I draw them up, / I'll think them every one an Antony.' Charmian recalls when Cleopatra and Antony wagered on their angling, and 'your diver / Did hang a salt fish on his hook, which he / With fervency drew up', and the allusion sends her into a fever of memory:

> That time,—O times!—
> I laugh'd him out of patience; and that night
> I laugh'd him into patience; and next morn,
> Ere the ninth hour, I drunk him to his bed.
>
> (II. v. 18-21)

The arrival of a messenger from Rome at this moment catches her in the rapture of relived passion, and she startles him with a spontaneous burst of erotic imagery that counterfeits sexual gratification itself: 'Ram thou thy fruitful tidings in mine ears, / That long time have been barren.' She sees bad news in the messenger's face and fears the worst: 'Antonius dead! If thou say so, villain, / Thou kill'st thy mistress.' She tries to bribe him to say that Antony is well, then threatens to melt the gold and pour it down his throat if he has equivocated; thus, alternately offering bribes and threats, stalling so as not to hear ill news of Antony, she prevents the messenger from divulging the truth through thirty-five lines before the murderous shaft: 'Madam, he's married to Octavia.' She then curses him, knocks him down, offers to stab him, drives him away in terror, then recalls him, puts the question again, and, on hearing the truth repeated, heaps curses on his head and again drives him out. That the news strikes deep appears in her subdued complaint: 'I am pale, Charmian.' Twice, plaintively, she tries to bribe the messenger to unsay the truth: 'Say 'tis not so, a province I will give thee.' Superficial cunning does not speak so; the blow has gone to her heart. At last the 'cunning queen' stands helpless as a confused child:

> Let him for ever go;—let him not—Charmian,
> Though he be painted one way like a Gorgon,
> The other way's a Mars. Bid you Alexas
> Bring me word how tall she is. Pity me, Charmian,
> But do not speak to me. Lead me to my chamber.
>
> (Ibid. 115-19)

The force of her words, and of the scene, is intensified by the fact that, in the scene just preceding, the Soothsayer has bluntly told us Antony's secret: he will indeed return to Egypt, but not for love.

In her final scene during Antony's absence, Cleopatra has regained composure enough to try self-deception again. Questioning the cringing messenger, whose discretion now bids him equivocate, she sucks comfort from each detail: his remark that Octavia is not as tall as she, and low-voiced, she converts to 'Dull of tongue and dwarfish'. As the interrogation continues, her soothing manner takes effect on the messenger, whose account of Octavia grows accordingly less complimentary: Octavia's face is 'Round even to faultiness', and her forehead is 'As low as she would wish it'—which details Cleopatra twists as she pleases: Octavia is merely foolish.

Thus this scene, like the others that show her in Antony's absence, exhibits a Cleopatra for whom time has stopped while her love is away. It is not, as Antony has made the world think, he who dotes, but Cleopatra. While she survives in Alexandria by feeding on the poison of self-deception, he, first in Rome and then in Athens, is too obsessed by his own problem to give her a thought. He will return to Egypt not because she is irresistible or Octavia dull, but because he needs to put space between himself and Caesar—and of course he returns the more precipitately because he learns that Caesar has eliminated both Pompey and Lepidus, leaving only two to grind each other.

And therefore it is that, on returning to Alexandria, he makes an immediate and colossal ostentation of the supposed passion that the world mistakes for real: 'Cleopatra / Hath nodded him to her', Caesar tells the poor pawn, his sister; and Maecenas echoes his master: '. . . th' adulterous Antony . . . gives his potent regiment to a trull'. But though the world so believes, the relationship of the pair continues as before—one-sided, with all the doting hers, not his. Agonizingly aware that his fall is imminent, Antony acts consistently in ways calculated to show the world that his defeat is the result of dotage, not Caesar's superiority. Hence, as we have noted, he insists on confronting Caesar by sea at Actium; and Canidius voices the very verdict that Antony wishes the world to hear: '. . . our leader's led, / And we are women's men'. Then, when Cleopatra flees the sea-battle at its height, he, recognizing an unprecedented opportunity to advertise his passion, flies after her 'like a doting mallard', and afterwards publicly chides her with words chosen for their effect: 'My heart was to thy rudder tied by the strings'. And even while, by word and action, he continues to practice on the world by embellishing the legend of his dotage, his blindness to Cleopatra's very real dotage continues. Hence it is that, when Thyreus has offered her all rights in Egypt if she will but turn Antony out, he says bitterly,

> To the boy Caesar send this grizzled head,
> And he will fill thy wishes to the brim
> With principalities.
>
> (III. xiii. 17–19)

Perhaps he does not really suppose that she will do so; and yet, so grossly does he misjudge her loyalty when she allows Thyreus to kiss her hand that it is entirely possible he could also think her capable of turning him out of Egypt. In any event, he is blind to the significance of her quiet reply: 'That head, my lord?' Then, moments later, after he has surprised her in the hand-kissing incident and has raged through four harsh consecutive speeches filled with vituperation and vileness, she again replies quietly, striking the deepest tragic tone of the play: 'Not know me yet?'

And indeed he does not know her; he plays a role throughout the Thyreus incident, using both the fury with which he addresses Thyreus and that with which he berates Cleopatra to serve his abiding public purpose. Playing his role, he remains insensitive to the tired despair of her words, which are as true as his own are histrionic. When at last he charges that she is cold-hearted towards him, she replies with yet another naked message from the heart:

> Ah, dear, if I be so,
> From my cold heart let heaven engender hail
> And poison it in the source, and the first stone
> Drop in my neck; as it determines, so
> Dissolve my life.
>
> (Ibid. 158–62)

But the force of this, too, is wasted on him; having made his public show, he abruptly breaks off with 'I am satisfied'—as if the forgiving should be his, not hers—and, without another word, changes the subject: 'Caesar sits down in Alexandria, where. . . .' To him the Thyreus incident has meant only another fortuitous occasion, like Cleopatra's flight at Actium, for burnishing the fiction of his enslavement. It is, after all, irrelevant to him whether Cleopatra's passion is real or feigned, so long as the world believes that his own for her is real. So, on the eve of battle with Caesar, they will have 'one other gaudy night', when he will show himself her slave, and she will seem, to himself as to the world, the cunning queen who holds him in thrall. It is just here that Enobarbus—who supposes that he understands the situation perfectly but in fact perceives it only as the world does—resolves to seek 'Some way to leave him'.

Antony's period of unnatural exhilaration, marked by Enobarbus's 'Now he'll outstare the lightning', begins at the end of the Thyreus episode and extends, with brief interruptions, through three of his scenes in Act IV. Shakespeare has taken care to ensure that here our own awareness that Antony's defeat is inevitable does not fail, however loudly Antony himself may boast of his expectations. We are first made to hear Enobarbus's declaration of intent to leave his master; then we hear Caesar laugh at Antony's challenge, remarking that his own army includes enough defectors from Antony's forces 'to fetch him in', and ending with his summation of Antony's condition: 'Poor Antony!' The whole of IV. iii, when night guards hear strange music and interpret it as signifying the desertion of Antony by Hercules, adds superfluous warning in the manner of the too-too solid supernatural manifestations that abound in *Julius Caesar*. And even Cleopatra, having just helped to arm her warrior for battle, ends IV. iv on a dismal note: 'That he and Caesar might / Determine this great war in single fight! / Then, Antony,—but now—Well, on.' Her wish that Antony might meet Caesar in single combat and so prevail is the sole revelation in the play of her own ignorance of Antony's deep, private terror.

At any earlier point in his career Shakespeare would doubtless have developed and then exploited a discrepancy between our own awareness, thus elaborately advised by trustworthy guides, and the unawareness of whatever hero was trapped in a situation as hopeless as Antony's. But here he exploits no gap, for none really exists: Antony is as aware as we that he is doomed. The exhilaration exhibited after the Thyreus scene, when he calls for 'one other gaudy night', is a new and fantastic form of the same old mask with which he has long disguised his soul's terror of Caesar. 'When one so great begins to rage,' Maecenas had told Caesar, '. . . Make boot of his distraction.' Antony's 'rage' takes the form of an exaggerated theatrical performance: Cleopatra is now, loudly, 'my Queen'; Antony grandly declares that he will return once more from the field 'To kiss these lips'; helping to dress him for battle, Cleopatra is saluted as 'the armourer of my heart', 'my Queen', 'my chuck'. Stimulated to such excesses by the approaching hour that must end his spirit's long, unequal contest with Caesar, his emotions relieve themselves in orgy, and in this condition it would be erroneous to regard the endearing terms showered on Cleopatra just before and just after his temporarily victorious skirmish as evidence of passion. Cleopatra is to him still an incidental in these final hours, as are the others about him—Eros, the 'sad captains', the servants, and, during and after the skirmish, even the heroic Scarus. 'I will be treble-sinew'd, hearted, breath'd, / And fight maliciously,' he boasts on the eve of battle, and calls his captains to a feast: 'I'll force / The wine peep through their scars.' 'By sea and land I'll fight. Or I will live / Or bathe my dying honour in the blood / Shall make it live again,' he tells Enobarbus, who, like Cleopatra, is a mere bystander to his distraction, ignorant of its root; so are the servants, whom Antony assembles and, in grotesque play, shakes by the hand one by one and compels all to weep by his own seeming overflow of emotion. 'What means this?' Cleopatra asks Enobarbus, who replies, ''Tis one of those odd tricks which sorrow shoots / Out of the mind.' Then, when Enobarbus has begged him to desist and bring an end to the universal embarrassment, Antony abruptly drops one actor's role for another no less histrionic: 'Ho, ho, ho! / Now the witch take me if I meant it thus!' and continues, 'I hope well of to-morrow, and will lead you / Where rather I'll expect victorious life / Than death and honour.' But this same outburst of joviality, put on with the same excess with which he had just made them weep, ends on a note that springs from the heart and echoes the ancient, bleak despair that harbours there: 'Let's to supper, come, / And drown consideration.' But, indeed, Antony's whole life in Egypt has been an effort, made in vain, to 'drown consideration'. Just so he had tried to drown it aboard Pompey's ship.

News of Enobarbus's desertion, brought him just as he sets forth to battle, doubtless moves him deeply: 'O, my fortunes have / Corrupted honest men!' Yet we cannot know in just what spirit he sends the deserter's treasure after him, together with 'gentle adieus and greetings'. Perhaps he is impelled by that same noble generosity for which he had once been famed and which is cited by the soldier who tells Enobarbus that the treasure has come: 'Your Emperor continues still a Jove.' But it was only the night before that Antony had deliberately wrung tears from his servants by overwhelming them with kindness—a performance, according to Enobarbus, that amounted to cruelty. Whatever was his purpose in sending the treasure, the result is of a kind with that which followed his display of generosity to the servants—but more severe: whereas the servants were moved only to shed tears, Enobarbus, on receiving the treasure, seeks the most ignominious of ends, death in a ditch.

Antony's brief triumph over Caesar, perhaps the only one in the history of their relationship, is, in our own perspective and Caesar's, insignificant; nor can Antony regard it as having importance, for the fact of his spirit's intimidation is too deep-seated ever to allow his imagining otherwise. Yet the immediate effect is his euphoric surge that dominates two brief scenes. Scarus, who, after Enobarbus's departure, appears for the first time in the play, a proxy Enobarbus, is the first recipient of Antony's burst of emotion: 'I will reward thee / Once for thy spritely comfort, and tenfold / For thy good valour.' The entire army is next:

> . . . you have shown all Hectors.
> Enter the city, clip your wives, your friends,
> Tell them your feats; whilst they with joyful tears
> Wash the congealment from your wounds and kiss
> The honour'd gashes whole.
>
> (IV. viii. 7–11)

While the same ecstatic fit continues, he greets his mistress:

> O thou day o' th' world,
> Chain mine arm'd neck; leap thou, attire and all,
> Through proof of harness to my heart, and there
> Ride on the pants triumphing!
>
> (Ibid. 13–16)

And with the rapture still upon him he presents Scarus to Cleopatra, demanding that she 'Commend unto his lips thy favouring hand', and commanding Scarus, 'Kiss it, my warrior.' The contrast of this incident with the Thyreus incident, when he had Thyreus whipped and gave his mistress a rude tongue-lashing for a similarly innocent kiss, is too obvious to be missed: Cleopatra is to him only what his mood of the moment dictates, only what he would have her seem to the world to be.

This scene of emotional orgy ends with Antony's boast that if Cleopatra's palace could but hold the numbers, all the army should dine with him 'And drink carouses'. 'I bind . . . the world to weet,' he had declared in the opening scene of the play, 'We stand up peerless'—that is to say, peerless lovers. But since that initial exhibition Antony has given us no sign that he is indeed enslaved by passion; his shows of passion have been public, calculated to enlarge the legend of his enslavement and to divert prying eyes that might, like the Soothsayer's, see into his heart. At first thought his rapturous salutation of Cleopatra on returning victorious from his skirmish with Caesar may appear to be an exception, sounding too true to be dismissed as public show. His words here do contrast with earlier ones; the ecstatic 'O thou day o' th' world' bears a ring of truth that Cleopatra has long yearned to hear, and she replies at the same high pitch, almost as though startled: 'Lord of lords! / O infinite virtue, com'st thou smiling from / The world's great snare uncaught?' but before concluding that Antony is, for once, over- come by feeling for his mistress, we should note the circumstances. He has just had the better of Caesar for the first time in his life; for a fleeting instant perhaps he feels free from the painful secret that gnaws his consciousness. It is, then, quite aside from Cleopatra, a moment for an outpouring of spontaneous emotion in which there is no feigning. It is appropriate to note, further, that Cleopatra is not the sole recipient of the outpouring: Scarus, a nobody of whom we have never heard before, is rewarded with a share not less extrava- gant than that accorded Cleopatra, and indeed the general army is recipient of the largest share of all: 'You have shown all Hectors', besides twenty additional lines of loving praise and the fanciful assurance that if space allowed all should dine at Cleopatra's table. In short, this emotional overflow is not selective, but engulfs all equally; had Mardian the eunuch confronted him, he would presumably have been saluted as exuberantly as Cleopatra. For Antony's euphoria does not derive from passion for Cleopatra, but from his one-time victory over the rival whose quails always beat his own, 'inhoop'd, at odds'.

If his euphoria is unselective, so is it extremely short-lived. Next day, when we hear him mention Cleopatra, his furious castigation begins with 'This foul Egyptian hath betrayed me' and worsens steadily thereafter: 'Triple-turn'd whore! 'tis thou / Hast sold me to this novice, and my heart / Makes only wars on thee.' His fleet has yielded without a fight; Caesar—whom, predictably, he here calls 'novice', as, elsewhere, in a rage, he calls him 'boy'—has beaten him again, and, again, even as after Actium, he seeks to set the blame on Cleopatra. Though Shakespeare otherwise follows Plutarch closely in representing the last details of Antony's life, he does not follow him here by so much as hinting that Cleopatra had in fact 'Pack'd cards with Caesar'. Shakespeare's Cleopatra—but not Plutarch's and not Antony's—is, as we have seen, so wholly given to Antony as to be incapable of betraying him. Repeating the charge loudly, categorically, Antony is determined to make the world, and himself, if possible, believe that his final defeat owes nothing to the 'novice', but all to Cleopatra. What is more, even while he rages of betrayal, he also insists—again just as after Actium—that she who has betrayed him is his heart's idol: 'This grave charm,' he cries, 'Whose eye beck'd forth my wars and call'd them home, / Whose bosom was my crownet, my chief end' has 'Beguil'd me to the very heart of loss'. In his next scene (IV. xiv) he speaks to the same tune: 'I made these wars for Egypt; and the Queen,— / Whose heart I thought I had, for she had mine . . . has / Pack'd cards with Caesar.' Thus by unceasing repetition during two scenes he strives both to set the blame on Cleopatra and to burnish the legend of his enslavement.

Next, as result of a practice that springs directly from Cleopatra's desperate passion, opportunity occurs for Antony to put a final sheen on the legend of his own passion. Desperately, not coyly, having witnessed his rage, Cleopatra sends false word of her death. For Antony, now completely beaten by his old rival, no word could be more fortuitous. Though he had long foreseen that Caesar would ultimately grind him under, and had prepared for this exigency by putting a solemn oath on Eros to end his misery with a sword-thrust, and, most recently, had recognized an immediate necessity for 'Ourselves to end ourselves', yet it is the arrival of the false news that conveniently precipitates his suicide. He quickly apprehends the magnitude of his opportunity, and his imagination supplies a vision of his earthly legend being perpetuated in the afterlife:

> Stay for me!
> Where souls do couch on flowers, we'll hand in hand
> And with our sprightly port make the ghosts gaze.
>
> (IV. xiv. 50–2)

Even in the next world, his 'passion' will amaze 'ghosts', and he and Cleopatra will steal the show from Dido and Aeneas.

It is noteworthy that, instantly after receiving word of her death, he ceases the vilification of Cleopatra in which he had engaged steadily since the first moment when he saw his fleet go over to Caesar. Here his pattern of conduct is identical to that which followed Cleopatra's flight and his ignominious pursuit, identical also to that which followed when he found Thyreus kissing her hand. On those occasions, after wrathful public castigation of Cleopatra together with extravagant protestations of his passion for her, he abruptly ended his tirade and resumed the role of passion's slave, having made his show for the world's consumption. So, now, on hearing of her death, already having raged enough to establish the point of her duplicity, he instantly ceases his tirade and resumes his doter's role: 'I will o'ertake thee, Cleopatra, and / Weep for my pardon.' 'Fall not a tear . . . Give me a kiss . . . Even this repays me,' he had said at the comparable point after Cleopatra's flight from Actium.

Antony, thus, gives himself a mortal blow in what seem to him ideal circumstances for the perpetuation of his legend: the world cannot fail to recognize that he has died for love; his act will supply definitive proof that the magnificent fiction he has built is true. In our own view, of course, the moment that he has chosen to kill himself is less than ideal, for even as he strikes we know that the report of Cleopatra's death is false. For a few moments, then, in his unawareness, Antony is in danger of wasting the best effect of his last ostentation.

But fortune reprieves him: he bungles his suicide and thus avoids dying under the false impression that his last grand gesture has succeeded. He lives to hear the truth—and to capitalize even on that, for by having himself borne to the monument he manages to die with his head in Cleopatra's lap. Bungling has thus contributed an unexpected dividend: not only will it appear to the world that he wounded himself mortally because of a trick perpetrated by the queen of tricks whose cunning has long enslaved him, but it will be seen that, even so, he forgave her and died in her arms who was at once his love and his destroyer.

Yet it is noteworthy that, having been brought to her, he speaks first not of love but of his true obsession:

> Not Caesar's valour hath o'erthrown Antony,
> But Antony's hath triumph'd on itself.
>
> (IV. XV. 14–15)

Then, recovering himself, still mindful of the world that watches the final tableau, he resumes the masquerade:

> I am dying, Egypt, dying; only
> I here importune death a while, until
> Of many thousand kisses the poor last
> I lay upon thy lips.
>
> (Ibid. 18–21)

But then again, having been heaved aloft to Cleopatra's arms, he lets the mask slip a little and dies with an oblique allusion to the unequal rivalry that has so long rebuked his spirit:

> ... and do now not basely die,
> Not cowardly put off my helmet to
> My countryman,—a Roman by a Roman
> Valiantly vanquish'd.

(Ibid. 55–8)

His last thought is thus not of his passion, or of concern for Cleopatra's fate, but of a claimed final victory over the scarce-bearded novice whose quails had always won.

So much for Antony's last hours. Two details remain to be discussed, and these reflect the least-flattering aspect of the portrait of Antony. The first is this: dying, still obsessed with the need to perpetuate his legend, he finds time for only an incidental word of advice for Cleopatra: 'One word, sweet queen: / Of Caesar seek your honour, with your safety.' How grossly he has mistaken her character throughout their relationship this insulting comment makes plain: he blindly assumes that, after his death, she will promptly use her wiles to make whatever adjustments her continuing personal comfort may require. At the same time, though assuming that she will ply her tricks, he cannot truly suppose that she will have any success with Caesar; his own experience has taught him the futility of any such expectation. Hence, even as he speaks, he must be aware that his advice is useless; but worse than the hollowness of his remark is its insult to Cleopatra, who, however, graciously replies, 'They do not go together.' The line matches her earlier responses, already noted, made when she observed Antony's insensitivity, his blindness to the reality of her affection: thus when Caesar had offered full amnesty if she would turn Antony out, and Antony had roared, 'To the boy Caesar send this grizzled head', she had replied quietly, 'That head, my lord?' and again, after Antony had caught Thyreus kissing her hand and had berated her, she expressed in one quiet question the tragedy of their relationship: 'Not know me yet?' Just so, in their last moment together, she sadly rejects the offensive suggestion that she seek her honour and her safety: 'They do not go together.'

But Antony's final act with respect to Cleopatra is worse than merely offensive. Either ignoring or not even hearing her 'Not know me yet?' he proceeds with the remainder of his advice: 'None about

Caesar trust but Proculeius.' The significance of this statement is too weighty to be passed over without examination.

Just a little later it is this very Proculeius whom Caesar sends as emissary to Cleopatra with explicit instructions:

> Go and say
> We purpose her no shame. Give her what comforts
> The quality of her passion shall require,
> Lest, in her greatness, by some mortal stroke
> She do defeat us; for her life in Rome
> Would be eternal in our triumph.
>
> (v. i. 61–6)

Proculeius easily grasps his master's meaning, and assents: 'Caesar, I shall.' He proves to be as good as his word, first telling Cleopatra to be of good cheer, for 'You're fallen into a princely hand; fear nothing'. And then, gratuitously, he proceeds with bland reassurances:

> Make your full reference freely to my lord,
> Who is so full of grace that it flows over
> On all that need. Let me report to him
> Your sweet dependency, and you shall find
> A conqueror that will pray in aid for kindness
> Where he for grace is kneel'd to.
>
> (v. ii. 23–8)

Even while Proculeius speaks thus, lying to lull her fears, Gallus and soldiers under Proculeius's command slip into the monument, catch Cleopatra off-guard, and seize her. When she tries vainly to stab herself, Proculeius prevents her and again seeks to lull her fears:

> Do not abuse my master's bounty by
> Th' undoing of yourself. Let the world see
> His nobleness well acted, which your death
> Will never let come forth.
>
> (Ibid. 43–6)

When she protests that she would rather be stretched naked on the Nile's mud and blown by water-flies than be exhibited in Rome, Proculeius seeks yet again to deceive her:

> You do extend
> These thoughts of horror further than you shall
> Find cause in Caesar.
>
> (Ibid. 62–4)

Cleopatra, certainly, is never deceived by these blandishments; when, in parting, Proculeius asks what word he should take to Caesar, she replies unequivocally: 'Say I would die.' But though he has failed to deceive her, Proculeius's intent to do so is unmistakable: mindful of Caesar's remark that 'her life in Rome / Would be eternal in our triumph', he has used every trick he knows in order to keep her alive for exhibition.

The quintessential question, then, remains: Why did Antony, with his last gasp, instruct Cleopatra to trust none about Caesar but Proculeius? Was it simply an honest error of judgement? If so, Shakespeare has left us wholly unprepared to recognize the fact; knowing Antony, we cannot believe him capable of misjudging the precise quality of 'any about Caesar'.

Or did Shakespeare, without intending one meaning more than another, simply follow Plutarch? Plutarch reports Antony as saying to Cleopatra that 'chiefly she should trust Proculeius above any man else about Caesar', and he reports Caesar as dispatching Proculeius to Cleopatra with the command 'to doe what he could possible to get Cleopatra alive, fearing least otherwise all the treasure would be lost; and furthermore, he thought that if he could take Cleopatra, and bring her alive to Rome, she would marvelously beautifie and sette out his triumphe.' Finally, in Plutarch as in Shakespeare, Proculeius assures Cleopatra 'that she should be of good cheere, and not be affrayed to referre all unto Caesar', then steals into the monument and takes her prisoner, foils her effort to slay herself, and insists that she should not deprive Caesar 'of the occasion ... to shew his bountie and mercie'. If Shakespeare differs from Plutarch, it is only by making the discrepancy between Antony's recommendation and Proculeius's actual performance more conspicuous. He does so by changing Plutarch's wording of Antony's statement that Proculeius should be trusted 'above any man else about Caesar' to 'None about Caesar trust but Proculeius.' The difference is slight but possibly significant: whereas Plutarch's Antony advises only that Proculeius is more to be trusted than any other about Caesar—a statement that allows for some degree of falseness in Proculeius as well as in the others—Shakespeare's Antony advises that Proculeius *is* to be trusted, though none else about Caesar is.

It is improbable that Shakespeare could have been following Plutarch so blindly as not to see that, when he has Antony recommend Proculeius unequivocally, and immediately afterwards shows this same Proculeius coolly and repeatedly lying to Cleopatra, the obvious contradiction must inevitably reflect on Antony and raise a terrible question about his motive: Did Antony himself, in his dying moment, while feigning concern for Cleopatra's fate, intend in fact to betray her in the worst way possible? If he knew that Proculeius was not to be trusted—and it is inconceivable that he should not have known—yet singled him out as trustworthy, the answer appears inescapable.

Harsh as it seems, this conclusion accords with the view of the pair's relationship that earlier incidents have suggested: Antony was never 'enslaved', but only nurtured the legend to disguise his true need to keep space between himself and Caesar; he remained blind to the reality of Cleopatra's own passion and her loyalty, undervaluing her except as a creature of many wiles; he always distrusted her motives, assuming that because she had turned and turned, she would turn again; at crucial moments his distrust, always lying just beneath the surface, erupted in outright hatred and foul words. At such moments he could kill her. Between the time of his fleet's surrender and the time that Mardian reported her death, he declared repeatedly that she must die for her treachery: 'The witch shall die . . . She dies for't . . . She hath betray'd me and shall die the death.' Up to the moment of Mardian's false news he was resolved to kill her immediately; then, hearing the false word, he changed abruptly and resolved to overtake her in death. But then, lying wounded and begging his guards to finish him, he learned that the word was false, that she had tricked him again. It can hardly be thought likely, whatever his words and actions during the final scene, that he came to think more kindly of her for the hoax.

It has earlier been suggested that his deepest purpose in having himself carried to die in Cleopatra's arms is the same that had long determined his course—advertisement to the world that he lived and died her slave, a martyr, finally, to his passion and her trickery. What is now suggested is an additional, and more reprehensible, motive: to take his threatened revenge on her, but revenge of a kind worse than the death he had first intended.

Shakespeare leaves no doubt that Antony knew exactly what fate awaited a live Cleopatra in Caesar's hands, or that he knew this to be, for her, the fate worse than death. Raging at her after the defection of his fleet, he threatened her thus:

> Vanish, or I shall give thee thy deserving
> And blemish Caesar's triumph. Let him take thee
> And hoist thee up to the shouting plebians!

Follow his chariot, like the greatest spot
Of all thy sex; most monster-like, be shown
For poor'st diminutives, for dolts; and let
Patient Octavia plough thy visage up
With her prepared nails.

(IV. xii. 32–9)

It is at this appalling prospect—not at his threat to kill her—that Cleopatra flees. Antony continues, ''Tis well thou'rt gone, / If it be well to live', and he concludes with a sentence of doom: 'The witch shall die.'

But then, having heard Mardian's report of her death and having responded by mortally wounding himself, he learns that the report was false and that she had tricked him again. It is likely that at this moment he reversed his earlier decision to kill her. Having been carried to her, he issues his solemn word of warning: 'None about Caesar trust but Proculeius.' And Proculeius, as we have noted, performs precisely as Antony knew he would: he smoothly lies and ends by capturing her alive. Perhaps, even beyond revenge, Antony was still conscious of an additional reason to wish her exhibited in Rome: displayed there, her charms would do more to verify the legend of his enslavement than any number of mere reports. Cleopatra in Rome would be as great a triumph for Antony as for Caesar.

Of course we have no explicit evidence for what Antony intended in his last moments; but, as we have repeatedly shown, Shakespeare's usual habits of overt exposition are mainly laid aside in this play: here no Hamlet announces his intent to put an antic disposition on and no Viola baldly tells us that she will masquerade as her lost brother. The new method is appropriate, for the deepest motives of the three high professionals can be best represented by being left obscure. The dramatist has rightly left it to us to seek their purposes by sifting and re-sifting their words and actions. Turning now from Antony, we must examine the highest of the high professionals, Caesar.

At the start of Act V, Antony has gone the way of Lepidus and Pompey. In eliminating his competitors, and more particularly the last, Caesar, like Antony, masks his deepest motives: whereas Antony built a legend to hide his shameful sceret, Caesar covers over his own purposes—which doubtless existed before the play begins and are coldly pursued throughout the action itself—with the implacable mask of justice, identified in his own mind with the good of Rome and indistinguishable from destiny itself. Like Antony, he is steadily concerned with his public face: after Antony's death, he says to his followers, 'Go with me to my tent, where you shall see / How hardly I was drawn into this war, / How calm and gentle I proceeded still / In all my writings.' Earlier, when the pawn Octavia arrived at Rome to attempt reconciliation, Caesar advised her that the rupture was final, and offered a word of comfort:

> Cheer your heart.
> Be you not troubled with the time, which drives
> O'er your content these strong necessities,
> But let determin'd things to destiny
> Hold unbewail'd their way.
>
> (III. vi. 81–5)

Between them, the two statements represent the face that Caesar turns to the world, the first blandly showing that he pursued Antony with reluctance, the second denying personal responsibility for past and future events in the march of destiny.

In exposing Caesar's private motives as opposed to his public ones, Shakespeare is hardly more explicit than in exposing Antony's; indeed, whereas Antony is given a single but deeply revealing soliloquy after the Soothsayer has diagnosed his secret, Caesar is given none at all. With respect to his true intent towards Pompey, Lepidus, and Antony we are given no direct statement, no private look into his mind; only the mask is shown us, the same that is seen by the world. Even so, surely we are not once in doubt about his ultimate intent. From our first sight of him straight on to his final speech, whether in conversation with lieutenants like Agrippa or in the presence of his fellow pillars, he is not merely Caesar but Rome and destiny. As an example of characterizing without characterizing, he is a Shakespearian masterpiece. But aside from the main consideration of what he is and aims to become, we might have expected the dramatist at least to advise us on the crucial details that are means to his end. The most critical of these, to which we now turn, is the marriage of Octavia to Antony. Though the handling of this marriage accords with the dramatist's method throughout the play, it stands as an especially conspicuous deviation from his expository method elsewhere.

What we are directly told is only that the marriage is suddenly conceived by Agrippa and then seconded by Caesar as means of cementing his alliance with Antony. If it were another play than this, we might safely assume that we have been told the whole truth. Enobarbus warns us that Caesar's plan will fail, for Antony 'will to his Egyptian dish again', and Antony himself, after agreeing to the marriage, confides that he will return to Cleopatra. But these fore-warnings do not compromise the essential fact of Caesar's intent as that is told us, and they give us no cause to suspect that this intent is other than what is told.

But of course what Caesar states is not truly his intent, for it runs counter to his sense of destiny, which requires the elimination of Antony along with Pompey and Lepidus. Caesar is to be 'Sole sir o' the world', and his every manœuvre is to this end; like the Fate of *Romeo and Juliet*, he wastes no motion. So Shakespeare's explicit representation of the marriage agreement conflicts with what is hidden even as Antony's much-advertised enslavement conflicts with his deep-lying secret.

Though Shakespeare gives no such soliloquy to Caesar as Iago's 'How, how?—Let's see.—After some time to abuse Othello's ear', the swift and shrewdly calculated events of the last two acts are inexplicable unless we recognize that Caesar coldly uses his sister as a pawn in his final move to destroy Antony. With Caesar as with Antony we are obliged to question all that appears to be laid open. What is here made explicit is only that the marriage is to cement the alliance; what is hidden is Caesar's will to breed a tremendous occasion that will justify, in the world's eye, a tremendous action; and afterwards, when Antony shall have been eliminated, Caesar will summon his lieutenants into his tent to see 'What I can show in this'. For even Caesar is mortal enough to covet justification of so monu-mental an act as the killing of an Antony.

But now, in the absence of Shakespeare's typically direct exposi-tion, what shreds and patches support this view of Caesar's pro-ceedings as a calculated practice on Octavia, Antony, and the world?

First is the way in which the idea of the marriage is initially intro-duced. During a lull in the angry quarrel that erupts as soon as Antony and Caesar sit down together, Agrippa tenders a proposal: 'To hold you in perpetual amity, / To make you brothers and to knit your hearts / With an unslipping knot, take Antony / Octavia to wife.' He speaks cautiously and at last offers an apology: 'Pardon what I have spoke, / For 'tis a studied, not a present, thought, / By duty ruminated.' But though Agrippa here implies that he is sole originator of the plan, we cannot properly ignore the fact that Caesar gave him his cue in the preceding speech: '. . . 't cannot be / We shall remain in friendship, our conditions / So diff'ring in their acts. Yet, if I knew / What hoop should hold us stanch, from edge to edge / O' th' world I would pursue it.' Agrippa picks up his cue and begins to speak: 'Give me leave, Caesar'—and proceeds to enunciate, as if it were exclusively his own, the marriage proposal. To be noted is the fact that Agrippa, though always talkative enough with his peers, speaks no words in the presence of the triumvirs except those that propose the marriage. Enobarbus once speaks out boldly in their presence and is harshly silenced by Antony: 'Thou art a soldier only; speak no more.' When he offers to speak again, he is again silenced: 'You wrong this presence; therefore speak no more.' Agrippa, who stands to Caesar as Enobarbus to Antony, is a more disciplined lieutenant than his counterpart; he would hardly speak in 'this presence' but by prearrangement.

Caesar's response after Agrippa has tendered his proposal is similarly suspect. He remains silent until Antony prompts him: 'Will Caesar speak?' He then replies, 'Not till he hears how Antony is touch'd / With what is spoke already.' Whatever preparation had preceded Agrippa's bold suggestion, it is evident that Caesar was involved in it, for he hears the proposal with no surprise; and when Antony asks 'What power is in Agrippa / If I would say, "Agrippa, be it so," / To make this good?' it is not Agrippa but Caesar who replies—and not tentatively, as if the idea were new, but firmly, as though the matter had been decided beforehand: 'The power of Caesar, and / His power unto Octavia.'

To suggest that Caesar himself had promoted Agrippa's proposal is not in itself to prove that he meant from the outset to sacrifice his sister in order to create a situation that would justify his eventual move against Antony. But, given his early assertion that he and Antony 'could not stall together / In the whole world', given the fact that Antony's chronic terror of Caesar becomes acute when he learns that Lepidus and Pompey have been eliminated—news that precipitates his flight to Egypt—and given Caesar's clinical pronouncement after Antony's death that '. . . we do lance / Diseases in our body'—given such clues, we cannot be expected to suppose that Caesar seconded the marriage in the naïve expectation of thus gaining the 'perpetual amity' that Agrippa promises and that he himself seems to applaud: 'Let her live / To join our kingdoms and our hearts, and never / Fly off our loves again.'

Further, the scene as the married couple leave Rome so heavily stresses Caesar's ostensible concern for his sister as to suggest that much more than this concern motivates him. 'Most noble Antony,' he begins,

> Let not the piece of virtue which is set
> Betwixt us as the cement of our love,
> To keep it builded, be the ram to batter
> The fortress of it. For better might we
> Have lov'd without this mean if on both parts
> This be not cherished.
>
> (III. ii. 27–32)

It is a thinly veiled warning that, if Antony should dishonour his marriage, Caesar will destroy him. That Antony understands is clear from his testy response: 'Make me not offended / In your distrust.' But to this Caesar replies curtly, 'I have said.' Plutarch reports once that Caesar 'dearly loved his sister', then says no more; but Shakespeare enormously enlarges the hint: Caesar loves his sister as no brother ever loved before; she is 'a great part of myself'. And more emphatic still is the parting scene itself, which is so tearful that it amazes and moves even Antony, who stands by while brother and sister whisper in each other's ear and weep. As they do so, Agrippa and Enobarbus converse aside: 'Will Caesar weep?' asks Enobarbus; and Agrippa continues,

> Why, Enobarbus,
> When Antony found Julius Caesar dead,
> He cried almost to roaring, and he wept
> When at Philippi he found Brutus slain.
>
> (Ibid. 53–6)

And to this Enobarbus replies pointedly, 'That year, indeed, he was troubled with a rheum; / What willingly he did confound he wail'd.' The line is spoken of Antony, but the conversation is occasioned by Caesar's weeping. Certainly Enobarbus possesses no special knowledge of what is in Caesar's mind, nor does he intend the remark to apply sarcastically to his secret purpose; but we are ourselves in position to apply it so.

Though once again nothing is made explicit, then, the parting scene adds cause to suspect that the marriage is Caesar's practice on Antony; and with this device Caesar has beaten him again—

though at the moment Antony seems unaware of the fact. It is not until later, in Athens, when Caesar's pressures have become intolerable, that he first begins to understand:

> ... he hath wag'd
> New wars 'gainst Pompey; made his will, and read it
> To public ear;
> Spoke scantly of me; when perforce he could not
> But pay me terms of honour, cold and sickly
> He vented them; most narrow measure lent me,
> When the best hint was given him, he not took 't,
> Or did it from his teeth.
>
> (III. iv. 3–10)

Caesar's sudden intensification of pressure directly contradicts his eager assent to the marriage 'as the cement of our love' and adds credence to the view that from the outset he forwarded the marriage only in order to disrupt it and to force Antony's flight to Egypt, then to use the desertion of Octavia as justification for the elimination of his remaining rival. Shakespeare read in Plutarch that when Octavia desired to join Antony in Athens, Caesar 'was willing unto it, not for his respect at all . . . as for that he might have an honest culler to make warre with Antonius if he did misuse her'. Enlarging upon this hint, he represents Caesar as taking outrageous measures to force the issue. It is in response to these measures that Octavia hastens back to Rome to attempt reconciliation and that Antony, on learning that Pompey and Lepidus have already been eliminated—blunt notice that his own turn is next—flees to Egypt.

Both Antony and Octavia thus react to Caesar's purpose as Caesar had desired, and it would be difficult to miss the exuberance that permeates his detailed account, previously cited, of Antony's new extravagances in Egypt (III. vi. 1–19). When Maecenas advises, 'Let Rome be thus inform'd,' Caesar's reply is blunt: 'The people knows it.' Antony could not have pleased him more than by his flight from Athens and his gaudy show in Alexandria, for these acts furnish argument for war—and Caesar has lost no time in spreading the word throughout Rome. When Octavia returns unheralded, all but unescorted, Caesar greets her with the loaded word 'castaway', then continues with one of the truly superb speeches in all Shakespeare, a speech that fairly bursts into a song of private jubilation even while it publicly exhibits the speaker's grief and commiseration:

> Why have you stol'n upon us thus? You come not
> Like Caesar's sister. The wife of Antony

> Should have an army for an usher, and
> The neighs of horse to tell of her approach
> Long ere she did appear. The trees by the way
> Should have borne men, and expectation fainted,
> Longing for what it had not; nay, the dust
> Should have ascended to the roof of heaven,
> Rais'd by your populous troops. But you are come
> A market-maid to Rome, and have prevented
> The ostentation of our love, which, left unshown,
> Is often left unlov'd. We should have met you
> By sea and land, supplying every stage
> With an augmented greeting.
>
> (III. vi. 42–55)

Caesar's practice on Antony, which began when Agrippa, as if unrehearsed, introduced the idea of marriage, has been carried through to the perfect conclusion: nothing could more gratify him than the unheralded arrival of his 'castaway' sister, whom all the world has been made to know he loves beyond measure, and whom he sacrificed in the interest of amity. Not elsewhere notably a singer, Caesar sings here, his private elation swelling and soaring in metaphor and hyperbole, all the while that he seems to lament the fate of the 'market-maid' and, by obvious implication, castigates the profligate Antony. After this incident the world will blame him for neither Octavia's grief nor its sequel, the swift annihilation of Antony; the world will embrace the verdict of Maecenas:

> Each heart in Rome does love and pity you;
> Only th' adulterous Antony, most large
> In his abominations, turns you off,
> And gives his potent regiment to a trull
> That noises it against us.
>
> (Ibid. 91–6)

Neither will the world suspect by what practice Caesar himself brought about the end over which he seems to weep.

The same method of expository inexplicitness that marks the representation of this practice continues through the final scenes, when, with Antony removed, the practices of Caesar and Cleopatra are pitted against each other.

It has been usual to say that the Cleopatra of Act V is a new Cleopatra, after Antony's death, who rises to great heights, casting off the physical and attaining a royalty of spirit not predictable earlier. But the unflattering question has also been raised whether, even so, she would not have made terms and continued to ply her mercenary wiles had she not instantly perceived the sheer impossibility of making terms with such a Caesar as Octavius. In treating her final hours, Shakespeare is even more inexplicit than in representing her earlier ones; he is, on the other hand, more explicit by far in dealing with Caesar's intentions during this period than at any other time. Thus we are advised not once but often that if Caesar can deceive her into placing her life in his hands he will exhibit her to the world as his rarest trophy: '. . . her life in Rome / Would be eternal in our triumph'. We are plainly told that Proculeius will seek to lull her with lies, seize her, and prevent her from killing herself; indeed, here the dramatist is explicit almost to a fault, so that we hardly need—as Cleopatra herself does not need—to hear Dolabella's blunt affirmation, in reply to her question. 'He'll lead me then in triumph?' Returns Dolabella, 'Madam, he will; I know't.' Nor do we need his later direct warning: 'Caesar through Syria / Intends his journey, and within three days / You with your children will he send before. / Make your best use of this.' This expository openness accords entirely with Shakespeare's earlier manner of making assurance doubly sure: not only are we told Caesar's intent, but we are made to know also that Cleopatra is as well aware of it as we. So, after Caesar has showered her with promises, she is made to hiss angrily to her girls, 'He words me, girls, he words me'.

She knows that she has no choice: she must kill herself or be a spectacle in Rome. But though both she and we have been elaborately advised on this point, the dramatist provides us with no similarly open answer to the more subtle question—whether she *had* hoped to make comfortable terms with Caesar and would in fact have made them had she found him amenable. The prevailing view of Cleopatra as the first four acts of the play expose it—that is to say, Philo's, Enobarbus's, Rome's, even Antony's—as a super-subtle courtesan argues that she would not act very differently after Antony's death: she would use all the tricks in her repertory to preserve the luxury of her former life. And this view the highly conspicuous incident involving Seleucus appears at first to confirm. For according to Seleucus she has secretly withheld from her inventory 'Enough to purchase what you have made known'. She appears to have made a final, desperate try at deceiving Caesar in order to live on luxuriously —and she has done so only to be trapped by her own practice. But the handling of the Seleucus incident is another remarkable example of Shakespeare's new fashion of expository inexplicitness, and we must not be satisfied to let our first view of it stand without examination.

If we accept the Roman view of Cleopatra's character, it follows that we must suppose her final decision for death to be made only after Seleucus has betrayed her and after Caesar's departure from the scene with the assurance that '. . . we remain your friend'—which she angrily rejects with 'He words me, girls, he words me, that I should not / Be noble to myself.' In short, if we take this view, we will conclude that, were she not sure that Caesar merely 'words' her, she would live on and trust to her wiles; but, being sure, she here resolves on death. But our analysis has shown that the Roman view is false, not at all in accord with that which the dramatist, for all his indirectness, has enabled us to gain. Antony, who from the outset until his death was too preoccupied with his own public image to question Cleopatra's, naturally assumed at last that she would 'seek your honour, with your safety', and paid no heed to her quiet demurrer, 'They do not go together.' But we see her in truer perspective than Antony: at his death she is ready to die, and says so: 'Come, we have no friend / But resolution and the briefest end.' When Proculeius introduces himself, she acknowledges that Antony 'Did tell me of you, bade me trust you, but'—she adds significantly for us—'I do not greatly care to be deceiv'd, / That have no use for trusting.' In plainest terms, she means that it does not matter whether she is or is not deceived, for she is resolved to die. When Proculeius surprises her in the monument, she struggles to stab herself then and there, despite her preference for some painless way of death. And all of this occurs before her first interview with Caesar, and, of course, well before the Seleucus incident.

But, then, how is this view of her intent to be reconciled with that incident? If she did not hope to live to enjoy more years of luxury, why did she hold back more treasures than were named in the account prepared for Caesar? Why did she call Seleucus to attest to the honesty of the inventory unless she counted on him to help her deceive Caesar? Had this incident appeared in an earlier play, and, as here, included no strategically placed forewarning to put us on our guard, we should have to take it to mean just what it seems to mean—that she intended to deceive Caesar, with her steward's backing, and was caught in her own practice. But we must recognize that the dramatist presents the Seleucus incident with the same expository inexplicitness with which he has presented other important matters in this play—indeed, has sometimes actually hidden the reality within walls of false evidence. So here: to all appearances her holding back more than half her treasure signifies that she hoped, after all, to live on in luxury. But we have been shown that she has no wish to live beyond Antony's death and that she has no illusions about her treatment in Caesar's hands. Therefore it would make no sense for her to hold back a great treasure.

We are left with no choice but to conclude that, with the aid of Seleucus, she staged the entire incident with the intent of deceiving Caesar, indeed—but for quite another reason than that of holding on to her treasure. By having her trick conveniently exposed by Seleucus, she expects to make him believe that she has no intention of killing herself—for, Caesar would reason, if she meant to kill herself, why would she have sought to withhold so much treasure? If she could make him believe that she meant to live to enjoy the treasure, she would doubtless be guarded less jealously and would thus be the freer to design means of killing herself. Proculeius had already prevented her once from ending her life; she could not count on a second chance unless Caesar's fears were somehow lulled. It is not difficult to imagine how Shakespeare would have prepared the Seleucus incident had he written it in the time of *Hamlet* or *Othello*: a brief dialogue of Cleopatra and Seleucus would have been set first, wherein she outlined her intent and gave him instructions for his own role in the practice. But though the earlier method would have had the merit of dealing openly with us, its use in *Antony and Cleopatra* would have clashed incongruously with the method of unexplicated means and motives that prevails elsewhere in the play.

Once we recognize that the Seleucus incident is a piece of acting staged, like many others, to serve Cleopatra's secret purpose, the conduct of the scene takes on its proper appearance as a performance shrewdly devised by its director. It begins abruptly with Cleopatra unexpectedly handing Caesar her 'brief', saying, 'Here, my lord.' It is significant that Caesar has demanded no such inventory: it comes as a surprise to him, as it does to us. Cleopatra declares the account complete, 'Not petty things admitted', and then calls out, 'Where's Seleucus?' On cue, the steward steps forth: 'Here, madam.' The performance continues, with just a shade of exaggerated emphasis that suggests its falsity; says Cleopatra, gratuitously, 'This is my treasurer; let him speak, my lord, / Upon his peril, that I have reserv'd / To myself nothing. Speak the truth, Seleucus.' We should bear in mind that, just as he had asked for no inventory in the first place, so Caesar has asked for no verification: the entire piece is gratuitously Cleopatra's. Playing his role correctly, Seleucus at first hesitates: 'Madam, / I had rather seal my lips than to my peril / Speak that which is not.' She gives him his next cue: 'What have I kept back?' At this invitation Seleucus delivers the studied line that carries the whole burden of the little drama: 'Enough to purchase what you have made known.' Urged by Caesar not to blush at the disclosure, Cleopatra first speaks a mere platitude—'O, behold, / How pomp is followed!'—then makes a show of rage at Seleucus; but what most marks the entire incident as contrived is the wording of her deeply ironical affectation of shame as she addresses Caesar:

> O Caesar, what a wounding shame is this,
> That thou, vouchsafing here to visit me,
> Doing the honour of thy lordliness
> To one so meek, that mine own servant should
> Parcel the sum of my disgraces by
> Addition of his envy!
>
> (v. ii. 158–64)

Her mock humility, irreconcilable with her mood in view of her awareness of what Caesar intends for her, is laid on so heavily that, from our vantage point, it must appear totally false and might be expected to strain Caesar's credulity as well. Next she shifts her tack and offers a broad hint of her hope to enjoy Caesar's favour despite the disgrace she has just suffered:

> Say, good Caesar,
> That I some lady trifles have reserv'd,
> Immoment toys, things of such dignity
> As we greet modern friends withal: and say
> Some nobler token I have kept apart
> For Livia and Octavia, to induce
> Their mediation; must I be unfolded
> With one that I have bred?
>
> (Ibid. 164–71)

After one final show of rage at Seleucus, her performance ends; and, at least seemingly, it has succeeded, for Caesar appears convinced of her intent to live on and, plying his own practice in turn, magnanimously offers her not only all the treasure that she had held back but all the rest as well: 'Still be't yours, / Bestow it at your pleasure.' He leaves, and it is precisely here that she hisses to her women, 'He words me, girls, he words me, that I should not / Be noble to myself.'

Has her performance lulled Caesar? Though he does not with-draw the guards that were set to prevent her death, yet these are sufficiently relaxed that her final ruse succeeds perfectly: the Clown is allowed to enter with what passes for an innocent basket of figs. Clearly, Caesar had not been completely put at ease by her staged incident, for at her death the Guard rushes in as though primed to prevent an expected catastrophe, and on the Guard's heels comes Dolabella, whose first words at sight of the dead queen imply that Caesar had had second thoughts immediately after he had seemed convinced by the brief comedy with Seleucus: 'Caesar, thy thoughts / Touch their effects in this; thyself art coming / To see perform'd the dreaded act which thou / So sought'st to hinder.' But she had managed to deceive the infallible Caesar long enough to win her death.

Finally, then, in this game of secret motives and marked cards dealt by high professionals, all three of the competitors both win and lose. Antony never learns what a prize he had in Cleopatra, and loses his life and his empire; but he successfully shields from the world his shameful secret and leaves behind him a magnificent fiction—that he had lost all for love. Cleopatra loses her kingdom and her life and never did secure the mind and heart of Antony; but she momentarily deceives Caesar, wins the right to be noble to herself, and in the next life can anticipate a second chance at an Antony who is no longer intimidated by Caesar. And Caesar loses his one crucial skirmish with Cleopatra, so that his triumphal procession at Rome will lack its prize trophy; but he has made himself sole sir of the world.

CHAPTER IX

Innocents at Home: Timon and Coriolanus

'. . . too noble for the world.'

WITH *Antony and Cleopatra* Shakespeare drastically revised his way
of making it easy for us to enjoy an unimpaired view of participants
as they engage in practices for sport or mildly malicious mischief,
as in the comedies, or with deadly purpose, as in the tragedies. The
change did not mean that *Antony and Cleopatra* should include no
practisers among the participants; on the contrary, that tragedy's
three principal figures are super-subtle practisers beside whom an
Iago or an Edmund seems a rank amateur. But Caesar, Antony, and
Cleopatra do not, like their predecessors, unfold their devices to us
in advance so that we can watch knowingly while unwary victims
blunder into them. Octavius never does tell us his real purpose in
forwarding his sister's marriage; Antony never directly advises us of
his true reason for retreating to Egypt; Cleopatra never tells us that
the famous wiles by which the world thinks she holds Antony serve
in fact to hide her sense of not holding him at all. In short, throughout the play, the dramatist never makes us privy to participants'
minds in the former way.

Timon of Athens and *Coriolanus* make yet another break with past
dramatic method. Not only does our interest depend less on exploitation of discrepant awarenesses in these than in earlier tragedies, but
these final ones confront us with protagonists who have no guile, no
secrets to hide, no capacity for conceiving or executing practices.
What Menenius says of one of them is apt for both Timon and
Coriolanus: 'His nature is too noble for the world.' Both are victimized by their worlds because they know no better than to wear their
minds and hearts on their sleeves; of all the tragic heroes they are
least equipped to survive in their worlds. Brutus and Othello are
their nearest kin in terms of innocence; yet even Brutus engages in
dark conspiracy and thus both countenances and participates in

deception, and Othello, though himself always an unsuspecting victim of deception, is quick to join his deceiver in conspiring against Cassio and Desdemona. Timon, it is true, deceives his former friends by inviting them to a banquet of tepid water, and Coriolanus conspires with Aufidius against Rome; but each of these is the act of a child striking back in frustration at his tormentors. Brutus and Othello have unique traits that blind them to limited segments of their environments; Timon and Coriolanus, alas, are blind not to such segments but to the full circle.

Not only are their tragedies unlike earlier ones in these and other respects shortly to be noted, but they also have particular resemblances to each other. These resemblances are not superficial—though many details that lie near the surface are also remarkably similar—but deep, lying at the core of the two actions; so much, indeed, are they alike that we might take the second, *Coriolanus*, for an attempt to correct, in a different setting, the manifest failure of the abortive first attempt. At the core, they are identical: they present and plead the cases of individuals whose natures are incompatible with the way of the world. As tragic heroes, Timon and Coriolanus differ from their predecessors in the completeness of their alienation from their societies. Hamlet, Othello, Lear, and Antony are at odds only with particular elements in their environments. If Hamlet's enemy seems to him, for a time, to be all Denmark, his enemy is in fact only that part that is momentarily infected by Claudius's evil; with this particular spot cut from the state, Denmark would offer a place where Hamlet could thrive. Similarly, Othello's enemy is not Venice or Cyprus or both, but only Iago; remove him, and Othello could survive with the same noble serenity that was his before Iago's poison infected him. Lear's enemies, which include Edmund and Cornwall besides his two wicked daughters, are more numerous than Hamlet's or Othello's; but only these, and not the world, are arrayed against him, and with them pared away Lear might live on, happily irresponsible, to a quiet end. And as for Antony, the sole irritant to his soul in the two worlds of Rome and Egypt is Octavius Caesar; remove Caesar, and Antony might revel long o' nights in either world or in one and then the other alternately.

In the other tragedies that precede *Timon* and *Coriolanus* the issue is equally clear-cut, for in these also the heroes are opposed by particular forces rather than by the total environment. The enemies of Titus Andronicus are only Tamora and Aaron, and not all Rome;

not all Verona, but only their fathers' feud, which brings Fate into the fray, destroys Romeo and Juliet; Brutus loves Rome and Rome loves him, and it is only a faction that defeats him; and finally is Macbeth, between whom and his total world an incompatibility does indeed develop, but in his unique case it is his own wickedness, not the wickedness of his world, or of any part of his world, that cuts him off. In contrast to these protagonists, Timon and Coriolanus are by nature at odds not with particular elements of their worlds but with the prevailing way of their worlds. It is this characteristic that both differentiates them from their predecessors and binds them together as brother sufferers.

1. *The Innocent of Athens: Timon*

It is only during the first two acts of *Timon* that Shakespeare's long-standing device of establishing and exploiting unequal awarenesses serves any useful purpose; thereafter, such discrepancies as occur are momentary and unproductive. Ponderously, repetitiously, inter-minably, the first half of the play proceeds with the business of mere exposition: again and again we are told and shown, told and shown what Timon is and what sort of men comprise his world. In the course of the tedious introduction two obvious advantages are forced upon us: we learn what he is incapable of guessing, that his wealth is exhausted and that the throng of supposed friends and admirers who frequent his house will coldly deny him once they know his need.

Throughout Act I and beyond, Shakespeare uses the method of illustration, or demonstration, that we noted in the early scenes of *Antony and Cleopatra*, where, for example, Philo bids us stand with Demetrius and observe the situation, after which the noble couple enter and perform exactly as foretold. But in *Timon* such illustration as in Antony's play takes up only the early scenes extends without interruption through nearly two full acts. Further, its emphasis is shifted from what, in the other play, was primarily situation-illustration to include, predominantly, character-illustration. Perhaps it is this very shift that initially differentiates *Timon* from earlier tragedies and at the same time relates it to *Coriolanus*, where also a disproportionate amount of space is consumed with repeated illustration of the hero's character. In other tragedies, protagonists are typically caught in some crucial situation that requires delinea-tion before we are introduced to any special propensities of the hero:

thus Brutus is caught in a crisis of threatened tyranny, Hamlet in the
need to rescue Denmark from a usurper and murderer, Lear in
the division of his kingdom, and so on. In these tragedies it is the
situation that requires the immediate exposition; but in *Timon* there
appears to be no special situation as such—for in a unique way it is
the very character of the protagonist that constitutes the situation.

Roughly a hundred lines of the opening scene partially introduce
us to Timon before he enters; we hear a veritable chorus of voices as
visitors file in: 'O, 'tis a worthy lord,' exclaims the Merchant; 'A
most incomparable man, breath'd, as it were,' he continues, 'To an
untirable and continuate goodness.' Less directly, the remarks of the
Poet and the Painter on their respective artistic pieces, wherein
Timon's quality is set forth in transparent allegory, contribute also
to the initial illustration. 'How this grace / Speaks his own standing,'
says the Poet of the Painter's piece: 'What a mental power / This eye
shoots forth!' 'How this lord is followed!' cries the Painter when
'certain Senators' enter and pass over. 'Happy man!' returns the
Poet, who then proceeds at length to interpret his own newest work:

> Sir, I have upon a high and pleasant hill
> Feign'd Fortune to be thron'd. The base o' the mount
> Is rank'd with all deserts, all kinds of natures
> That labour on the bosom of this sphere
> To propagate their states. Amongst them all,
> Whose eyes are on this sovereign lady fix'd,
> One do I personate of Lord Timon's frame,
> Whom Fortune with her ivory hand wafts to her,
> Whose present grace to present slaves and servants
> Translates his rivals.
>
> (I. i. 63–72)

The Poet continues to praise Timon, who seems indeed the chosen
one of Fortune; but then, just before Timon's entrance, an ominous
note sounds:

> When Fortune in her shift and change of mood
> Spurns down her late beloved, all his dependants
> Which labour'd after him to the mountain's top
> Even on their knees and hands, let him slip down,
> Not one accompanying his declining foot.
>
> (Ibid. 84–8)

After commenting that painting is able to depict 'these quick blows
of Fortune' more forcefully than poetry, the Painter praises his
companion's effort: '. . . you do well / To show Lord Timon that

mean eyes have seen / The foot above the head'. It is precisely upon this cue, his approach signalled by a blare of trumpets, that Timon makes his first appearance, 'addressing himself courteously to every suitor'.

This uncommonly repetitive and static introductory portion is notably inferior to the corresponding portion of any other play of Shakespeare's. Interesting only for the novelty of presenting the hero and his world through the 'feigned' works of the Poet and the Painter, it appears to over-accomplish its purpose, for it leaves us with nothing more to learn in the rest of the play: Lord Timon is a noble, generous giver; he will inevitably fall; and when he falls, no hand will support him. Thus when we first see him, our advantage is already complete, like his own unawareness of the condition in which he stands. Still, perhaps the novel device of an allegorical or 'feigned' representation of the hero serves a special purpose that could not have been as well served by the usual, more direct method. By prefiguring Timon's fall through the Poet's piece, stilted and obvious though that is, Shakespeare confers upon it the quality of moral fable: Timon is an example—neither the first nor the last of men whom Fortune has first lifted high and then let slip. ''Tis common,' remarks the Painter: 'A thousand moral paintings I can show / That shall demonstrate these quick blows of Fortune's / More pregnantly than words.' Since Timon's house is peopled exclusively with visitors who come to sing his praises, dine royally, and bear away rich gifts—for indeed these Lords, Senators, Merchants appear to be without exception sycophants and parasites— who is there among them whose estimate of Timon we can truly credit? Unanimously they sing his praises, marvel at his bounty, vie in describing his goodness, his nobility; but all of them have wares to sell—and gifts to be harvested that are valuable far beyond those they bring: 'If I want gold,' says a Senator, 'steal but a beggar's dog / And give it Timon; why, the dog coins gold.' The Merchant boasts of Timon's 'untirable and continuate goodness'. Excepting only the churl Apemantus, all 'this confluence, this great flood of visitors' speaks with a single voice: 'A most incomparable man'.

A pack of sycophants would be as quick to flatter an ignoble man as a noble one, so long as his bounty continued; for anything that the visitors say directly, then, we cannot tell whether Timon is or is not the paragon that he is represented to be by the chorus of voices. Perhaps, indeed, surfeited with these incessant paeans addressed to his goodness, we may idly wish that he may turn out to be another 'honest' Iago, or at the least a Macbeth; but, alas for both the hero and the tragedy, Timon proves even as virtuous as his flatterers assert. When both the Poet and the Painter depict his nobility in their respective works of art, then—even though these same artists intend to claim treasures in return for their idealized portraits—we seem, at least, to be in the presence of some higher truth, from which we gain a view of Timon that transcends the patently mercenary expressions of the general visitors and leaves no doubt that he deserves the accolades: he is indeed a paragon, 'A most incomparable man'. And thus we are made to understand from the outset that the play which follows these scenes of character-illustration will present a classic case of one who is wholly good and admirable and who is betrayed not by individuals who comprise a segment of his world, but by his entire world. We are to understand that Timon is literally too good for the world.

This first impression the dramatist promptly confirms by showing his hero in two actions that directly demonstrate the fact. Timon's first line spoken in our hearing refers to a friend's misfortune: 'Imprison'd is he, say you?' On being advised that Ventidius's debt is five talents, he neither hesitates nor asks assurances: 'I am not of that feather to shake off / My friend when he must need me.' He will pay the 'ransom' immediately, and, afterwards,

> . . . being enfranchis'd, bid him come to me.
> 'Tis not enough to help the feeble up,
> But to support him afterwards.
>
> (Ibid. 106–8)

This business is barely finished when an Old Athenian complains against a servant of Timon's whose courtship of the Athenian's daughter is unwelcome; Timon pauses only to make sure that the young people are in love; and then,

> This gentleman of mine hath serv'd me long;
> To build his fortune I will strain a little,
> For 'tis a bond in men. Give him thy daughter;
> What you bestow, in him I'll counterpoise,
> And make him weigh with her.
>
> (Ibid. 142–6)

Neither act bears a hint of ostentation or is performed with expectation of gain; both are acts of simple kindness, undertaken because— or so Timon's nature requires him to suppose—there 'is a bond in

286

men', and they evince that 'untirable and continuate goodness' of which his nature is composed.

With virtually the first half of the play given exclusively to verbal praise and illustrative incidents, it is plain that Shakespeare had in mind to establish Timon's character beyond mistaking, and this fact must be remembered when a problem of interpretation later arises. We are soon to see the hero transformed from a selfless lover of men to a hater of men whose detestation exceeds all bounds. So intemperate is Timon's hatred during the latter half of the play, so bestial does his former nobility turn, that temptation is then strong to forget what he had been—or, worse, to review and misconstrue the evidence of the earlier characterization. It has sometimes been erroneously argued that from the outset an element of impurity existed in Timon's nature that caused him to respond as he does to his misfortune—that bestiality lurked in his nature all the while and is merely bared by his misfortune. But to this view Shakespeare's unprecedented early emphasis on the precise quality of his hero's character lends no support; the dramatist's care to establish Timon's character advises us that what we are to witness afterwards is the change of a purely noble, selfless nature to its diametric opposite. The force of the tragedy—such as it is—depends on our recognizing that what occurs in *Timon* is a perfect reversal, and that this reversal occurs not, indeed, because of some lurking measle that was always present but because of the shock given his nature at the instant of his discovery that the way of the world is not remotely as, in his innocence, he had assumed it to be. It is because Timon's nature is pure, even as the first half of the play has represented it, that the shock is peculiarly intense and therefore ruinous.

At the same time that the first two acts establish Timon's character, they also, as we have noted, lift our awareness above his with respect to two facts—that his resources are exhausted and that none of his friends will open a purse to rescue him. Timon remains oblivious of the first until the middle of Act II and does not fully apprehend the second until about the middle of Act III—the midpoint of the tragedy. We begin to gain our advantage very early and easily; we scarcely need the Poet's pointed warning of the 'dependants' who let Fortune's chosen one 'slip down, / Not one accompanying his declining foot', or the Painter's boast that he can show 'A thousand moral paintings' that celebrate the same theme. Neither do we really need Apemantus's coarse admonitions: 'I should fear

those that dance before me now / Would one day stamp upon me.'
The extraordinary amount of emphasis that the dramatist gives to
the establishment of Timon's character provides us with all the
forewarning that our advantage really needs: even without much
dramatic sophistication we would expect a hero so selflessly noble
and so bounteous to beggar himself eventually; further, our own
very ordinary good sense enables us to recognize at once that the
way of the world is not at all as Timon thinks and that, once down,
he will be deserted by those he has befriended. But Shakespeare takes
no chances with our sophistication or our good sense; hence, from
the first lines onward he prods us with ominous warnings, some
direct, like Apemantus's barbs, some indirect, like the 'feignings' of
the two artists.

At the same time that we recognize the certainty of Timon's
friends' denying aid, we are also enabled to recognize that those who
frequent his house and enjoy his bounty are not practisers on him in
the usual sense—not such deliberate deceivers as we have met elsewhere before *Timon*. What Shakespeare means to convey through
the scenes that precede his hero's fall is that his visitors represent the
mill-run of common humanity, and not a special, limited class: the
Merchants, Lords, Senators who attend on him include 'all conditions, . . . all minds / As well of glib and slipp'ry creatures as / Of
grave and austere quality'. They number 'all kinds of natures / That
labour on the bosom of this sphere'. If the dramatist cannot bring all
Athens upon his stage, yet he can and does insist that those he does
bring on represent the entire spectrum; his 'free drift', like the Poet's,
'Halts not particularly', but 'flies an eagle flight'. We are to recognize
that though the flood includes some 'glib and slipp'ry creatures' and
'glass-fac'd flatterers' among the rest, none is bent on ruining Timon
or even bears him any ill will. When the Merchant rhapsodizes that
'''Tis a worthy lord!' and the Jeweller concurs with 'Nay, that's most
fix'd', we have no cause to suspect that either's thought and speech
are at odds. The throng includes no Iago to Timon's Othello, no
Edmund to his Edgar. The Jeweller has a stone to sell Timon 'If he
will touch the estimate', but there is no hint that the stone is worth
less than the figure that the Jeweller has in mind or that the Jeweller
intends any manner of deceit in offering it. Ventidius's appeal for
five talents to free himself from prison is an honest one and in no
sense a practice meant to hoodwink Timon; the Old Athenian whose
daughter loves a serving-man has not devised a plot to extort money,

but truly has a daughter and is truly concerned for her welfare. Though the Third Lord spoke admiringly of Timon's bay courser 'the other day' and was awarded the horse simply 'because you lik'd it', yet certainly the expression of admiration was honest, and Timon's gift springs from his free choice and not from his having been deceived. The Senator whose ominous word implies that Timon is often the victim of designing villains—'If I want gold, steal but a beggar's dog / And give it Timon; why, the dog coins gold'—makes the fact clear that men do indeed take advantage of Timon's generosity; yet this speech, too, characterizes the actions of men in general and does not single out the practices of aberrant individuals.

For Timon, as the entire first half of the play exists to advise us, mistakes the way not of such individuals—as Othello mistakes the way of Iago—but of the world, the general run of mankind. It is his inability to see what men in general are, not his failure to see through the devices of particular deviates from the general, that ruins him. When, too late, he recognizes his error, the shock to his nature is correspondingly severe. It is one thing to find that one has mistaken the character of a villain or two, and quite another to find that one has been wrong about the world of men.

The structure of awarenesses that comes into being during the opening speeches, simultaneously with the characterization of the pair of incompatibles, the hero and his world, remains essentially intact until the middle of Act III, when Timon, now aware both that his wealth is gone and that his friends are false, enters 'in a rage'. He had glimpsed the first truth midway through II. ii, when, after being surprised by the demands of Caphis and the servants of Isidore and Varro for money owed their masters, he retired from the scene and was then presumably set straight about his affairs by Flavius; for when he re-enters it is to demand, 'Wherefore, ere this time, / Had you not fully laid my state before me, / That I might have rated my expense / As I had leave of means?' Just here we abruptly lose half of our advantage over him. But the other—and much more important—half we retain until III. iv, when, not in our hearing, Timon receives reports of his servants' failure to raise even a single solidare by their pleas on his behalf.

The dramatist's opportunity to exploit his two discrepancies for dramatic effect inevitably arose with creation of our advantages over Timon, and thereafter exploitation has continued steadily until the moment that we lose all advantage when the hero recognizes that his pleas for aid have failed. Throughout this long period, the effects of exploitation have flowered without the dramatist's special effort. Each time a visitor utters a breath of extravagant praise, each time either Poet or Painter expounds upon his work with its obvious application to Timon, and of course each time Apemantus snarls a churlish warning, the gulf between our awareness and Timon's blindness to his perilous situation is illuminated—and virtually the entire first half of the play is composed of these words of praise, 'feignings', and warnings. Especially in the case of Apemantus's remarks—'. . . what a number of men eat Timon, and he sees 'em not' and '. . . my friends, if I should need 'em'—both the intent to exploit and the effects of exploitation are obvious even to excess.

After Timon has learned that he is bankrupt and we have lost one advantage, exploitation of the more important gap continues with rising intensity. Laments the loyal Flavius,

> Great Timon! noble, worthy, royal Timon!
> Ah, when the means are gone that buy this praise,
> The breath is gone whereof this praise is made.
> Feast-won, fast-lost; one cloud of winter showers,
> These flies are couch'd.
>
> (II. ii. 177–81)

Flavius's statement is actually the first direct assertion that Timon's friends will fail him, though of course we have known from the first speeches onward what the end must be; hence, for our own advisement, the statement is not significant. But it is vital to Timon, for it marks the first occasion on which his naïve assumption that all his admirers are loyal has been challenged, and he responds exactly as we should expect:

> Come, sermon me no further.
> No villanous bounty yet hath pass'd my heart.
> Unwisely, not ignobly, have I given.
> Why dost thou weep? Canst thou the conscience lack
> To think I shall lack friends? Secure thy heart;
> If I would broach the vessels of my love,
> And try the argument of hearts by borrowing,
> Men and men's fortunes could I frankly use
> As I can bid thee speak.
>
> (Ibid. 181–9)

Timon's speech takes its place in a long succession, running throughout the Shakespeare canon, of utterances made by heroes, heroines, and a wide range of other participants who, at a critical moment, totally mistake their situations. Here Shakespeare makes his hero's unawareness yield a maximum of effect by having him, in his excess of confidence, express pleasure at his misfortune:

> And, in some sort, these wants of mine are crown'd,
> That I account them blessings; for by these
> Shall I try friends. You shall perceive how you
> Mistake my fortunes; I am wealthy in my friends.
>
> (Ibid. 190–3)

His faith remains firm as he dispatches servants to Lucius, Lucullus, and Sempronius; but Flavius, for our benefit, mutters 'Lords Lucius and Lucullus! Hum!' Airily, Timon orders that a thousand talents be requested of the Senators, 'Of whom, even to the state's best health, I have / Deserv'd this hearing'. But his faith is jolted at the end of the scene when, having ordered the Senators to be solicited, he hears Flavius report that they have already been asked and have refused. In an earlier play Shakespeare would likely have advised us of this fact in advance and thus have opened a gap to be exploited throughout the passage in which Timon so confidently orders them solicited. In any event, Flavius's report of the denial shakes him—'Is't true? Can't be?'—though he makes an effort to laugh off the Senators' rebuff: 'Prithee, man, look cheerly. These old fellows / Have their ingratitude in them hereditary. / Their blood is cak'd, 'tis cold, it seldom flows.' Next he dispatches a servant to Ventidius, saying, 'When he was poor, / Imprison'd, and in scarcity of friends, / I clear'd him with five talents.' Again he speaks airily to Flavius: 'Ne'er speak or think / That Timon's fortune 'mong his friends can sink.' Earlier, he had dispatched servants to Lucius, Lucullus, and Sempronius with casual requests for fifty talents each, and of the Senators he had demanded a thousand. That he should now, at last, think of Ventidius, the only one of the friends we know to be directly indebted, and the paltry sum of five talents, suggests how near he is to apprehending the full truth. Then, in quick succession, we see the servants rebuffed by those to whom they have been sent, and in III. iv, Timon enters 'in a rage'. The gap between his awareness and ours has closed abruptly, with a roar.

The gap has been of long duration, was central, and has been exploited steadily, with rising intensity; it has, in short, the three characteristics of such productive discrepancies as we noted in

Romeo and Juliet, *Hamlet*, and *Othello*. Accordingly, we should have expected it to produce dramatic effects comparable in force to those of the earlier plays, and to involve our emotions as profoundly. But if Timon actually claims our sympathy for himself and his plight, it is a curiously uninvolved, academic sort of sympathy the intensity of which falls far below that we feel for the earlier protagonists. Even Macbeth, who proves himself a conscienceless villain bent on murder for selfish gain, makes, in the end, a stronger bid for our compassion than Timon, who is shown to be wholly noble, selfless, and good, and who is betrayed in a classic instance of ingratitude.

Whatever were his intentions in the elaborate and insistent characterization of Timon and his incompatibility with the way of the world, what Shakespeare achieved is only the representation of a hypothetical 'case'. Romeo, Brutus, Hamlet, Othello, Lear, Macbeth, and Antony affect us not as cases but as men; they do not illustrate, exemplify, or 'stand for', but directly *are*. But Timon, though one admirer after another voices praise of him during two acts and though he shows himself to be truly noble and good, emerges for us only as an illustration of an argument, more abstraction than man.

In *King Lear* we are introduced to the protagonist by a succession of individuals who have identities of their own—Kent, Gloucester, Edmund, and, shortly afterwards, Cordelia, Regan, Goneril; we are immediately caught up in a conflict of human personalities and contradictory motives, and after only ninety lines, when Cordelia shatters the formalized proceedings with her startling 'Nothing, my lord', we are already inextricably involved with beings that have flesh, blood, and vitality and appear capable of rushing in any direction and carrying us along involuntarily. But in *Timon* we are introduced to the hero by a succession of nameless and faceless examples of the way of the world: a Painter, a Poet, a Merchant, a Jeweller, sundry Lords, Senators, an Old Athenian; none has any real being, and all serve the common function of marking the contrast between Timon's world, of which they are collectively the symbol, and Timon's character. Timon, too, when he enters to stand, speak, and perform before us, similarly lacks being and serves to complete this same contrast. When we see Lear mistaken in supposing that Regan and Goneril love him and that Cordelia and Kent do not, our emotions become agitated, and we feel for the protagonist; but when we see Timon grossly mistaken in and cruelly cheated by the world that he has thought as noble and selfless as himself, we remain curiously unaffected. It appears, then, that an advantage in awareness, in itself, carries no guarantee of emotional involvement, but functions effectively only when we have been made to care about the participants over whom we hold advantage. We care deeply about Romeo and Juliet, Othello, and Lear, and therefore our sense of their unawareness in the midst of peril burdens us with anxiety and distress; but if we care about Timon at all, it is only as an abstraction.

During the brief period just before Timon learns that the world is false, Shakespeare does make an effort to stir our sympathies. In succession, he shows us the fury of the servants who solicit their master's friends and are rejected, as though a display of their emotions might awaken ours. The first servant hurls back a proffered gratuity and adds a curse that modestly anticipates the curses of Timon that fill the second half of the play; the second servant turns away with only a bitter remark or two; but the third, after being denied, first castigates Sempronius and then extracts a pointed and heavy moral from the case. Besides the servants' reactions, the dramatist shows us those of three Strangers who hear of Timon's wrongs. Being strangers, unacquainted with Timon, these three must be thought of as impartial; hence their remarks should especially affect us. Among them they speak about twenty-five lines the tenor of which is singular: '... O, see the monstrousness of man / When he looks out in an ungrateful shape!' cries the first, and continues,

> I never tasted Timon in my life,
> Nor came any of his bounties over me
> To mark me for his friend; yet, I protest,
> For his right noble mind, illustrious virtue,
> And honourable carriage,
> Had his necessity made use of me,
> I would have put my wealth into donation,
> And the best half should have return'd to him,
> So much I love his heart.
>
> (III. ii. 84–92)

We cannot know, certainly, whether the Third Stranger, if put to the test, would in fact have responded differently from the Lords and Senators who turned their backs; but the dramatist's effort to stir our own sympathy by showing that even disinterested passers-by are moved by Timon's wrongs is obvious; yet, alas, this direct bid also fails.

Shakespeare's efforts to stir our sympathy do not stop with the outraged expressions of servants and strangers. It was surely not by chance that the scene in which the Senators deny the plea of Alcibiades for the life of a nameless friend is set to follow that in which the ugly fact of man's ingratitude at last penetrates Timon's consciousness. Alcibiades' plea is among the few moving passages in the play; it is filled with passion and looks more finished than other prominent passages that were apparently left in rough cast. The case for the friend, who had killed another man 'in hot blood' because he saw that his own reputation was 'touch'd to death', is forcefully made. Alcibiades speaks 'like a captain' on behalf of a valiant soldier; he acknowledges, but does not seek to extenuate, the man's fault: 'O my lords, / As you are great, be pitifully good. / Who cannot condemn rashness in cold blood?' The nameless soldier had done noble service for the State: 'His service done / At Lacedaemon and Byzantium / Were a sufficient briber for his life.' Further, when the most that he can say for his friend fails utterly, Alcibiades places his own accomplishments in the balance:

> My lords, if not for any parts in him—
> Though his right arm might purchase his own time
> And be in debt to none—yet, more to move you,
> Take my deserts to his and join 'em both.
>
> (III. v. 76–9)

But this plea, too, fails, and the Senators condemn the man. Alcibiades continues to argue with great passion, and at last, for his pains, gains a penalty as severe as that given his friend: 'We banish thee for ever.' Alcibiades' rage at this verdict anticipates the intensity of Timon's fury that dominates the rest of the play.

By introducing the parallel case of Alcibiades, Shakespeare clearly meant to heighten our experience of Timon's wrong, even as he meant Gloucester's ordeal to heighten our experience of Lear's. Three cases of gross ingratitude figure in the play: that of Timon, that of Alcibiades, and that of the nameless friend who had done noble service to the State. At the same time, the outraged reactions of Timon's servants and the three Strangers to Timon's wrongs are meant to serve a similar function by prodding us into feeling for Timon's case; further, these reactions may be intended to suggest that, however extreme it will prove to be, Timon's subsequent reaction to the wrong done him is not excessive but appropriate. Though he does not, like Timon, rage throughout two and a half acts, yet Alcibiades rages as violently, and, besides, has means to strike back at Athens whereas Timon has no vent but words for his passion.

The most potent exhibition of others' concern for Timon's wrong is reserved to IV. ii, when Flavius and the other servants part from Timon's house and one another. Reminiscent of those scenes in *Antony and Cleopatra* when Enobarbus, soldiers, and servants weep for the fallen Antony, this scene appears calculated by its very placement—Shakespeare sets it just after we have heard Timon's first sustained and most horrendous curse—to prod our recollection of what Timon was before his reversal and thus to compel our sympathy. 'So noble a master fall'n!' cries one servant, and 'Yet do our hearts wear Timon's livery', cries another. Timon's goodness has permeated and infected them all: 'Wherever we shall meet,' says Flavius as they embrace, 'for Timon's sake / Let's yet be fellows.' Flavius's function here resembles Enobarbus's after Antony has sent his treasure after him; it is Flavius who is most affected and who best recapitulates and points Timon's tragedy:

> Poor honest lord, brought low by his own heart,
> Undone by goodness! Strange, unusual blood,
> When man's worst sin is he does too much good!
> (IV. ii. 37–9)

So he continues at length, and ends with 'Alas, kind lord! / He's flung in rage from this ingrateful seat / Of monstrous friends.' This baldly instructive scene, set between Timon's first terrible outburst and his second, serves to remind us, at a moment when we are tempted to forget, of Timon's former nobility. Subsequent scenes that show Timon raging, though they ring several variations on the theme of misanthropy and exhibit numerous ingenious forms of vituperation, never show him in a more unflattering light than those that initiate his long spell of cursing; it is when we first see him in his new savagery that we most urgently need Flavius's reminders of his former nobility.

The remaining scenes of the play, when successively Alcibiades and his whores, Apemantus, Banditti, Flavius, Poet and Painter, and the Senators visit Timon in the woods, seem calculated also to prevent us from being simply repelled by the protagonist as one whose hatred is excessive. Had he wished, Shakespeare could have made the world that Timon reviles begin to take on a less odious look after the fall; but then our sympathy would surely have been all the more difficult to win, for we would have been quick to judge the hero's misanthropy as exceeding due measure. The dramatist avoids that mishap by making all the former acquaintances, and thus the world they represent, look even worse than before. Even Alcibiades, the first and best of the visitors, appears tarnished in the company of his two whores, one of whom he calls 'sweet' Timandra. Though the warrior is solicitous of Timon's welfare and offers him gold, it is clear that his real concern is not for Timon but for his own revenge on Athens; if Timon is avenged, it will be only incidentally. As for the whores, who are totally foul, their manner caricatures the way of the world that Timon has abandoned: like his former friends, they accept his new-found gold, ask for more, boast that 'we'll do anything for gold', and flatter him with 'bounteous Timon'. Apemantus, next to appear, engages in an intolerably long and ludicrous exchange of abuses with Timon the effect of which is to make both speakers look as disgusting as the world they revile. The Banditti, who next enter, make the world look worse by showing themselves better than it; but Flavius, the next to arrive with all his goodness still shining, magnifies by contrast the world's vileness.

Yet it is the visit of the Poet and the Painter, followed by that of the Senators that most vividly assures us of the world's abiding vileness. In the first scene of the play the hypocrisy of this pair was implied but not directly asserted; now, as they approach Timon's cave, they are made to express their conviction that their former benefactor still has vast wealth and 'Therefore 'tis not amiss we tender our loves to him in this supposed distress of his'. This scene alone in the second half of the play exploits a situation of unequal awarenesses. Timon, overhearing their plan to promise but not produce gifts of new artistic works, gains for once, and perhaps even relishes holding, an advantage. Having led his sycophants on to a point of great expectations, he turns suddenly upon them, berates them as villains and dogs, beats them out, and retreats to his cave. This situation of the tables turned is one that we might normally expect to be productive of rich dramatic effect, but, curiously, the effect is minimal, as in the earlier climactic turn when Timon served his flatterers a banquet of tepid water; like the earlier failure, this one again suggests that we simply do not care enough about Timon to applaud his momentary triumph. Timon's victory over his final visitors, the Senators who come to offer 'heaps and sums of love and wealth' if he will accept the captainship of Athens to repel the 'approaches wild' of Alcibiades, follows the same tables-turned pattern, and it, too, yields less dramatic satisfaction than we should have expected—especially since Timon's victory here culminates in a glorious speech in which he invites all his friends, all Athens, to use his tree to hang themselves.

In his reading of Plutarch for *Antony and Cleopatra* Shakespeare would certainly have seen and was no doubt struck by the half-dozen lines in which Timon, standing in the pulpit of the market-place, graciously invites the citizens of Athens to come and hang themselves on his fig tree before he cuts it down. It is tempting to suppose that this incident alone attracted him to the subject and that the entire tragedy was constructed as a frame for Timon's monumentally ironic invitation. What in any event is evident is that little else in the play shows a sign of having seriously engaged the dramatist's mind. From his seeming lack of interest in Timon and his friends as human individuals involved in a real situation, as opposed to mere abstractions figuring in a 'case', follows all that is dramatically unsatisfactory. For it is not only the chaotic condition of the text, with its unaccountable mixtures of prose and verse, its wild tossing-about of the value of talents, its confusion of entrances and exits—not only such gross examples of negligence as these that evince a lack of commitment to the undertaking; it is also the nameless Lords, Senators, Strangers, Poet, Painter, Merchant, Jeweller, old Athenian, and even the named figures—Ventidius, Lucullus, Sempronius, Lucius—who, functioning as mere illustrations of the way of the world, betray the dramatist's indifference. Even Alcibiades—who would have been a better choice for tragic hero than Timon—emerges as just another item of evidence in Timon's case against the world.

And, finally, there is Apemantus, who participates in segments that total about four hundred lines—a number hugely disproportionate to his dramatic usefulness. Twice Apemantus engages in abusive dialogue that extends to eighty lines, and his final session of trading insults with Timon at the cave extends to two hundred. At his first appearance Apemantus aids in creating our initial advantage over Timon by assuring us that his wealth will soon be gone and that his friends will then abandon him. Further, he serves as a constant against which we can measure Timon's transformation: during the first half of the play his churlish misanthropy heightens by contrast our sense of Timon's goodness; during the latter half this same churlishness appears actually moderate and restrained beside the impassioned loathing that eats Timon and makes him, too, loath-some. So much for his legitimate dramatic service; but in addition we must suspect that Apemantus and his tirades serve mainly by supplying so much sheer padding for a play in which there is simply not enough of dramatic substance or action to fill up five acts. The eighty-line exchanges of Apemantus with Timon and others in I. i, and with the Fool and the creditors' servants in II. ii, are mere inter-ludes that take time out from such action as is in progress. Most obnoxious of all is the two-hundred-line exchange of abuses presented in IV. iii, after Timon's fall, when the chronic and the acute misanthropes vie in venting hatred for man in general and each other in particular. Whatever effect this episode was meant to produce, it contributes not at all to tragic effect. Sometimes merely trading epithets—'Beast!' . . . 'Slave!' . . . 'Toad!' . . . 'Rogue, rogue, rogue!'—and sometimes exchanging elaborately contrived insults, the rival haters strive to outdo each other as though they were comedians. On the face of it, we might expect that bringing together two similar yet dissimilar haters, the chronic carper and the sudden convert, would accomplish something remarkable, like bringing together the antic Edgar and the mad Lear. But the effort fails, and midway of the long exchange the context of the play itself tends to slip from our consciousness, leaving but the spectacle of two buffoons striving to entertain us with foul words. Perhaps nothing more strongly suggests Shakespeare's lack of commitment than the mere presence of this endless exchange late in Act IV, when, in all the earlier tragedies, the final march to the catastrophe has begun.

2. *The Innocent of Rome: Coriolanus*

The decline in Shakespeare's use of unequal awarenesses to create dramatic effect has been noted in both *Antony and Cleopatra* and *Timon of Athens*; what had been a basic principle of his method earlier becomes, in these plays, a relatively unimportant element in a few scenes, a primary element in none. With *Coriolanus* the decline continues: if *Romeo and Juliet*, *Hamlet*, and *Othello* represent the dramatist's fullest exploitation of gaps between awarenesses, *Coriolanus* represents his near abandonment of it. And if it is apt to call *Romeo and Juliet* and *Othello*, in particular, tragedies of un-awareness, it is also appropriate to call *Coriolanus* a tragedy of awareness. In the earlier tragedies, *Hamlet* mainly excepted, catastrophe strikes heroes who are unknowing and does so *because* they are unknowing. But Coriolanus is well aware of the destructive forces both within and without his own nature, and disaster strikes despite the fact.

To say so much is not to say that *Coriolanus* totally lacks scenes in which discrepant awarenesses occur; we do, in fact, hold secrets that are unknown to certain participants in nine scenes. We are not, however, as in earlier tragedies, possessed of a great central secret; nothing here is comparable to our private knowledge that Fate is bent on sacrificing Romeo and Juliet or to our knowledge that 'honest' Iago is dishonest. Further, in this play neither the course of action nor its outcome would be essentially changed if the few details privately known to us were known also to the protagonist— for indeed virtually all that is known to us is also known to him. Though three active practisers are at work against the hero—two of whom, Sicinius and Brutus, are incessantly busy—Coriolanus is never really deceived by these enemies. Though Timon and Coriolanus are alike in being innocents in worlds with which their natures are incompatible, they differ diametrically in that whereas Timon, during the first half of the play, is wholly blind to his situation, Coriolanus is keenly aware of his: whereas Timon fails to recognize the reality, Coriolanus sees it clearly but will not alter his conduct to accommodate it.

Before attacking the central problem of the hero's relation to his world, we should briefly survey the scenes in which we hold advantage over participants, acknowledging in advance that these are relatively few and unimportant. First is I. vi. In I. v, we have seen Marcius and Lartius triumph in Corioli; but now a messenger advises Cominius that the Roman forces have been driven back to their trenches, so that, for about thirty-five lines, he remains ignorant of the truth known to us. Except that the interval of uncertainty prepares for the jubilant reunion of Cominius and Coriolanus directly afterwards, it contributes negligible effect. No further instance occurs until III. i, when again the exploitable interval is brief and hardly productive. In II. iii, we have seen Brutus and Sicinius prevail with the citizens, who, earlier, had given their voices for Coriolanus's consulship; now they agree to confront the new consul at the Capitol and revoke their approval. Thus, in III. i, for a space of twenty-five lines, Coriolanus and Menenius, with others of the nobility, remain ignorant of the reversal that the tribunes have effected. Marching confidently towards the Capitol and speaking of the Volsces' defeat, they are suddenly barred by Sicinius and the commoners with the command to 'Pass no further', and the brief space of their unawareness comes to an end. Though the tribunes have worked covertly to manipulate the citizens' minds, they have not deceived Coriolanus, who instantly charges them with having 'set them on', and Cominius, too, flatly asserts that 'The people are abus'd'. The same pattern is repeated in the next instance of our advantage over Coriolanus, which occurs in III. iii. Here the scene opens with the tribunes directing the Aedile how to prime the citizens to demand the hero's death just as soon as Sicinius and Brutus, with charges of treason, shall have 'put him to choler'. Coming innocently to this meeting, unaware that a trap has been set and determined to bear himself humbly in order to regain the consulship—as he would easily succeed in doing but for the tribunes' prior coaching of their mindless constituents—Coriolanus falls quickly into the trap, losing his temper at the first shout of 'traitor'. But he would doubtless have reacted exactly as he does had he known of the trap in advance; thus, though the tribunes' practice achieves its purpose, no one has been fooled: says Volumnia immediately afterwards, ''Twas you incens'd the rabble.'

In all the play the one scene that approximates the effects created by exploitation of unequal awarenesses in earlier plays—and that would be elaborately recreated in the dramatic romances that were to follow *Coriolanus*—is IV. v, when the banished hero, muffled and incognito, banters with servants who deny him entrance to Aufidius's house. Here, briefly, the hero commands a situation like that enjoyed by a succession of disguised heroines—Portia, Rosalind, Viola—and by, among others, the hooded Duke in *Measure for Measure*, the King, masquerading among his troops as a plain soldier in *Henry V*, and Pericles, his royalty hidden in rusty armour at the court of Simonides. Ignorant who it is that confronts them, the servants threaten by turns to thrust the mighty warrior out of doors, and though Coriolanus is himself of no humour to relish his advantage, for us the situation generates a welcome interlude of comic relief at the same time as it gives the unique sort of thrill that arises from the spectacle of greatness unrecognized and mistreated by persons who would cringe and stammer if they but knew. Near the end of his encounter with the unknowing servants, even the grim and humourless 'lonely dragon' seems almost to enjoy the banter:

3 Serv. How, sir! do you meddle with my master?
Cor. Ay; 'tis honester service than to meddle with thy mistress.

(IV. V. 51–3)

But the fun is brief, for in the next moment Aufidius enters and Coriolanus breaks off the lone episode that recalls many on which the dramatist formerly lavished great care in setting up and exploiting and which he clearly regarded as worthy of large space in comedy, history, and tragedy alike. But in *Coriolanus* it is apparent that graver matters were on his mind.

The balance of this scene, showing Coriolanus's meeting with Aufidius and ending with the servants' advice that the two leaders will immediately move against Rome, prepares us to enjoy the next exploitable discrepancy, which occurs in the following scene. Here, during some thirty-five lines, the tribunes and their constituents are exhibited gleefully congratulating themselves on the hero's banishment and the 'happier and more comely time' in which Rome now basks. On this occasion the universal ignorance of what we know— that Coriolanus is even now advancing on Rome with the intent of burning it—is repeatedly exploited. Sicinius boasts of 'the present peace', taunting Menenius with 'Your Coriolanus is not much missed' and 'The commonwealth doth stand / And so would do were he more angry at it.' A citizen, blissfully ignorant of the imminent disaster that the tribunes have invited by banishing Coriolanus, asserts that 'Ourselves, our wives, and children, on our knees, / Are bound to pray for you both.' An instant before an Aedile enters with word that the Volsces have entered Roman territory, Brutus boasts to Menenius that 'Rome / Sits safe and still' without Coriolanus. At first news of the invasion, the tribunes refuse to believe, and Brutus orders 'this rumourer' whipped. Their reluctance to acknowledge the truth is extended through another twenty lines, until Cominius comes to confirm the worst: 'He'll shake your Rome about your ears.' Well prepared in advance and thoroughly exploited, these brief moments during which we hold a strong and gratifying advantage over the treacherous tribunes are among the most agreeable in the play.

Briefly during the closing scenes exploitation again produces potent, but this time painful, effects. But in order to set these moments in perspective we must first resolve certain problems about the state of Coriolanus's awareness. These are complicated by two heavily weighted statements made early by Aufidius, the first of which occurs at the end of Act I, after the Volsces' defeat; says Aufidius,

> If e'er again I meet him beard to beard,
> He's mine, or I am his. Mine emulation
> Hath not that honour in't it had; for where
> I thought to crush him in an equal force,
> True sword to sword, I'll potch at him some way;
> Or wrath or craft may get him.
>
> (I. X. 11–16)

He next vows that no circumstance can diminish his hatred, but,

> Where I find him, were it
> At home, upon my brother's guard, even there,
> Against the hospitable canon, would I
> Wash my fierce hand in's heart.
>
> (Ibid. 24–7)

This unequivocal vow, which the dramatist clearly means to fix securely in our consciousness, warns us that at whatever time he shall find opportunity to pay Coriolanus home, Aufidius will do so. But subsequent events appear to compromise the forewarning; for when Coriolanus meets Aufidius after being banished and offers himself to the Volsces for an attack on Rome, Aufidius seems genuinely moved:

> O Marcius, Marcius!
> Each word thou hast spoke hath weeded from my heart
> A root of ancient envy.
>
> (IV. V. 107–9)

He embraces his old rival, calls him 'Worthy Marcius', insists that he will henceforth 'contest / As hotly and as nobly with thy love / As ever in ambitious strength I did / Contend against thy valour', asserts that he would gladly fight against Rome if there were no other cause 'but that / Thou art thence banish'd', and concludes by offering 'The one half of my commission' to Marcius for the assault. Later a servant tells us that 'Our General himself makes a mistress of him, sanctifies himself with 's hand, and turns up the white o' the eye to his discourse.' If Aufidius is not perfectly sincere in these protestations, yet we are given no hint of the fact; and if he is indeed sincere, then, it appears, we are obliged to disregard his earlier vow to kill Coriolanus 'Where I find him'.

In that case, we hold no advantage over Coriolanus when, with characteristic innocence and faith, he embraces the offer to share the

Volscian leadership. We cannot know, at this moment, whether or not he stands in peril. But in Aufidius's very next scene we find cause to question his sincerity, and to do so the more intently because Shakespeare gives an entire scene to the new warning. A lieutenant reports that 'Your soldiers now use him as the grace 'fore meat, / Their talk at table, and their thanks at end, / And you are dark'ned in this action, sir, / Even by your own.' Aufidius acknowledges that Coriolanus 'bears himself more proudlier, / Even to my person, than I thought he would / When first I did embrace him', and then goes farther:

> When he shall come to his account, he knows not
> What I can urge against him. Although it seems,
> And so he thinks, and is no less apparent
> To the vulgar eye, that he bears all things fairly,
> And shows good husbandry for the Volscian state,
> Fights dragon-like, and does achieve as soon
> As draw his sword, yet he hath left undone
> That which shall break his neck, or hazard mine,
> Whene'er we come to our account.
>
> (IV. vii. 17–26)

Speech and scene close on a dismal promise: 'When, Caius, Rome is thine, / Thou art poor'st of all; then shortly art thou mine.' This is plain talk that gives us immediate advantage over Coriolanus, who is doomed either way: if he falters in his assault on Rome, he is doomed; and if he faithfully burns Rome, yet Aufidius will destroy him.

The closing scene of Act IV thus casts a chilling shadow over all six scenes of Act V—over the hero, his family, and his friends. First Cominius, newly returned from the Volsces' camp, reports failure to win mercy for Rome and describes the seemingly godlike place of Coriolanus among the enemy forces:

> I tell you, he does sit in gold, his eye
> Red as 'twould burn Rome; and his injury
> The gaoler to his pity. I kneel'd before him;
> 'Twas very faintly he said, 'Rise'; dismiss'd me
> Thus, with his speechless hand.
>
> (v. i. 63–7)

'... his eye / Red as 'twould burn Rome'; perhaps in part the dreadful sound of this is meant for the ears of the tribunes, whose discomfiture at the words we might relish immoderately but that the direr sound of Aufidius's warning at the end of the preceding scene still rings in our ears; for while the advantage appears all Coriolanus's and the despair all Rome's, the hero's fate is certain, whereas Rome has still the possibility of salvation, for Menenius, Volumnia, and Virgilia have yet to try their pleas for mercy. But for Coriolanus there is no escape.

In v, ii we are shown the doomed hero as, imperiously, he dismisses Menenius's plea: '. . . be gone. / Mine ears against your suits are stronger than / Your gates against my force.' Turning, he bids Aufidius note well his fidelity to the Volscian cause: 'This man, Aufidius, / Was my belov'd in Rome. Yet thou behold'st.' Betraying nothing of what we know lies in his heart, Aufidius replies, 'You keep a constant temper.' Next, in v. iii, our advantage over the hero still holds while he first begins to weaken at the sight of his family, then gives way utterly to their entreaties. Yielding, he recognizes that the success of his mother's plea has doomed him:

> You have won a happy victory to Rome;
> But, for your son,—believe it, O believe it—,
> Most dangerously you have with him prevail'd,
> If not most mortal to him.
>
> (v. iii. 186–9)

Here the hero's unawareness of Aufidius's intent to destroy him whether or not he burned Rome strongly reflects the nobility of his decision not to do so: had he known that he was doomed either way, his decision would have been only indifferently noble; but in deciding not to burn Rome when he believes that the decision will prove fatal to him whereas a decision to do so would secure his place with Aufidius and the Volsces, he reveals a true greatness of soul and a selflessness that relate him to Timon. Further, having made the fatal decision, he again shows the purity of his honour by giving no thought to the chance of saving himself by abandoning the Volsces: 'I'll not to Rome, I'll back with you,' he tells Aufidius. And here again the dramatist prods our awareness of the hero's peril by having Aufidius speak aside: 'I am glad thou hast set thy mercy and thy honour / At difference in thee. Out of that I'll work / Myself a former fortune.' At the end of the scene Coriolanus remains ignorant that Aufidius meant to destroy him even if he had burned Rome, whereas Volumnia, happily returning to Rome, remains ignorant that her success in persuading him has sealed his doom—yet, ironically, unaware also that it was sealed in any event by Aufidius's prior vow.

In v. iv, Menenius, unaware that Volumnia has triumphed and that Rome is no longer in danger, continues, with obvious relish, to twist the knife in the flesh of the tribunes; thus of Coriolanus:

The tartness of his face sours ripe grapes; when he walks, he moves like an engine, and the ground shrinks before his treading. He is able to pierce a corslet with his eye; talks like a knell, and his hum is a battery. He sits in his state, as a thing made for Alexander. What he bids be done is finished with his bidding. He wants nothing of a god but eternity, and a heaven to throne in.

(v. iv. 18–26)

Volumnia, he asserts, has no more chance of moving her son than Sicinius has of moving 'yond coign o' th' Capitol' with his little finger. The scene involves our own awareness in a conflict of forces. Though we enjoy, with Menenius, the discomfort of the vicious tribunes and perhaps rejoice that Rome will not, after all, be burned, we cannot easily forget that Coriolanus, who, so far as Menenius knows, 'wants nothing of a god but eternity', is in fact at this moment marching back towards Corioli and his death. Similarly, in the next scene of only seven lines, when all Rome rejoices as the triumphant ladies pass over the stage, our own awareness attends the doomed hero.

The details of the final practice are arranged by Aufidius and the conspirators during the first sixty lines of the closing scene. The practice is reminiscent of that devised earlier by the tribunes, who won the hero's banishment from Rome by first putting him into a passion with taunts of 'traitor' and then trusting to his own fury and the shouts of the well-coached citizens to do the rest. Since the final trap is set just before Coriolanus's entrance, with the commoners crowding about and hailing him as he marches with drum and colours, we hold the most distressing advantage that has been ours at any time in the action. The hero knows that he will have to justify his decision to spare Rome, but, in his innocence, assumes that he can count on Aufidius's support. Shakespeare makes the fact apparent that, but for the plot devised by Aufidius, Coriolanus's rigidly honest report made to the Volscian lords, with the commoners already wholly his, would clearly salvage his honour, his high place, and his life: 'Our spoils we have brought home / Doth more than counterpoise a full third part / The charges of the action. We have made peace / With no less honour to the Antiates / Than shame to the Romans.' As always, what he speaks is perfect truth, and it is noteworthy that, even after the crooked Aufidius has said the worst of him and has set on his accomplices to demand his instant death, the First and Second Lords demand peace and a fair hearing; says the Second Lord,

> Peace, ho! no outrage: peace!
> The man is noble and his fame folds in
> This orb o' th' earth. His last offences to us
> Shall have judicious hearing.

(v. vi. 125–8)

Coriolanus himself, though he reacts with a flash of rage, as in Rome, at the hated word 'traitor', nevertheless manages to speak with remarkable restraint: 'Pardon me, lords, 'tis the first time that ever / I was forc'd to scold. Your judgements, my grave lords, / Must give this cur the lie.' It is evident that his manner and the expressed disposition of the Lords would even yet win him at least a hearing but for Aufidius's sudden, calculated, and—by Coriolanus—wholly unexpected attack, which immediately ensues and to which the citizens of Corioli—who, moments earlier, had wildly greeted Coriolanus as their returning hero—respond with the same mindless surge with which, earlier, their Roman counterparts responded to the tribunes: 'Kill, kill, kill, kill, kill him!'

The foregoing survey of incidents during which we hold advantage over certain participants reveals that, though the dramatist obviously wished us at times to be cognizant of our advantage, he took no special pains, as in earlier plays, to manipulate situations for the primary purpose of developing exploitable gaps. The time had been —indeed, had endured from the beginning until *Antony and Cleopatra*, and would return with the last plays, the dramatic romances—that, for Shakespeare, the ideal dramatic situation was that which offered the richest exploitable gap. In many of the most memorable scenes of earlier plays the gap itself literally *is* the dramatic situation: Malvolio strutting cross-gartered before Olivia under the practice-induced illusion that she dotes on the fashion; Romeo drinking poison above Juliet's just-waking form; Othello executing Desdemona as an act of 'justice'. Such scenes mesmerize us not by their action merely but by the force of our private knowledge upon our view of the action.

But in *Coriolanus*, instead of seeking to devise such situations, the dramatist appears sometimes even to shun the former method, as if that way meant theatrical trickery and was inferior to and incompatible with some newer purpose. Though he shows the tribunes, and later Aufidius, plotting Iago-like, yet he does not do so for the primary purpose of creating exploitable discrepancies. With the exception of the brief scene in which the muffled Coriolanus banters with Aufidius's unwitting servants, he not only develops such situations sparingly but makes negligible exploitation for dramatic effect when they do occur. Though the hero is briefly ignorant first at Rome and then at Corioli that his enemies have devised practices against his life, yet the dramatist neither exploits the discrepancy for passing effect nor represents it as an important factor in the course of the tragedy: Coriolanus, being what and as he is, would have confronted the tribunes and their mob on the final occasion in Rome even had he known in advance what trap was laid for him; and, just so, had he known in advance what we know of Aufidius's scheme to kill him upon his return to Corioli, he would still have proceeded just as he does proceed, spurning the chance to escape by returning to Rome.

In short, then, it appears that the deliberate creation of exploitable conditions was farther from Shakespeare's mind in *Coriolanus* than at any previous time in his career. Something else, and very much different, was evidently on his mind; hence the questions most appropriate to ask of this play differ in kind from those we ask of other tragedies, with the partial exception of *Timon*, the unfinished and chaotic condition of which leaves uncertainty about the dramatist's intention. For in both these tragedies it appears that Shakespeare undertook to construct a kind of 'case', and to present it, perhaps, without bias: thus what we have is not the case for Coriolanus, or the case against Coriolanus, but the case *of* Coriolanus. To this end, instead of a succession of exploitable situations, the dramatist provides us with a relentless flow of evidence about the hero's character and his deeds, rather as though it were a legal matter on which we are finally to render a verdict.

In earlier tragedies, too, of course, a considerable amount of discussion of the hero's character takes place. We hear of Romeo, for example, and his love-sickness, before we meet him, and there-after at intervals we hear him discussed by all the principal persons of the play—Juliet, Mercutio, Benvolio, Tybalt, the Nurse, Capulet, the Friar. The same is true of Brutus, Hamlet, Othello, Lear, Macbeth, Antony, none of whom escapes being talked of, sometimes briefly, sometimes at length, sometimes generally, sometimes particularly: Lear's daughters discuss their father's bad temper and slight understanding; Claudius and Gertrude, with Polonius, Ophelia, and others, fairly anatomize Hamlet, or try to; Duncan, Lady Macbeth, Banquo, the Witches, Lennox, and others all have their say on the character of Macbeth; Cassius soliloquizes on Brutus, Iago on Othello. In all the tragedies such discussions occupy space and are among our major sources of information on the hero; they are, in short, the serviceable, indispensable devices by which the dramatist establishes our view of the protagonist.

But in *Coriolanus* discussion of the hero by other participants assumes an unprecedented proportion and serves a purpose that goes far beyond the usual. Whereas in the earlier tragedies its obvious function is characterization of the hero so as to explain and make credible his part in the action, in *Coriolanus* the prodigiously elaborate and nearly incessant discussion of the hero appears rather to be an end in itself. In this tragedy Shakespeare is intent on representing the case of Coriolanus, of which the heart is the nature of the man himself; hence not only elaborate discussion of the man by others but the very action of the play serves to illustrate his nature. The proportion of scenes directly given to the character-ization of this hero, who is, in fact, steadily subjected to analysis by others or placed on exhibit by illustrative action, will best be represented by statistics. The tragedy is composed of twenty-nine scenes. Of the ten in Act I, five—i, iii, v, vi, and ix—are dominated by discussion of the hero's character, and of the remaining five, all but Scene vii include passages and individual lines that bear directly on it. In Act II, all three scenes are weighted heavily with such comment. In Act III, again, all three scenes are given to discussion of the hero. In Act IV, only Scenes i and iv—which the hero's own presence dominates—include nothing said about him by others, while the remaining five scenes present both extended and passing com-ment on him—most notably Scene vii, where Aufidius offers the finest single assessment to be found in the play. In Act V, only Scene v (containing but seven lines) has no discussion of Coriolanus; of the remaining five, the first four are dominated by such discussion. Thus, of the full twenty-nine scenes, only four—all brief and special in their functions—include no discussion of Coriolanus's character; of the twenty-five that present such discussion, twenty may be said to be dominated by it. When compared with corresponding statistics for earlier tragedies (*Timon* again excepted), these proportions are startling, and they reflect the drastic change of both dramatic method and purpose that makes *Coriolanus* unique.

Impressive, too, is the sheer number of participants who discuss the hero's character. Not only all the principal persons, who speak repeatedly and often lengthily about him, but literally every person who speaks in the play at all speaks of him. Citizens, Senators, Soldiers, Messengers, Sentinels, Servants, both in Rome and in Corioli—all sum up his parts and render their verdicts as though, for the dramatist's special new purpose, every facet of the hero's nature must be illuminated, and from every angle. The mass of testimony thus offered in the case of Coriolanus is enormous; a count of individual lines reveals it to be approximately fifty times that given to the character of Hamlet. What is more, this mass represents a vastly broader spectrum of personal assessments than is brought to bear on Hamlet, who is directly characterized by only five persons, and then briefly and incidentally. Further, whereas Hamlet is a hero of formidable complexity, whose elements amount to a veritable universe of human qualities, attitudes, feelings, ideas, Coriolanus is a man of remarkable simplicity; only Timon, whose whole sum of parts consists of unqualified goodness that turns to unqualified sourness, is less complex. With Hamlet, after all evidence of every sort has been tallied, it is still necessary to suppose that the whole man, if we could come to know him wholly, would prove even more complex; but with Coriolanus it is plain that no more exists than is exhibited: this hero does not appear simple because the dramatist neglected to show all of him, but because he is a simple man.

Necessarily, then, though many persons representing many views speak almost incessantly about him, all say essentially the same things; it is only their own attitudes that vary and conflict: the citizens and the tribunes hate and fear him; his mother, Cominius, and Menenius love and admire him; Aufidius hates and admires him; and various soldiers, servants, and others reflect diverse attitudes. Yet all recognize what Coriolanus essentially is, and all point straight to the same narrowly concentrated aggregate of qualities that comprise his nature. The character of the hero thus stands as an absolute, a fixed mark; when successive speakers appraise him, they do not alter the portrait itself, but betray their own biases—as, indeed, critics of the play have betrayed theirs for two centuries: *Coriolanus* has inspired more bad criticism than any other play of Shakespeare's.

The 'case of Coriolanus' begins with the first evidence offered by the citizens, who assert that Caius Marcius is 'chief enemy to the people . . . a very dog to the commonalty'. He has done great service to his country, but he 'pays himself by being proud'. Nothing spoken later, nothing in all the mass of testimony, ever enlarges much on the First Citizen's summary:

> I say unto you, what he hath done famously, he did it to that end. Though soft-conscienced men can be content to say it was for his country, he did it to please his mother, and to be partly proud; which he is, even to the altitude of his virtue.

(I. i. 36–41)

It is this fusion of uncompromising valour and uncompromising pride that makes Coriolanus's way incompatible with the world's way, even as uncompromising goodness and naïve assumption that all men are as selfless as he make Timon's way incompatible with the world's. But though all persons agree in identifying the hero's qualities, the opposing camps differ diametrically in their attitudes to them; and as for Coriolanus himself, in his own view no particle of a fault exists in his nature: for him the marriage of pride with valour marks the perfection of virtue, and to be other than he is by so much as a hair would be to be imperfect. To the citizens and the tribunes his valour is welcome—though certainly not to be emulated by themselves—but his pride is a capital offence that merits condemnation: 'Let us kill him. . . . To the rock, to the rock!' Between this attitude and the hero's own stands that of his family and friends, who hold his valour dear but deplore the pride that loses him the consulship and gets him banished. However, they see his pride not as a genuine defect but only as an obstacle: they would not have him purge his system of it, but only hide it. 'Why did you wish me milder?' the hero asks his mother, who replies: 'Lesser had been / The thwartings of your dispositions if / You had not show'd them how ye were dispos'd / Ere they lack'd power to cross you.' And Menenius agrees: 'Before he should thus stoop to the herd, but that / The violent fit o' th' time craves it as physic / For the whole state, I would put mine armour on, / Which I can scarcely bear.' Volumnia urges her son to use the same craftiness in dealing with the citizens that he uses in confronting his enemies in war. She advocates a course of studied hypocrisy, commanding her son to speak words 'that are but roted in / Your tongue' and to win his election 'with this bonnet in thy hand . . . Thy knee bussing the stones'. And

310

Cominius, too, alas—closest to Coriolanus in spirit and the best of men in his world—sides with Menenius and Volumnia in urging him to frame his spirit to the role of hypocrite, and adds, 'We'll prompt you.'

As his friends see it, then, the hero's fault is not that he possesses a pride so absolute that it must be disguised if he is to survive, but that he *will* not disguise it. But of course his very refusal to play the hypocrite is an integral part of his nature. Having temporarily been persuaded—for the good of others, not himself—to try dissembling, he makes the attempt, but not without words that reveal how abhorrent to him is the role:

> Away, my disposition, and possess me
> Some harlot's spirit! My throat of war be turn'd,
> Which choir'd with my drum, into a pipe
> Small as an eunuch's, or the virgin voice
> That babies lull asleep! The smiles of knaves
> Tent in my cheeks, and schoolboys' tears take up
> The glasses of my sight! A beggar's tongue
> Make motion through my lips, and my arm'd knees,
> Who bow'd but in my stirrup, bend like his
> That hath receiv'd an alms!
>
> (III. ii. 111–20)

Here revulsion overwhelms him, and he breaks off: 'I will not do't, / Lest I surcease to honour mine own truth / And by my body's action teach my mind / A most inherent baseness.' Even so, yielding to the pleas of mother and friends, he twice essays the dissembler's part in the market-place. On the first occasion, when he solicits 'voices' for the consulship, his sturdiest try at hypocrisy produces only a mocking caricature of humility that barely wins approval: 'To my poor unworthy notice,' remarks the Second Citizen soon afterwards, 'He mock'd us when he begg'd our voices.' On the second occasion, when his very life is at stake after the citizens have revoked their voices and threatened to hurl him from the Rock, his promise to 'answer mildly' is shattered by Sicinius's first utterance of the word 'traitor'—the word that is truly anathema to him, irreconcilable with his nature: 'The fires i' th' lowest hell fold in the people! / Call me their traitor!'

In posing the case of one whose nature is incompatible with the world's, Shakespeare appears less concerned with the usual dramatic effects of action and outcome than with placing upon us the burden of a verdict. Are we to accept the one extreme view, that of the tribunes and their constituents, that Coriolanus is a monster who deserves quick death? Or the middle view of Volumnia, Menenius, and Cominius, that not his nature itself is reprehensible, but only his refusal to dissemble when it is expedient to do so? Or the other extreme, that of Coriolanus himself, who finds the steep Tarpeian death preferable to any falsification of his nature? Does the dramatist himself, in packing our awareness with an enormous body of evidence, set one view above the other? It is, after all, the dramatist's play, not ours; and if we are to deal with it on its own terms, as we should, then it is the latter question to which we should seek an answer.

To suppose that Shakespeare moves us to see his hero as the tribunes and their constituents see him is obviously to err badly. If, in the end, we side with them, we must do so because our personal prejudices have directed us despite the dramatist's strictures. Even ignoring all the unflattering remarks made about them by the hero and his faction, we must yet concede that the citizens are a thoroughly unreliable lot. That they do require corn to survive goes without saying, and that Coriolanus is an outspoken opponent of their demands is amply apparent; but the elimination of the hero would not increase the short store: the gods, says Menenius, and not the patricians, have made the dearth. The wild and irresponsible economic fantasies of the citizens are epitomized by the First Citizen's second inflammatory exhortation: 'Let us kill him, and we'll have corn at our own price.' That the others loudly applaud the logic of this mad *non sequitur* confirms in advance the truth of the most caustic remarks later made by the hero and his friends about the people and their powers of reason; says Menenius, '. . . though abundantly they lack discretion, / Yet are they passing cowardly'. The Third Citizen himself acknowledges the limitations of his fellows' intellects: 'I think if all our wits were to issue out of one skull, they would fly east, west, north, south, and their consent of one direct way should be at once to all the points o' th' compass.' Nor are their limits only intellectual. When Marcius invites those who have been demanding corn to fight the Volsces and 'gnaw their garners', the citizens swiftly 'steal away'. When he storms the gates of Corioli, his soldiers hang back and leave him to be shut inside alone; when he fights his way out again, against appalling odds, and then, with Titus Lartius's help, re-enters the city, the citizens follow after—and immediately begin looting 'Ere yet the fight be done'. The citizens throw their caps in the air to welcome the hero Marcius home after victory; they lend him their voices for the consulship—and abruptly take them back again and shout their desire to hurl him from the Tarpeian cliff. Their minds and emotions are switched to and fro at a word from the tribunes. So much for the Romans. The Volscian citizens and soldiers are cut from the same cloth: like sheep, they desert their former master Aufidius to bleat after Coriolanus, and, when he returns from Rome's gates to Corioli, on the heels of Aufidius, it is he who gains the ovation: 'Your native town you enter'd like a post, / And had no welcomes home,' a Conspirator tells Aufidius; and he continues: 'But he returns, / Splitting the air with noise.' But a few moments later, responding to the prompting of Aufidius, even as the Romans to that of the tribunes, the people cry in chorus: 'Tear him to pieces! Do it presently!' And so they do.

If the dramatist has thoroughly discredited the citizens' views of the hero by repeatedly exhibiting the instability of their minds and emotions, he has doubly and triply discredited those of the tribunes. In the entire range of Shakespeare's plays it is impossible to find another pair of individuals, or even one individual, whose words, deeds, and personalities are so unrelievedly repugnant and whose total image is so destitute of a redeeming quality. The tribunes are villainous wretches of a unique species, worse in kind and degree than the dramatist's usual villains because they are evidently loathed by the dramatist himself, who elsewhere confers a measure of attractiveness on even his wickedest creations. Aaron the Moor is a spectacular theatrical villain the very audacity of whose crimes gives him a fascination; moreover, Shakespeare saw fit to glamourize him briefly, and to sentimentalize his instincts, in the incident of the black baby. Cassius, the sly misleader of noble Brutus, comes in the last scenes to claim as much sympathy as the hero himself. Claudius, the brother-murderer, throne-usurper, and queen-corrupter, grips and squeezes spectators' hearts during his agonized efforts to pray, and achieves a degree of grandeur in his dying moments. Iago, diabolical and mad as he is, is a masterful comic ironist, always magnetic, capable at any moment of diverting all eyes from Othello to himself. Edmund of *King Lear* shines with an authentic Renaissance spirit; his Machiavellian boldness 'sticks fiery off' and makes his wholly good brother Edgar seem a clod by comparison. The best (and worst) of these villains, Iago and Edmund, are darkly romantic figures who, like Satan, simultaneously fascinate and repel. No less is to be said of the villains of the comedies, most of whom exude charm and win admiration despite their taints of evil: Shylock is as romantic a figure as Hamlet, and commands as strong sympathy as dislike; Don John is but a warped prankster whose very delight in doing ill for its own sake is charming within the comic frame of *Much Ado;* Duke Frederick is merely villainy 'as you like it', and, besides, is converted at last to pure goodness; Angelo, too, is converted, but from angel to devil, and he is converted not by lust for evil but by lust for beauty and goodness, and for his transgression suffers more anguish than he inflicts; Iachimo of *Cymbeline,* the finest of Shakespeare's non-tragic villains, bears an artist's eye and a poet's tongue and is the dramatist's greatest connoisseur of beauty; but, indeed, even the brute Caliban is irresistibly gifted with a child's capacity for wonder and delight.

But Sicinius Velutus and Junius Brutus are simply little, mean, vicious, contemptible, calculated only to repel; and the dramatist takes pains to discredit them anew in every scene in which they appear. In their first appearance, after the citizens have slunk away at Coriolanus's suggestion that they follow him to the wars and gnaw the Volsces' corn, the tribunes, who have not yet spoken, are left alone, and, during the closing lines of the scene, discuss Coriolanus's pride. 'But I do wonder / His insolence can brook to be commanded / Under Cominius,' remarks Sicinius. Quickly, then, they agree on a theory that accords with their own mean little minds: Marcius is content to serve under Cominius because thus, either way, he is bound to preserve his fame, for if the wars prove disastrous the blame will fall on Cominius, and if they are successful the victory will be credited to Marcius. Here, were we to believe them, we should have to find Coriolanus not only proud and insolent, but shrewd, guileful, and purely self-serving—in short, the exact image of the tribunes themselves, who have interpreted his motives according to their own.

But what chiefly runs throughout their first appearance is the vice of envy. The tribunes are little men, little and crabbed, and theirs is the jaundiced view that such pygmy souls take of stalwart youthful heroes. Many times they have seen Marcius return from the wars to loud acclaim, and now they have heard him called for in the newest crisis; as his fame has soared, envy has gnawed them raw, and they vent their frustration in outcries: 'The present wars devour him! He is grown / Too proud to be so valiant!' He has been 'Tickled with good success', is 'already . . . well grac'd' with fame, and 'Opinion . . . sticks on' him. By the end of their first scene, through a conversation of only twenty-five lines, the dramatist has exposed their characters: they are crooked little men, gnawed by envy, who have just now been made acutely vicious by the likelihood that Coriolanus will shortly achieve yet higher honours. We should misread the dramatist grossly if we allowed a spot of their venom to taint our view of the hero.

In II. i, at their second appearance, Shakespeare introduces one of the few trustworthy voices in the play to expose their meanness just as sharply as their own words expose it elsewhere. Between I. i and II. i we have seen the hero in action and heard him wildly acclaimed by Cominius and Titus Lartius. Having now seen his greatness demonstrated by his deeds, we are in good position to recognize the validity of Menenius's remarks, when, in II. i, he defends Coriolanus and delivers a comprehensive and scathing castigation of the tribunes. It is they who begin the argument by charging the hero with pride and with boasting. Menenius replies by observing that the tribunes are themselves '. . . a brace of unmeriting, proud, violent, testy magistrates, alias fools'. The name-calling continues at length, and there can be no doubt whom Shakespeare intends as the victor. Whenever the tribunes open their mouths, Menenius tells them, he finds 'the ass in compound with the major part of your syllables'. They are ignorant men, he says, who 'know neither me, yourselves, nor anything', but they are ambitious 'for poor knaves' caps and legs', and though they spend much of their time in hearing quarrels over trivial affairs, the only peace they manage to make 'is calling both the parties knaves'. When they speak most wisely, what they say is 'not worth the wagging' of their beards, and the beards deserve no fate so grand as 'to stuff a botcher's cushion or to be entomb'd in an ass's pack-saddle'. And it is just such fellows as they, he concludes, who 'must be saying Marcius is proud, who . . . is worth all your predecessors since Deucalion'. At last he ends the exchange because 'More of your conversation would infect my brain'.

And thus it goes throughout the play: no figure in Shakespeare, not even the worthless Parolles of *All's Well*, who is at last verbally flayed by the old lord Lafeu, is more systematically reduced to a lump of pure repulsiveness. If the Monk of the *Canterbury Tales* is, as has been said, the sole person in the total canon whom Chaucer himself could not bear, the tribunes clearly have best claim to the title in the Shakespeare canon.

Besides permanently fixing the view that we are to take of the tribunes, that portion of II. i in which Menenius castigates them prepares us for their reappearance just after a conversation among Coriolanus's family in which all that is said of the hero is laudatory. The climactic moment of this scene is reached with the triumphant return of the hero himself, fresh from the wars. Flanked by Cominius and Titus Lartius and wearing the oaken garland, he enters to a flourish of trumpets and is greeted with an ovation: 'Welcome to Rome, renowned Coriolanus!' Here the dramatist gives the hero two lines that grace him as well as any in the play: 'Wouldst thou have laugh'd had I come coffin'd home,' he gently chides his tearful wife, 'That weep'st to see me triumph? Ah, my dear'. Moved by this tender interlude, Menenius speaks for all—or nearly all—of Rome:

> A hundred thousand welcomes! I could weep
> And I could laugh; I am light and heavy. Welcome!
> A curse begin at very root on's heart
> That is not glad to see thee!

> (II. i. 200–3)

The closing line is sharply pointed: no sooner have they marched off—the jubilant crowd of generals, captains, soldiers, family, friends, both those who have returned in triumph and those who have come to greet the victors—than the dramatist again brings on his two jackals. In such a situation, just following the outpouring of joy that we have seen Coriolanus accorded, even the mildest derogatory sounds that the tribunes might make would be enough to win them our loathing; but Shakespeare does not spare them by making their utterances mild: he fills their mouths with venom that beslubbers each word they speak. Brutus thus describes the ovation given the hero: the 'bleared sights' of citizens are 'spectacled to see him'; the 'prattling nurse' neglects her baby while she 'chats' him; the 'kitchen malkin' wraps her finest linen collar around her 'reechy neck' while 'clamb'ring the walls' to view the hero; priests themselves push their way through the throng and 'puff / To win a vulgar station'; veiled ladies subject their 'nicely gawded cheeks' to the sun's burning rays to have a look—and, all told, there is such a 'pother' as though not a man but a god inhabited the hero's form.

Nowhere else in Shakespeare is such another succession of nasty epithets so obviously employed to degrade the speakers themselves.

Set just to follow the happy reception accorded the returning hero, the tribunes' remarks reek of envy, spite, meanness, with each loaded word picked for its power to make the speaker offensive. The scene enlarges on an important fact: the tribunes do not hate Coriolanus alone; they hate everyone and despise even their constituents who keep them in power. Yet they hate the hero most because in his rising power they see a threat to their own; if Coriolanus is made consul, says Brutus, 'Then our office may / During his power, go sleep'. It is this motive that prompts their repeated practices against Coriolanus, who must, at any cost, be represented as the citizens' mortal foe:

> We must suggest the people in what hatred
> He still hath held them; that to 's power he would
> Have made them mules, silenc'd their pleaders, and
> Dispropertied their freedoms, holding them,
> In human action and capacity,
> Of no more soul nor fitness for the world
> Than camels in the war . . .
>
> (Ibid. 261–9)

Further, this image must be 'suggested' at times when the hero's 'soaring insolence' is at its height; hence the tribunes must breed occasion for him to be 'put upon 't'—kindled to anger that can be made to incense the people against him. The scene ends when a messenger calls the tribunes to the Capitol, where Marcius is to be made consul. 'Let's to the Capitol,' says Brutus, 'And carry with us ears and eyes for th' time, / But hearts for the event.' Such direct confessions of their hypocrisy Shakespeare repeatedly makes the tribunes speak openly to each other, for neither has any illusion about what both are.

Perhaps by this point—the middle of Act II—the dramatist might safely have trusted us, too, to remember what they are without further demonstration. He does not cease, however, but continues in scene after scene to expose their meanness. In II. iii, after Coriolanus has won the necessary voices, the pair again remain before us after the hero and his friends have departed. First they berate their constituents for having failed to rebuff Coriolanus 'As you were lesson'd', in Brutus's words, and 'As you were foreadvis'd' in Sicinius's. Here the very repetition of their wording makes the dramatist's intent plain: he will not risk, even yet, our failure to apprehend that the tribunes did indeed coach their constituents prior to the hero's plea for their voices. Next, predictably, the pair are shown again infecting the citizens' minds with the idea that giving Coriolanus their voices was a monstrous act: '... tell those friends / They have chose a consul that will from them take / Their liberties, make them of no more voice / Than dogs. ...'

Nor is Shakespeare yet finished with his portrait of the people's misleaders; he continues to strive for means of showing them as even more devious and contemptible than he had yet succeeded in showing them to be. Having ordered the citizens to revoke their voices, Brutus specifies their procedure:

> Lay
> A fault on us, your Tribunes, that we labour'd,
> No impediment between, but that you must
> Cast your election on him.

> (II. iii. 234–7)

'Lay the fault on us', Sicinius repeats; and Brutus, 'Ay, spare us not. Say we read lectures to you.' To the end of the scene they continue to harp on the one string: 'Say you ne'er had done 't— / Harp on that still—but by our putting on.' Thus their practice is to ingratiate themselves simultaneously with the nobility, by making it appear that they advised the citizens to accept Coriolanus, and with the citizens, by grandly offering to take the blame upon themselves. Having so devised against both sides, they scurry on to the Capitol, there to direct the conspiracy against the hero.

That the tribunes, through crooked counselling both before and during this confrontation, are to blame for the reversal of the hero's fortunes, Shakespeare makes unmistakable. In II. iii we heard them instruct the citizens to put Coriolanus into a rage; now, in III. i, we see their practice executed exactly as planned: blocking the hero's way, the tribunes announce that the people have revoked their voices, and then, with calculated insults climaxed by the key word 'traitor', quickly enrage him; right on cue, Sicinius cries out 'Bear him to the rock Tarpeian, and from thence / Into destruction cast him.' Though certainly the hero, once angered, responds with epithets hardly calculated to win friends—'Hence, rotten thing! or I shall shake thy bones / Out of thy garments'—yet Shakespeare has so pointedly assigned all blame to the tribunes that perhaps nothing Coriolanus might say or do could shift our sympathy from him to his enemies. But in fact, even at the height of his fury, he speaks with clear logic and even with restraint; what is more, the Senators,

Cominius, and Menenius direct our sympathies by supporting our view of him as the innocent victim of a crooked practice. Needing no debate, they stand solidly against his detractors; both Menenius and Cominius would do battle against the tribunes and the mob but that, in the latter's trained judgement, ''tis odds beyond arithmetic'. When the hero, for the others' safety, not his own, has retired, the First Patrician states an obvious fact: 'This man has marr'd his fortune.' But the wise Menenius's assessment of the hero's conduct, far from being censorious, precisely labels the deepest cause of his tragedy: 'His nature is too noble for the world.'

Even after Coriolanus has gone, the tribunes continue their hate campaign with blatantly loaded words: 'Where is this viper?' cries Sicinius; again, 'This viperous traitor'; and yet again, 'He's a disease that must be cut away'—echoing an earlier image of the hero as an 'infection' that must be destroyed before it spreads. In short, throughout the climactic scene of confrontation, the dramatist has so relentlessly scored the baseness of the tribunes that their victim demands our full sympathy.

Before we next see the tribunes, we watch a scene in which the hero, yielding once more to his mother and his friends, agrees to try what is most abhorrent to his nature—play the hypocrite in order, even yet, to appease the enemy. 'Arm yourself / To answer mildly', urges Cominius; and Coriolanus replies, 'The word is "mildly" . . . Let them accuse me by invention, I / Will answer in mine honour.' With these modest sentiments still sounding in our ears, we hear the first lines of the next scene, when Brutus enjoins his fellow practiser,

> In this point charge him home, that he affects
> Tyrannical power. If he evade us there,
> Enforce him with his envy to the people.
>
> (III. iii. 1–3)

Sicinius then instructs the Aedile precisely in details of their plot:

> Assemble presently the people hither;
> And when they hear me say, 'It shall be so
> I' th' right and strength o' th' commons,' be it either
> For death, for fine, or banishment, then let them,
> If I say fine, cry 'Fine!' if death, cry 'Death!'
>
> (Ibid. 12–16)

'Put him in choler straight', Brutus orders—and out of what the hero will then surely speak will arise occasion 'to break his neck'. With clear deliberateness, directly after these most vicious remarks in a series wherein each utterance of the tribunes reeks of malice, the dramatist then brings on the hero and puts into his mouth this wholly selfless sentiment:

> Th' honour'd gods
> Keep Rome in safety, and the chairs of justice
> Supplied with worthy men! plant love among's!
> Throng our large temples with the shows of peace,
> And not our streets with war!
>
> (Ibid. 33–7)

'A noble wish', says Menenius; but then Sicinius cries, 'Draw near, ye people!' and the rest proceeds just as the tribunes had planned: Coriolanus is 'put to choler' on being charged with contriving to 'wind' himself into tyrannical power and thus proving himself 'a traitor to the people'. Predictably, he flares with anger at the word, and the citizens, taking their cue from their masters, scream for his banishment. When the hero withdraws, they shout, throw their caps, and, with the tribunes, hoot him out of the city. The scene ends with a jubilant chorus that touches off a glaring flash of irony: 'The gods preserve our noble tribunes!'

During the first three acts, thus, the tribunes' role has consisted exclusively of plotting to rid themselves of the chief threat to their own power; each scene has shown them either devising practices, executing them, or gloating over their success. Shakespeare's method with these practices is of a kind with that in earlier plays, with one notable difference: except for those whom they supposedly serve— the citizens of Rome—the tribunes actually deceive nobody, for the hero and his side have repeatedly expressed their awareness; Volumnia's angry assertion after her son has been hooted from Rome is typical: ''Twas you incens'd the rabble.' Elsewhere the wickedness of Shakespeare's villains is often relieved by touches that claim a moment of sympathy or admiration: Iago's effusive wit and humour, Edmund's early show of brilliance, Claudius's pangs of conscience; but the dramatist allows Sicinius and Brutus no redeeming word or act.

Nor does he allow them to look any better after they have driven their rival from Rome. In IV. ii, just after the hero's tender parting— 'Come, my sweet wife, my dearest mother, and / My friends of noble touch; when I am forth, / Bid me farewell and smile'—the tribunes are brought on again, reeking of hypocrisy: 'Now we have shown our power,' says Brutus, 'Let us seem humbler after it is done / Than when it was a-doing.' Meeting the returning family, they are berated not only by the formidable Volumnia but also by the mild and gentle Virgilia—and to these stricken women they turn faces of mock humility: 'I would he had continued to his country / As he began, and not unknit himself / The noble knot he made,' Sicinius whines; and Brutus echoes him, 'I would he had.' Perhaps at no point does Shakespeare bare them to fiercer feelings of revulsion than here, when they feign humility and seem to turn the other cheek; yet their show of concern does not even last out the scene: 'Why stay we to be baited / With one that wants her wits?' snarls Sicinius, turning his back on the ladies and skulking off.

Shakespeare is kind enough to spare us further sight of this pair during the next three scenes, while Coriolanus joins Aufidius and wins command of half the Volscian forces bent on burning Rome. In IV. v, announcing these developments, the Third Servant comments ominously, 'I would not be a Roman, of all nations. I had as lieve be a condemned man.' Coriolanus, he asserts, will go 'and sowl the porter of Rome by the ears'. Directly after hearing this warning of Rome's imminent destruction, we are again shown the tribunes as they gloat over the city's seeming prosperity in Coriolanus's absence; says Sicinius, 'Here do we make his friends / Blush that the world goes well.' When Menenius enters, they taunt him repeatedly: 'Your Coriolanus is not much miss'd,' says Sicinius; and Brutus sneers, 'Rome / Sits safe and still without him.' These hubristic mouthings are ended shortly after with word of the Volsces' approach, Coriolanus at their head. With the tables turned, Menenius and Cominius—who are clearly used to voice our own exultation—have their share of fun at intimidating the cowering tribunes: 'O, you have made good work!' cries Cominius; and Menenius, 'You have made good work, / You and your apron-men; you that stood so much / Upon the voice of occupation and / The breath of garlic-eaters!' In all, Menenius and Cominius bait the snivelling culprits during a period of fifty lines. Elsewhere, when fortunes suddenly are reversed and those who have wielded and abused great power have become underdogs—Richard III, Shylock, even Macbeth—Shakespeare's way is to point the scene so that we are moved to sympathy, however intense has been our former dislike. But it is no such thing here: when the tribunes are suddenly thrown from their exalted state into ignominy and peril, the dramatist uses an entire scene to twist the knife in their wounds. Not only the taunts of Menenius and Cominius, but their own words are turned against them and require our contempt: 'Say not we wrought it', they cry in unison when Cominius lays to their charge the imminent destruction of the city.

Discredited as the tribunes are throughout the play, and rendered utterly despicable by the dramatist's unrelenting exposure of their littleness, their views of Coriolanus are obviously not to weigh against the hero in the mass of evidence from which we are to arrive at a verdict. More reliable—at least on first consideration—is the evidence supplied by the hero's friends and family, who stand between the extremes of the tribunes, on the one hand, and the evidence of Coriolanus's own words and deeds, on the other. From this group it is fair to select a single spokesman, Menenius, whose explicit attitudes best represent all. If his view is not quite identical to Volumnia's or Virgilia's, yet he may be said to mediate the extremes within this small group very much as, at the same time, he mediates the polar opposites of tribunes and hero. But to recognize that the sweep of his vision is the broadest in the play is not to suggest also that his is the point of view of the tragedy—the attitude towards persons and events that the dramatist means us to adopt.

Menenius's view of the citizens and their tribunes does, however, appear the same as that which the dramatist has laboured to make our own; further, this view is no harsher than the hero's own. Menenius spends half again as many lines as the hero in castigating the tribunes and their constituents, and if, unlike Coriolanus, he avoids such direct invectives as 'dissentious rogues', 'curs', and 'rats' in addressing them, he does so only because his superior grace of phrasing enables him to be more devastating otherwise. His slashes cut deeper than do the hero's: '. . . though abundantly they lack discretion', he says of the citizens, 'Yet are they passing cowardly'. The tribunes, he insists, are 'the herdsmen of the beastly plebians', and the citizens are 'Your multiplying spawn . . . That's thousand to one good one'. After the hero has first won and then lost their voices, the citizens are momentarily 'beat in' by Coriolanus, and while the patricians await a second encounter with the mob, the Second Patrician remarks, 'I would they were abed.' 'I would they were in Tiber!' replies Menenius. Once the hero has been banished and word comes that he has joined with Aufidius to destroy Rome, Menenius's contempt for the tribunes and their followers is expressed in ever more contemptuous terms: the citizens are the tribunes' 'apron-men' and 'garlic-eaters', the 'clusters' who cast their 'stinking greasy caps'.

Thus no appreciable distinction exists between the attitudes of Menenius and Coriolanus towards the hero's enemies, or between their attitude and that which the dramatist urges on us. But there is a sharp distinction between Menenius's view of the hero and that which the dramatist would have us take. Though Menenius approves wholly of Coriolanus's pride and valour, he disapproves of his unwillingness to dissemble when it is politic to do so and scolds him for allowing his dislike of his political foes to show at impolitic moments. Should we, then, conclude that Menenius provides the dramatist's directive for our own point of view, we should have to characterize the play as a case for hypocrisy. If Menenius, Volumnia, Titus Lartius, and the otherwise wholly admirable Cominius were but once to scold the hero for *his attitude itself* towards the people, we might be justified in taking their view as the dramatist's directive to ourselves. But they never do so; they demand only that the hero mask his attitude. In the final assessment of the body of evidence, then, their testimony deserves hardly more weight than that of the tribunes and the citizens.

Menenius and the rest deal just as crookedly with the hero when they admonish him to be false to his nature as do the tribunes when they direct the citizens how to deal with him. Surely it cannot be to the hero's discredit that—though in deference to his mother and his friends he makes a valiant attempt—he fails utterly at the hypocrite's role. Coriolanus is caught between two equally reprehensible elements in his world—the tribunes who manipulate the citizens against him, and his mother and friends who would manipulate him against the citizens, and against himself. He is deceived by neither set of practisers; innocent though he is, he is not even self-deceived. He knows as well as others, as well as we, that the course urged on him by Menenius is the one that he must take if he would be consul, or even survive in his world. His fault, if there be one, is only that he is by nature incapable of dishonesty, and therefore he is incompatible with the world in which the dramatist has placed him.

Having weighed the great body of evidence spread before us, we surely cannot reasonably conclude that the hero is more to be censured for not following Menenius's counsel to a prosperous end than for somehow corrupting his nature in order to follow it. The single fault on which citizens, tribunes, family, and friends agree in blaming him is pride; but in the circumstances of his world as the dramatist has represented that world, his lone alternative is hypocrisy. Pride is the quality in his nature that makes dissembling abhorrent to him. His pride is unfortunate in that it destroys him, and, viewed in that perspective, is a fault; but because it is also what prevents him from lying even to those he despises, it is a solid virtue.

Index

(The Index omits reference to characters in the chapters on their own plays.)

325